Science and Religion

Science and Religion

A HISTORICAL INTRODUCTION

Edited by

Gary B. Ferngren

The Johns Hopkins University Press
Baltimore and London

Printed in the United States of America on acid-free paper
9 8 7 6 5 4 3 2

The Johns Hopkins University Press
2715 North Charles Street
Baltimore, Maryland 21218-4363
www.press.jhu.edu

Library of Congress Cataloging-in-Publication Data

Science and religion: a historical introduction / edited by Gary B. Ferngren.
p. cm.
Includes bibliographical references and index.
ISBN 0-8018-7038-0 (pbk. : alk. paper)
1. Religion and science—History. I. Ferngren, Gary B.
BL245 .S37 2002
291.1'754—dc21 2002016042

A catalog record for this book is available from the British Library.

Contents

Introduction ix

Part I : : Science and Religion: Conflict or Complexity? 1

1 The Conflict of Science and Religion 3
COLIN A. RUSSELL

2 The Historiography of Science and Religion 13
DAVID B. WILSON

Part II : : The Premodern Period 31

3 Aristotle and Aristotelianism 33
EDWARD GRANT

4 Early Christian Attitudes toward Nature 47
DAVID C. LINDBERG

5 Medieval Science and Religion 57
DAVID C. LINDBERG

6 Islam 73
ALNOOR DHANANI

Part III : : The Scientific Revolution 93

7 The Copernican Revolution 95
OWEN GINGERICH

8 Galileo Galilei 105
RICHARD J. BLACKWELL

9 Early Modern Protestantism 117
EDWARD B. DAVIS AND MICHAEL P. WINSHIP

10 Causation 130
JOHN HENRY

11 Mechanical Philosophy 143
MARGARET J. OSLER

12 Isaac Newton 153
RICHARD S. WESTFALL

13 Natural Theology 163
JOHN HEDLEY BROOKE

Part IV : : Transformations in Geology, Biology, and Cosmology, 1650–1900 177

14 Geology and Paleontology 179
NICOLAAS A. RUPKE

15 Natural History 195
PETER M. HESS

16 Charles Darwin 208
JAMES MOORE

17 Evolution 219
PETER J. BOWLER

18 Cosmogonies 234
RONALD L. NUMBERS

Part V : : The Response of Religious Traditions 245

19 Roman Catholicism since Trent 247
STEVEN J. HARRIS

20 Evangelicalism and Fundamentalism 261
MARK A. NOLL

21 Creationism since 1859 277
RONALD L. NUMBERS

22 The Scopes Trial 289
EDWARD J. LARSON

Part VI : : The Theological Implications of Modern Science 299

23 Physics 301
RICHARD OLSON

24 Twentieth-Century Cosmologies 314
CRAIG SEAN MCCONNELL

25 Scientific Naturalism 322
EDWARD B. DAVIS AND ROBIN COLLINS

26 The Design Argument 335
WILLIAM A. DEMBSKI

27 Ecology and the Environment 345
DAVID N. LIVINGSTONE

Part VII : : Current Historiographical Issues 357

28 Gender 359
SARA MILES AND JOHN HENRY

29 The Social Construction of Science 374
STEPHEN P. WELDON

30 Postmodernism 381
STEPHEN P. WELDON

Index 389

Introduction

ANDREW DICKSON WHITE'S *A History of the Warfare of Science with Theology in Christendom* (1896) was published more than a century ago. In it White argued that Christian theologians had a long history of opposing scientific progress in the interest of dogmatic theology. The charge was not new. It grew, in fact, out of the view of the eighteenth-century philosophes that the church was an institution whose ignorance and intolerance had hindered human progress, while science was a force of cultural liberation. An argument similar to that of White had been made by John William Draper in his *History of the Conflict between Religion and Science* (1874), and it struck a responsive chord in American thought, which was at the turn of the century adopting an increasingly secular outlook as it came to recognize the central role that science played in modern society. The Draper-White thesis, as it has come to be known, was enormously influential. For the past century it has been the predominant view of the relationship of science and religion among scientists and laymen alike. It wedded a triumphalist view of science with a patronizing view of religion. Popular misconceptions doubtless underlay the widespread presumption that religion was opposed to science. Grounded in faith, religion seemed bound to suffer when confronted by science, which was, of course, based on fact.

While some historians had always regarded the Draper-White thesis as oversimplifying and distorting a complex relationship, in the late twentieth century it underwent a more systematic reevaluation. The result is the growing recognition among historians of science that the relationship of religion and science has been much more positive than is sometimes thought. Although popular images of controversy continue to exemplify the supposed hostility of Christianity to new scientific theories, studies have shown that Christianity has often nurtured and encouraged scientific endeavor, while at other times the two have co-existed without either tension or attempts at harmonization. If Galileo and the Scopes trial come to mind as examples of conflict, they were exceptions rather than the rule. In the words of David Lindberg, writing on medieval science and religion for this volume, "There was no warfare between science and the church. The story of science and Christianity in the Middle Ages is not a story of suppression nor one of its polar opposite, support and encouragement. What we find is an interaction exhibiting all of the variety and complexity with which we are familiar in other realms of human endeavor: conflict, compromise, understanding, misunderstanding, accommodation, dialogue, alienation, the making of common cause, and the going of separate ways" (pp. 70–71). What Lindberg writes of medieval Europe can be said to describe much of Western history. Evidence that the relationship of science and religion has exhibited a multiplicity of attitudes, reflecting local conditions and particular historical circumstances, has led John Brooke to speak of a "complexity thesis" as a more accurate model than the familiar "conflict thesis." But while Brooke's view has gained widespread acceptance among professional historians of science, the traditional view remains strong elsewhere, not least in the popular mind.

The purpose of this volume is to provide an introduction to the historical relationship of the Western religious traditions and science. With one exception, the 30 essays assembled here are drawn from *The History of Science and Religion in the Western Tradition: An Encyclopedia,* Gary B. Ferngren, General Editor (New York: Garland, 2000), whose 103 articles provide a comprehensive survey of the field. Edward J. Larson's essay, "The Scopes Trial," was commissioned specifically for this volume. Richard J. Blackwell's "Galileo Galilei" and William A. Dembski's "The Design Argument" are expanded versions of their original articles. All other essays remain the same except for minor stylistic changes and additions to the bibliographies. The essays chosen for inclusion here were selected because they deal with major sites of the intersection of science and religion or with historical periods or events in which that intersection was prominent. The decision to limit the volume's coverage to the West is based on the belief that, underlying the diversity of the several streams that have fed Western civilization, there exists a basic substratum, formed by the West's dual heritage from the classical world of Greece and Rome and the monotheistic traditions of Judaism and Christianity. The focus on Christian theology reflects the dominance of that religious tradition in European scientific and philosophical thought.

In the past generation, historical scholarship has devoted a good deal of attention to the relationship of science and religion. Two historiographical perspectives that have influenced, even revolutionized, that scholarship provide the orientation of many of the essays in this volume. The first is the reaction against what is commonly called the Whig interpretation of history, which dominated the historiography of the nineteenth century and the first half of the twentieth. The phrase, coined by the English historian Sir Herbert Butterfield, denotes a perspective that views the past through the lens of the present and sees history as moving progressively toward the ideas and institutions of a later age, particularly one's own. "Whiggish" historians have been accused of distorting the past to affirm the values of the present by dividing historical figures into the friends and enemies of progress.

Most historians of science writing today attempt to avoid presentist and essentialist approaches, which have often influenced our historical understanding of both religion and science. *Presentism* is the tendency to shape the past by employing modern definitions and understandings. *Essentialism* assumes that an idea or a discipline is basically the same in all ages. In fact, however, words like *science* and *religion* have had many different meanings throughout history. *Natural philosophy* (like *scientia*) had a very different connotation in the seventeenth century from that of *science* in the twentieth. It contorts the history of science to impose modern definitions on the study of the past and to admit as truly scientific only what a modern scientist would regard as within his or her purview. What are termed the occult or marginal sciences (e.g., astrology, alchemy) would not fall under the rubric of science today. Yet many of the most distinguished practitioners of science in the early modern period (such as Isaac Newton) embraced one or more of them.

Studies of the interaction of science and religion by historians of science during the past generation have consciously sought to place that interaction in its social and historical context. Such studies have enriched our understanding of some of the prominent controversies involving religion and science. Thus the traditional picture of Galileo as a martyr to intellectual freedom and a victim of the Catholic Church's opposition to science has been demonstrated to be little more than a caricature. Viewed against the backdrop of its age, the trial of Galileo was an episode that involved multiple issues and included powerful personalities as well as vested professional and ecclesiastical interests. Similarly, Isaac Newton, who was at one time presented solely as a great natural philosopher, is now known to have had a deep interest in theology and biblical prophecy and to have practiced his science within a theological context. Finally, Charles Darwin has been portrayed in recent historical studies not merely as the discoverer of a scientific theory that revolutionized natural history but as a man who struggled with personal religious issues and was long influenced by the tradition of natural theology that he had imbibed from the writings of William Paley.

Social constructionism provides the second modern perspective that informs much recent writing in the history of science. *Social constructionism* is a methodological orientation that views the pursuit of scientific knowledge not as an autonomous but as a human endeavor that reflects social backgrounds and conditions as well as the spirit of the age. It marks a reaction to the dominance of positivistic science. In the Western world science has long enjoyed a privileged position of authority because it has been regarded as empirically based and therefore rigorously objective. Much of contemporary science operates (often unconsciously) on the positivist principle that the highest form of knowledge is the description of sensory phenomena. Hence, the term *positivism* is commonly associated with any system that limits knowledge to empirical data and excludes theological or metaphysical claims.

Social constructionism as an intellectual movement was inspired by Thomas Kuhn's *The Structure of Scientific Revolutions* (1962), which accounted for major transformations in science (Kuhn termed them "paradigm shifts") as the result of social and psychological factors. Kuhn's approach owed much to antirationalistic and relativistic notions and marked a reaction to the longstanding emphasis on the autonomy and rationality of traditional science. It unintentionally created an agenda for the next generation of scholars, who pursued what came to be called *science studies.* Their research program emphasized the sociology of knowledge and the contextual and social-cultural setting in which science developed. Positivistic scientists and intellectual historians of science reacted vigorously in defense of older views, however, and the resulting "science wars" have characterized the last two decades of the twentieth century. Adding to the ferment was the emergence of feminism and postmodernism. Feminism became a major force in the scholarship of many academic disciplines, including science studies, with its study of gender as a factor in the construction of scientific knowledge and its exploration of the extent of women's participation in science. Postmodernism injected a more radical influence. Postmodernists rejected all "totalizing" systems, arguing that no human endeavor, not even science, could claim to be based on universal truths or assumptions, such as rationality. Hence, they denied the claims of both science and religion, which found themselves deprivileged and attacked for their authoritarian pretensions. The science wars have raged for a generation, with many (but by no means all) scholars accepting, in at least some form, the legitimacy of a program of social construction of science. Radical Kuhnians, however, including many feminists and postmodernists, have gained considerably less acceptance. Nevertheless, in attempting to appreciate the way in which modern scholars have reconstructed the historical intersection of science and religion, it is important to understand these intellectual movements, which have created the historiographical agenda for contemporary science studies.

It should not be surprising that some scientists have been uncomfortable in seeing the rationality and autonomy of their discipline attacked and their epistemological credentials undercut by critical approaches that are rooted in assumptions of cultural relativism. By contrast, both scholars and believers have long recognized that religious traditions are neither monolithic nor static—that they have developed over time and reflect the diverse circumstances of their geography and culture. Yet "science" (in the words of Alfred North Whitehead) "is even more changeable than theology." If the historical landscape is littered with discarded theological ideas, it is equally littered with discarded scientific ones. Failure to understand this historical reality has led those who see the march of science as one of inexorable progress to view controversies between science and religion as disputes in which (to quote Whitehead again) "religion was always wrong, and . . . science was always right. The true facts of the case are very much more complex, and refuse to be summarized in these simple terms" (*Science and the Modern World* [New York: New American Library, 1960], 163).

Recognition that both science and religion have developed historically does not necessitate belief in cultural relativism. Nor need it deny the claims of Christianity or other religions to special (i.e., divine) revelation. It does, however, require an awareness of the cultural limitations that are imposed on all societies, ideas, and disciplines, including, of course, our own. Perhaps it is hardest for us (as it is for any generation) to accept the fact that the modern age is a historical period like any other, limited in its perspectives by space and time and subject to the constraints of its own *Zeitgeist*. Understanding that we, too, have historical and cultural limitations forces us to view the past in a manner that is neither patronizing nor disparaging, but capable of appreciating the power of ideas that we do not share or that have fallen out of fashion in our own day. If the study of the intersection of religion and science demonstrates anything, it is the enduring vitality and influence of some of the basic traditions of the Western world—religious, philosophical, and scientific—which still retain their ability to shape ideas and inform our culture in the twenty-first century.

Gary B. Ferngren

Part I :: Science and Religion: Conflict or Complexity?

1 The Conflict of Science and Religion

Colin A. Russell

The most common view of the relationship between science and religion is that they are, and have been historically, in mortal conflict. While it cannot be denied that isolated cases of real conflict have existed, as in the cases of Galileo and Darwin, recent historiography suggests that it is wrong to extrapolate from these examples to the view that science and religion are necessarily hostile. Studies of the social history of science in the nineteenth century have, moreover, suggested that other factors account for the emergence of the conflict theory.

Colin A. Russell is emeritus professor of history of science and technology at the Open University. He holds both the Ph.D. and D.Sc. from London University. His recent books include *Edward Frankland: Chemistry, Controversy, and Conspiracy in Victorian England* (Cambridge: Cambridge University Press, 1996); *Michael Faraday: Physics and Faith* (New York: Oxford University Press, 2000); and (as editor) *Chemistry, Society, and Environment: A New History of the British Chemical Industry* (Cambridge: Royal Society of Chemistry, 2000).

The Conflict Thesis

The history of science has often been regarded as a series of conflicts between science and religion (usually Christianity), of which the cases of Galileo Galilei (1564–1642) and Charles Darwin (1809–82) are merely the most celebrated examples. Some would go further and argue that such conflict is endemic in the historical process, seeing these and other confrontations as occasional eruptions of a deep-seated inclination that is always present, if not always quite so spectacularly visible. There is usually the additional assumption, implicit or explicit, that the outcome of such conflict will always and inevitably be the victory of science, even if only in the long term. Such a view of the relations between science and religion has been variously described as a "conflict thesis," a "military metaphor," or simply a "warfare model."

The considerable literature on this subject began with two famous works of the nineteenth century: John William Draper's *History of the Conflict between Religion and Science* (1874) and Andrew Dickson White's *A History of the Warfare of Science with Theology in Christendom* (1896). A more mature work of the twentieth century, J. Y. Simpson's *Landmarks in the Struggle between Science and Religion* (1925), adds to the vocabulary of metaphors by positing a struggle between science and religion. The first two books achieved a wide circulation and have been repeatedly reprinted. They were written at a time when science seemed triumphant at home and abroad, and each author had his particular reasons for settling old scores with organized religion. Draper, a professor of chemistry and physics in a medical school in New York, feared the power wielded by the Roman Catholic Church and was worried by the promulgation of the dogma of papal infallibility of 1870. White, professor of history at the University of Michigan and later president of Cornell (the first private non-sectarian university in the United States), was not surprisingly opposed by the advocates of sectarian theology. White's book thus became a manifesto directed (in the last version) not so much against religion as against dogmatic theology.

For nearly a century, the notion of mutual hostility (the Draper-White thesis) has been routinely employed in popular-science writing, by the media, and in a few older histories of science. Deeply embedded in the culture of the West, it has proven extremely hard to dislodge. Only in the last thirty years of the twentieth century did historians of science mount a sustained attack on the thesis, and only gradually has a wider public begun to recognize its deficiencies.

Issues of Contention

First, it may be helpful to spell out briefly the chief issues of contention around which the real or imagined conflict revolves. Initially these issues were in the area of epistemology: Could what we know about the world through science be integrated with what we learn about it from religion? If not, a situation of permanent conflict seemed probable. Such epistemological issues were first raised on a large scale by the Copernican displacement of the earth from the center of the solar system, which was clearly incompatible with what seemed to be the biblical world picture of a geocentric universe. The question, though posed by Copernicus (1473–1543) himself, caused little public stir until the apparent conflict became inextricably intertwined with other clericopolitical disagreements at the time of Galileo. With hindsight, it is truly remarkable that, as early as the sixteenth century, Copernicus and his disciple Georg Joachim Rheticus (1514–74) resolved the issue to their satisfaction by invoking the patristic distinction between the Bible's teaching on spiritual and eternal realities and its descriptions of the natural world in the language of ordinary people. Rheticus specifically appealed to Augustine's doctrine of "accommodation," asserting that the Holy Spirit accommodated himself on the pages of Scripture to the everyday language and terminology of appearances. What began to emerge was what later became the distinction between world picture and worldview, the former being mechanistic, tentative, and expendable, while the latter concerned values and principles that were likely to endure. This same principle imbued the work both of Galileo and his followers and of Johannes Kepler (1571–1630) and effectively defused the issue for a majority of Christian believers. If they were right, there was an absence of conflict not only over the specific case of cosmology but, in principle, over anything else in which scientific and biblical statements appeared to be in contradiction. A "conflict thesis" would have seemed untenable because there was nothing to fight about. However, the historical realities were such that these lessons were not quickly learned.

Despite the advent in the late eighteenth century of evidence for a much older Earth than had been imagined on the basis of the Mosaic account in Genesis, little opposition arose until the emergence in early-Victorian England of a disparate but vocal group of "scriptural geologists." They were not, as is often claimed, a group of naive scientific incompetents but, indeed, were often rather able men who saw a distinction between biblical descriptions of the present natural world and of events in the past, respectively corresponding to their understandings of physical science and history. While for the most part happy to accept "accommodation" over biblical references to the sun and Earth, they were not prepared to extend it to what appeared to be descriptions of history, including chronology. The potential for conflict was greatest where science had a historical content (as in geology or biology). The war cries of the "scriptural geologists" were echoed by those who, in due course, assailed Darwinian evolution on the same grounds.

A second, and related, area of contention has been in the realm of methodology. Here we find the age-old polarization between a science based on "facts" and a theology derived from "faith," or between a naturalistic and a religious worldview. Naturalism has had a long history, going back to the early Middle Ages and beyond, with a spectacular revival in nineteenth-century England that was dignified by the title of "scientific naturalism." It was a view that denied the right of the church to "interfere" in the progress of science by introducing theological considerations into scientific debates. By the same token, any appeal to divine purpose as an explanation of otherwise inexplicable phenomena has been a famous hostage to fortune. This philosophy of "God of the gaps" has generated special heat when one of the "gaps" has later been filled naturalistically. In these cases, conflict has certainly appeared, though whether it is really about methodological issues may be doubted. It has also been argued in a veritable torrent of informed and scholarly works that the methodologies of science and of religion are complementary rather than contradictory, and local instances of dispute have been assigned to other causes. Yet, this confusion still penetrates popular thinking, and the conflict thesis has been thereby sustained.

The third potential for conflict has been in the field of ethics. Most recently this has been realized in questions about genetic engineering, nuclear power, and proliferation of insecticides. Past debates on the propriety of such medical procedures as vaccination and anesthesia have been replaced by impassioned conflict over abortion and the value of fetal life. In Victorian times, one of the more serious reasons for opposing Darwin was the fear that his theories would lead to the law of the jungle, the abandonment of ethical constraints in society. In nearly all of these cases, however, it is not so much science as its application (often by nonscientists) that has been under judgment.

Fourth, some opposition between science and religion has arisen from issues of social power. In Catholic cultures in continental Europe, the polarity between sacred and secular was often much sharper than in Britain and the United States, with the result that progressive science-based ideologies were more frequently in explicit contention with conservative political and ecclesiastical forces. In early-nineteenth-century Britain, certain high-church Anglicans turned on science for threatening their dominant role in society. While this debate was formally about the authority of Scripture, in reality it was about the growing spirit of liberalism within the universities. Not surprisingly, the community of science resented such attacks and, in due course, turned the table on the enemy.

Their response came in the form of a concerted effort by certain scientific naturalists in Victorian England, most notably those associated with Thomas Henry Huxley (1825–95), to overthrow the hegemony of the English church. The movement, which was accompanied by bitter conflict, generated a flood of articles, lay "sermons," and verbal attacks on the clergy and included conspiratorial attempts to get the "right men" into key positions in the scientific establishment. It involved lectures, secular Sunday schools, and even a successful lobby to have Charles Darwin's body interred in Westminster Abbey, yet it was not a battle between science and religion except in the narrowest sense. Unlike White, who averred that he opposed not religion but dogmatic theology, Huxley sought to undermine organized religion, though his rhetoric frequently sought to convey the impression of a disinterested defense of truth. One recent writer identifies the driving force behind at least the Victorian struggles as "the effort by scientists to improve the position of science. They wanted nothing less than to move science from the periphery to the centre of English life" (Heyck 1982, 87). It was at this time that science became professionalized, with the world's first professional institute for science, the Institute of Chemistry, established in 1877. In Europe, this was also the period when scientific leadership began to slip from Britain to Germany, generating a fierce rearguard reaction by some British scientists against anything that could diminish their public standing. If the church was seen to be in their way, it must be opposed by all means, including the fostering of a conflict myth, in which religion routinely suffered defeat at the hands of triumphalist science.

The Weaknesses of the Conflict Thesis

The conflict thesis, at least in its simple form, is now widely perceived as a wholly inadequate intellectual framework within which to construct a sensitive and realistic historiography of Western science. Nor was it merely a case of British controversy. Ronald L. Numbers has suggested that "the war between science and theology in colonial America has existed primarily in the cliché-bound minds of historians." He regards the polemically attractive warfare thesis as "historically bankrupt" (Numbers 1985, 64, 80). In the composite volume *God and Nature: Historical Essays on the Encounter between Christianity and Science* (1986), edited by Numbers and his colleague at the University of Wisconsin, David C. Lindberg, an effort is made to correct the stereotypical view of conflict between Christianity and science.

The shortcomings of the conflict thesis arise from a multiplicity of reasons, some of which may be briefly summarized as follows. First, the conflict thesis hinders the recognition of other relationships between science and religion. At different phases of their history, science and religion were not so much at war as largely independent, mutually encouraging, or even symbiotic. Certainly there are well-documented cases, such as those of Galileo and Darwin, in which science and religion seemed to wage open war with each other. But recent scholarship has demonstrated the complexity of the issues at stake in even these cases, with ecclesiastical politics, social change, and personal circumstances as relevant as questions of science and religion. Quite apart from those considerations, such cases have been too often taken as typical, and, consequently, a generalized conflict thesis has been erected on insubstantial foundations. As a historical tool, the conflict thesis is so blunt that it is more damaging than serviceable. One has only to consider the "two books" of Francis Bacon (1561–1626)—nature and Scripture—each of which had a role complementary to that of the other. They were held not to be at odds with each other because they dealt with different subjects. Again, for many scientific figures in the seventeenth and eighteenth centuries, Christianity played a central role in fostering and even shaping their scientific endeavors: The instances of Kepler, Robert Boyle (1627–91), Isaac Newton (1642–1727), and René Descartes (1596–1650) are the most conspicuous. The historical relations between religion and science are certainly more rich and complex than a simple conflict thesis suggests.

Second, and more specific, the conflict thesis ignores the many documented examples of science and religion operating in close alliance. This was most obviously true of the seventeenth and eighteenth centuries, as evidenced by the names of Boyle, Newton, Blaise Pascal (1623–62), Marin Mersenne (1588–1648), Pierre Gassendi (1592–1655), and Isaac Beeckman (1588–1637). Since then, a continuous history of noted individuals making strenuous efforts to integrate their science and religion has testified to the poverty of a conflict model. This was particularly true in Britain, where representatives in the nineteenth century included most famously Michael Faraday (1791–1867), James Joule (1818–89), James Clark Maxwell (1831–79), William Thomson (Lord Kelvin [1824–1907]), and George Gabriel Stokes (1819–1903). In the next century, a number of distinguished scientists of religious persuasion were ready to join societies like the Victoria Institute in London or its successors in Britain and the United States, which were dedicated to bringing together religious and scientific ideas. The English-speaking world was not unique in this quest for integration but has certainly been the most subject to historical scrutiny.

Third, the conflict thesis enshrines a flawed view of history in which "progress" or (in this case) "victory" has been portrayed as inevitable. There appears to be no inherent reason why this should be so, though it is readily understandable why some should wish it to be the case. This approach represents and embraces a long-demolished tradition of positivist, Whiggish historiography.

Fourth, the conflict thesis obscures the rich diversity of ideas in both science and religion. Neither of these has ever been monolithic, and there was seldom a unified reaction from either. Thus, in the case of Galileo, it was the Roman Catholic, not the Protestant, wing of Christianity that appeared to be at odds with science. In the Darwinian controversy, a uniform response was lacking even within one branch of Protestantism, for Anglicans of low-, high-, or broad-church persuasion tended to respond to Darwin's theories in different ways. Moreover, the scientific community was deeply divided over religion in Victorian England, the mathematical physicists being far more sympathetic than the scientific naturalists. The conflict thesis fails to recognize such variety.

Fifth, the conflict thesis engenders a distorted view of disputes resulting from causes other than those of religion versus science. Given this expectation, conflict is not difficult to find in every circumstance, whether or not justified by the available historical evidence. A classic case is that of the alleged opposition to James Young Simpson (1811–70) for his introduction of chloroform anesthesia in midwifery. Despite repeated claims of clerical harassment, the evidence is almost nonexistent. Insofar as there was any conflict, it was between the London and Edinburgh medical establishments or between obstetricians and surgeons. The origins of that myth may be located in an inadequately documented footnote in White (1896, 2.63).

Finally, the conflict thesis exalts minor squabbles, or even differences of opinion, to the status of major conflicts. The confrontation between Samuel Wilberforce (1805–73) and Huxley in 1860 has been so frequently paraded as a one-sided battle on a vast scale that one is liable to forget that, in fact, it was nothing of the kind. Such exaggeration is an almost inevitable accompaniment to the exposition of a conflict theory. It is excellent drama but impoverished history, made credible only by a prior belief that such conflict is inevitable. Of such material are legends made, and it has been well observed that "the dependence of the conflict thesis on legends that, on closer examination, prove misleading is a more general defect than isolated examples might suggest" (Brooke 1991, 40).

Reasons for Its Endurance

Given, then, that the warfare model is so inaccurate, one may wonder why it has lasted so long. This is, indeed, a major question for historians. The explanation may lie at least partly in the celebrated controversy of Huxley and his friends with the Anglican and Roman Catholic churches. In addition to the strategies mentioned above, they had another tactic, more subtle and yet more bold than anything else they accomplished. By establishing the conflict thesis, they could perpetuate a myth as part of their strategy to enhance the public appreciation of science. Thus, Huxley could write, with a fine disregard for what history records: "Extinguished theologians lie about the cradle of every science as the strangled snakes beside that of Hercules; and history records that wherever science and orthodoxy have been fairly opposed, the latter have been forced to retire from the lists, bleeding and crushed if not annihilated; scotched if not slain" (Moore 1979, 60). The Huxleyite warriors were outstandingly successful in this respect, and their ideals were enshrined in the works of Draper and White, best understood as polemical tracts that advanced the same cause. Draper takes such liberty with history, perpetuating legends as fact, that he is rightly avoided today in serious historical study. The same is nearly as true of White, though his prominent apparatus of prolific footnotes may create a misleading impression of meticulous scholarship. With an astonishing breadth of canvas, his writing exudes confidence in his thesis and conveys a sense of truly comprehensive analysis. Yet, with his personal polemic agenda, selectivity was inevitable. His selectivity exposed him to the criticism that he was trapped by his own presuppositions of an inherent antagonism between the theological and the scientific views of the universe. His book, which he commenced writing in the 1870s, is no longer regarded as even a reliable secondary source for historical study. It is, however, an accurate reflection of how certain liberal-minded men of his day perceived the relationship between religion and science and of how "history" (or a version of it) was pressed into service for their cause. The remarkable thing about the whole conflict thesis is how readily the Victorian propaganda in all of its varied forms has become unconsciously assimilated as part of the received wisdom of our own day. However, it is salutary to note that serious historical scholarship has revealed the conflict thesis as, at best, an oversimplification and, at worst, a deception. As a rare example of the interface between contemporary public opinion and historical scholarship, it is high time for a robust exposure of its true character.

BIBLIOGRAPHY

Brooke, John Hedley. *Science and Religion: Some Historical Perspectives.* Cambridge: Cambridge University Press, 1991.

Corsi, P. *Science and Religion: Baden Powell and the Anglican Debate, 1800–1860.* Cambridge: Cambridge University Press, 1988.

Draper, John William. *History of the Conflict between Religion and Science.* London, 1874.

Farr, A. D. "Religious Opposition to the Obstetric Anaesthesia: A Myth?" *Annals of Science* 40 (1983): 159–77.

Gilley, S., and A. Loades. "Thomas Henry Huxley: The War between Science and Religion." *Journal of Religion* 61 (1981): 285–308.

Heyck, T. W. *The Transformation of Intellectual Life in Victorian England.* London: Croom Helm, 1982.

Hooykaas, R. *Religion and the Rise of Modern Science.* Edinburgh: Scottish Academic Press, 1972.

———. *G. J. Rheticus' Treatise on Holy Scripture and the Motion of the Earth.* Amsterdam: North-Holland, 1984.

Jensen, J. V. "Return to the Wilberforce-Huxley Debate." *British Journal for the History of Science* 21 (1988): 161–79.

Lindberg, David C., and Ronald L. Numbers, eds. *God and Nature: Historical Essays on the Encounter between Christianity and Science.* Berkeley: University of California Press, 1986.

———. "Beyond War and Peace: A Reappraisal of the Encounter between Christianity and Science." *Perspectives on Science and Christian Faith* 39 (1987): 140–45

Livingstone, David N. *Darwin's Forgotten Defenders: The Encounter between Evangelical Theology and Evolutionary Thought.* Edinburgh: Scottish Academic Press, 1987.

Livingstone, David N., D. G. Hart, and Mark A. Noll. *Evangelicals and Science in Historical Perspective.* New York: Oxford University Press, 1999.

Lucas, J. R. "Wilberforce and Huxley: A Legendary Encounter." *Historical Journal* 22 (1979): 313–30

Moore, James R. *The Post-Darwinian Controversies: A Study of the Protestant Struggle to Come to Terms with Darwin in Great Britain and America, 1870–1900.* Cambridge: Cambridge University Press, 1979, 20–49.

Numbers, Ronald L. "Science and Religion." *Osiris* 2d ser. (1985): 58–80.

Rupke, N. A. *The Great Chain of History: William Buckland and the English School of Geology (1814–1849).* Oxford: Clarendon, 1983

Russell, C. A. "Some Approaches to the History of Science." Unit 1 of undergraduate course *AMST 283, Science and Belief: From Copernicus to Darwin.* Milton Keynes, U.K.: Open University Press, 1974, 30–49.

———. *Cross-Currents: Interactions between Science and Faith.* 1985. Reprint. London: Christian Impact, 1995.

———. "The Conflict Metaphor and Its Social Origins." *Science and Christian Belief* 1 (1989): 3–26.

———. "Objections to Anaesthesia: The Case of James Young Simpson." In *Gases in Medicine: Anaesthesia,* ed. E. B. Smith and S. Daniels. Cambridge: Royal Society of Chemistry, 1998, 173–87.

Russell, C. A., N. G. Coley, and G. K. Roberts. *Chemists by Profession.* Milton Keynes, U.K.: Royal Institute of Chemistry / Open University Press, 1977.

Simpson, James Y. *Landmarks in the Struggle between Science and Religion.* London: Hodder and Stoughton, 1925.

Turner, Frank M. "The Victorian Conflict between Science and Religion: A Professional Dimension." *Isis* 69 (1978): 356–76.

White, A. D. *A History of the Warfare of Science with Theology in Christendom.* 2 vols. New York: Appleton, 1896.

Wilson, David B. "Galileo's Religion versus the Church's Science? Rethinking the History of Science and Religion." *Physics in Perspective* 1 (1999): 65–84.

2 The Historiography of Science and Religion

David B. Wilson

A historiographical revolution in recent decades has resulted from studies that have argued that science and religion have related to each other in many ways, including various harmonious combinations. These studies have pointed out that the very meanings of the words *science* and *religion* have changed considerably over time. Historians now examine historical figures on their own terms rather than from the perspective of the present, eschewing approaches that press the past into the service of present concerns. Their goal is to deepen our comprehension of the discussions themselves instead of justifying particular views involved in current debates. As a result, a complexity thesis has replaced the traditional conflict thesis in historians' understanding of the relationship of science and religion.

David B. Wilson is professor of history, mechanical engineering, and philosophy at Iowa State University. He received his Ph.D. in the history of science from the Johns Hopkins University. His publications include *Kelvin and Stokes: A Comparative Study in Victorian Physics* (Bristol: Adam Hilger, 1987). His current research focuses on the relationship of physics, philosophy, and theology during the Scottish Enlightenment.

THE HISTORY of science and religion has been a contentious subject. In addition to the usual scholarly disputes present in any academic area, this historical subject has been enmeshed in more general historiographical debates and influenced by the religious or antireligious beliefs of some historians. After considering some basic issues, this essay discusses several works written during the previous century and a half, while focusing on the last fifty years. Recent decades have seen a radical shift in point of view among historians of science.

Although historians have espoused various approaches to the past, it will make our subject more manageable if we concentrate on the polar opposites around which views have tended to cluster. One approach has been to examine past ideas as much as possible in their own context, without either judging their long-term validity or making the discussion directly relevant to present issues. Another approach has been to study past ideas from the perspective of the present, taking full advantage of the hindsight provided by later knowledge to judge which ideas have proven to be valid. The second approach has apparent advantages. It does not exclude current knowledge that can assist us in the historical task. It also keeps present issues to the fore by insisting that historians draw lessons from the past that are relevant to current issues. However, historians have tended to regard the second approach as precariously likely to lead to distortion of the past in the service of present concerns. Dismissing this as "presentism," therefore, historians of science have come to favor the first, or contextualist, approach.

Whichever method historians use, they might reach one of several possible conclusions about the historical relationship between science and religion. Conflict, mutual support, and total separation are three obvious candidates. One of these models might long have predominated, or the relationship might have changed from time to time and place to place. The discovery of conflict might raise the further questions of which side emerged victorious and which side ought to have done so. The discovery of mutual support might lead to the question of whether either science or religion contributed to the other's continued validity or even to its origin.

The Conflict Thesis

The most prominent view among both historians and scientists in the twentieth century has been a presentist conflict thesis that argues as follows. To engage in the history of science, one must first know what science is. It is certainly not religion, and, indeed, it is quite separate from religion, as can clearly be seen in science as practiced in the modern world. The historian of science, then, should properly examine the internal development of the scientific ideas that made modern science possible (i.e., to the exclusion of such external factors as religion). The proponents of some ideas in the past were closer to the right track in this process than were others. Those who expanded the realm of religion too far were on the wrong track, so that religion improperly intruded on the realm of science. In such instances, conflict ensued between science and religion, with scientific advances eventually making the truth clear to all and invariably (and rightly) emerging victorious. The historical process need not have occurred in this way, but it so often did that conflict has been the primary relationship between science and religion. Science's best-known victories were those of Copernicanism and Darwinism. Presentism, internalism, and the conflict thesis coalesced into a de facto alliance, with the result that the conflict model is still widely accepted by academics (historians and scientists alike), though generally no longer by historians of science. A gulf in point of view thus marks the immediate setting of any scholarly treatment of the subject for a popular audience.

That this alliance was not a necessary one can be seen in the work of William Whewell (1794–1866), the most prominent historian of science during the first half of the nineteenth century. Known today primarily as a historian and philosopher of science, Whewell was, first of all, a mathematical physicist, but also an Anglican clergyman and a moral theorist. His philosophy of science featured a series of what he called "fundamental ideas" (like the idea of space) that, as part of man's mind created in the image of God, figured crucially in scientific knowledge of God's other creation, nature. Moral knowledge was structured similarly. Both moral and scientific knowledge were progressive. Scientists, for example, gradually became aware of the existence and implications of fundamental ideas. The study of history, that is, disclosed (a sometimes lurching) progress toward the present or, at any rate, Whewell's particular version of the present. Great scientists, such as Isaac Newton (1642–1727), were both intellectually strong and morally good.

Whewell did not think that conflict between science and religion had been especially significant historically, nor, indeed, was it in Whewell's own day. From his vantage point, he could give medieval science the uncomplimentary epithet "stationary" for several reasons that did not particularly include religious repression. The Roman Catholic Church had acted against Galileo (1564–1642), to be sure, but, for Whewell, that episode was an aberration. A tightly knit, biblical-historical-philosophical-moral-scientific-theological unity was manifested in Whewell's major, mutually reinforcing books: *History of the Inductive Sciences* (1837), *Foundation of Morals* (1837), *Philosophy of the Inductive Sciences* (1840), and *Elements of Morality* (1845).

John William Draper (1811–82), author of *History of the Conflict between Religion and Science* (1874), and Andrew Dickson White (1832–1918), author of *The Warfare of Science* (1876) and *A History of the Warfare of Science with Theology in Christendom* (1896), lived in an age that was different from Whewell's. While the Darwinian debates of the 1860s preceded Draper's book, what really alarmed him during that decade was the formulation of the doctrine of papal infallibility and the Roman Catholic Church's pronouncement that public institutions teaching science were not exempt from its authority. In his *History,* Draper depicted these developments as merely the latest phase in a long history of "the expansive force of the human intellect" in conflict not with religion generally, but with that "compression" inflicted by Catholicism. White developed and first published his views at about the same time as Draper. White's insights stemmed from his presidency of the new Cornell University, which was founded as a secular institution that stood in sharp contrast to the traditional religious sponsorship of colleges and universities. The withering criticism and innuendo directed at him personally by some religious figures led eventually to the writing of his books. Like those of Draper, White's books did not condemn all religion. They attacked what White called "that same old mistaken conception of rigid Scriptural interpretation" (White 1876, 75). White proclaimed that, whenever such religion sought to constrain science, science eventually won but with harm to both religion and science in the process. Science and "true religion," however, were not at odds.

Had Whewell still been alive, White and Draper might have told him how their circumstances had helped them improve on his writing of history. Unlike Whewell, they believed that they had stood in the shoes, as it were, of those who had been persecuted. White seemed especially to identify with Galileo. Their improved awareness had, they thought, enabled them to observe factors that he had overlooked. In any case, their books were highly influential. Moreover, it was not their whispered qualifications but their screaming titles that were to thunder through the decades, remaining audible more than a century later.

Differences of opinion did not seem to alter what was to become the widely current views of Draper and White. In *Metaphysical Foundations of Modern Physical Science* (1924), E. A. Burtt argued that the foundations of science were often theological. Galileo's God, for example, labored as a geometrician in creating the world, with the result that man, who knew some mathematics as well God did, was capable of grasping nature's essential mathematical logic. In *Science and the Modern World* (1926), Alfred North Whitehead maintained that the origin of modern science depended upon medieval theology, which had long insisted on God's rationality and hence also the rationality of his creation. Yet, in the 1930s, when his research suggested that seventeenth-century English Puritanism had fostered science, Robert K. Merton found that prevailing scholarly opinion, which had been shaped by the books of Draper and White, held that science and religion were inherently opposed and necessarily in conflict. Of course, the 1920s were the decade not only of Burtt and Whitehead, but also of the Scopes trial, which was generally interpreted as yet another in a long series of confrontations between science and religion. Also, during the 1920s and 1930s (and for some time afterward), the still undeveloped discipline of the history of science was pursued mainly by men trained in the sciences, who found presentist internalism a natural point of view.

Reaction to the Conflict Thesis

The Whig Interpretation of History (1931), written by the young general historian Herbert Butterfield, was eventually to influence the history of science deeply. Butterfield argued that historians had tended to be Protestant in religion and Whig in politics. They liked to divide the world into friends and enemies of progress—progress, that is, toward their own point of view. History was thus peopled by progressives and reactionaries, Whigs and Tories, Protestants and Catholics. Whig historians made the mistake of seeing Martin Luther, for example, as similar to modern Protestants rather than, as was actually the case, closer to sixteenth-century Catholics. By reading the present into the past in this way, Whig historians ratified the present, but only by misshaping the past. A better way was to assume that the sixteenth century was quite different from the twentieth and to explore the sixteenth century on its own terms, letting any similarities emerge from historical research rather than from prior assumption.

Butterfield's *Origins of Modern Science* (1949) applied this methodology to the history of science, including the relationship between science and religion during the scientific revolution. By not viewing scientists of the past as necessarily similar to modern scientists, it was possible to reach historical insights quite different from those of, say, Whewell or White. Overall, the scientific revolution resulted not from accumulating new observations or experimental results, but from looking at the same evidence in a new way: It was a "transposition" in the minds of the scientists. The alleged revolutionary Copernicus (1473–1543) could now be understood as a "conservative," much akin to the Greek astronomers with whom he disagreed. Religion was not necessarily either opposed to or separate from science in the modern sense but could, in principle, be viewed in any relationship, depending on the historical evidence.

Reading the evidence in a non-Whiggish way, Butterfield saw variety. There was, to be sure, theological opposition to the Copernican system, but it would not have been very important if there had not also been considerable scientific opposition. Even Galileo did not actually prove the earth's motion, and his favorite argument in support of it, that of the tides, was a "great mistake." Christianity favored the new mechanical worldview because it allowed a precise definition of miracles as events contrary to the usual mechanical regularity. Newton's gravitational theory required God's continued intervention in the universe he created, and one of Newton's possible explanations of gravity "made the existence of God logically necessary" (Butterfield 1949a, 157). Butterfield's *Christianity and History* (1949) made his own Christian faith explicit, but his religious views did not turn *Origins of Modern Science* into a Christian tract, though they guaranteed that Christian factors received a fair hearing.

Whatever the exact influence of Butterfield on them, three books published during the 1950s revealed the progress of non-Whiggish studies of science and religion during the scientific revolution. Alexandre Koyré, influenced by Burtt, had already published such studies as "Galileo and Plato" (1943) a few years before Butterfield's *Origins of Modern Science*. In *From the Closed World to the Infinite Universe* (1957), Koyré argued that the revolution involved philosophy and theology as well as science and that all three dimensions of thought usually existed in "the very same men," such as Johannes Kepler (1571–1630), René Descartes (1596–1650), Isaac Newton, and Gottfried Wilhelm Leibniz (1646–1716). Koyré thus portrayed the conflict between Newton and Leibniz, one that involved Leibniz's stiff opposition to Newton's gravitational theory, as primarily a theological conflict. He contrasted Newton's "work-day God" (who caringly involved himself in the operation of his universe) with Leibniz's "God of the sabbath" (who created the world skillfully enough for it to run by itself). In his *Copernican Revolution* (1957), Thomas Kuhn adopted the "unusual" approach of treating astronomers' philosophical and religious views as "equally fundamental" to their scientific ones. For the early Copernicans, at the center of the universe resided the sun, "the Neoplatonic symbol of the Deity" (Kuhn 1957, 231). Unlike Koyré's and Kuhn's books, Richard Westfall's *Science and Religion in Seventeenth-Century England* (1958) examined a variety of better- and lesser-known men of science (virtuosi) in a particular national context. In general, the virtuosi regarded their scientific discoveries as confirmation of their religious views, thus answering charges that studying nature both led man to value reason over revelation and made it difficult to know the nonmaterial side of existence. While there existed in the seventeenth century a multiplicity of ways to dovetail science and religion, there was a general movement from revealed religion to a natural theology that prepared the way for the deism of the next century.

The 1950s witnessed non-Whiggish studies of science and religion, not only in the century of Galileo and Newton but in Darwin's century, too. In his "second look" in *Isis* at Charles Gillispie's *Genesis and Geology* (1951), Nicolaas Rupke credited Gillispie with transforming the historiography of geology by going beyond the great ideas of great men as defined by modern geology to the actual religious-political-scientific context of British geology in the decades before Darwin's *Origin of Species* (1859). Explicitly rejecting the conflict thesis of Draper and White, Gillispie saw "the difficulty between science and Protestant Christianity . . . to be one of religion (in a crude sense) *in* science rather than one of religion *versus* science" (Gillispie 1951, ix). Writing about a period in which geologists were often themselves clergymen, Gillispie thought "that the issues discussed arose from a quasi-theological frame of mind within science" (Gillispie 1951, x). At the end of the decade, John Greene published *The Death of Adam* (1959), an examination of the shift from the "static creationism" of Newton's day to the evolutionary views of Darwin's. Without making any particular point of rejecting the Draper-White conflict thesis, Greene nevertheless did so implicitly, calling attention "to the religious aspect of scientific thought" (Greene 1959, vi) and infusing his book with examples of a variety of connections between religion and science. Thus, Georges Louis Leclerc, Comte de Buffon (1707–88), was forced to fit his science to the religious views of the day but found evolution contrary to Scripture, reason, and experience. William Whiston (1667–1752) employed science to explain scriptural events, rejecting alternative biblical views that were either too literal or too allegorical. Charles Darwin (1809–82) jousted with fellow scientists Charles Lyell (1797–1875) and Asa Gray (1810–88) about the sufficiency of natural selection as opposed to God's guidance and design in evolutionary processes.

Christian Foundations of Modern Science

If these notable books of the 1950s rejected the conflict thesis in various ways, two books from the early 1970s went even further, turning the thesis on its head to declare (echoing Whitehead) that Christianity had made science possible. The first was Reijer Hooykaas's *Religion and the Rise of Modern Science*

(1972). The Protestant historian Hooykaas (1906–94) had explored the relations between science and religion for several years. His *Natural Law and Divine Miracle* (1959), for example, showed the compatibility of what he called "a Biblical concept of nature" with nineteenth-century biology and geology. In 1972, he went further by arguing for a Christian, especially Calvinistic, origin of science itself. After discussing Greek concepts of nature, Hooykaas concluded that, in the Bible, "in total contradiction to pagan religion, nature is not a deity to be feared and worshipped, but a work of God to be admired, studied and managed" (Hooykaas 1972, 9). Not only did the Bible "de-deify" nature, Calvinism encouraged science through such principles as voluntaristic theology, a "positive appreciation" of manual work, and an "accommodation" theory of the Bible. Voluntarism emphasized that God could choose to create nature in any way he wanted and that man, therefore, had to experience nature to discover God's choice. This stimulus to experimental science was reinforced by the high value that Christianity placed on manual labor. The view that, in biblical revelation, God had accommodated himself to ordinary human understanding in matters of science meant that Calvinists generally did not employ biblical literalism to reject scientific findings, particularly Copernican astronomy.

Stanley L. Jaki's *Science and Creation* (1974) also expanded themes that were present in his earlier chapter "Physics and Theology" in his *Relevance of Physics* (1966). Jaki was a Benedictine priest with doctorates in both theology and physics. His *Science and Creation,* a book of breathtaking scope, examined several non-Western cultures before focusing on the origin of science within the Judeo-Christian framework. Jaki argued that two barriers to science pervaded other cultures: a cyclic view of history and an organic view of nature. Endless cycles of human history made men too apathetic to study nature. Even when they did, their concept of a living, willful nature precluded discovery of those unvarying patterns that science labels natural laws. The Judeo-Christian view, in contrast, historically regarded nature as the nonliving creation of a rational God, not cyclic but with a definite beginning and end. In this conceptual context (and only in this context), modern science emerged, from the thirteenth through the seventeenth centuries. Earlier adumbrations of science were pale, short-lived imitations, doomed by hostile environments. Unfortunately, Jaki thought, amidst attacks on Christianity in the twentieth century, there had arisen the theory of an oscillating universe, which was another unwarranted, unscientific, cyclic view of nature. Hence, consideration of both past and present disclosed the same truth: "the indispensability of a firm faith in the only lasting source of rationality and confidence, the Maker of heaven and earth, of all things visible and invisible" (Jaki 1974, 357).

The Continuing Influence of the Conflict Thesis

Despite the growing number of scholarly modifications and rejections of the conflict model from the 1950s on, the Draper-White thesis proved to be tenacious, though it is probably true that it had been more successfully dispelled for the seventeenth century than for the nineteenth. At any rate, in the 1970s leading historians of the nineteenth century still felt required to attack it. In the second volume of his *Victorian Church* (1970), Owen Chadwick viewed the conflict thesis as a misconception that many Victorians had about themselves. His *Secularization of the European Mind* (1975) presented Draper's antithesis as the view to attack by way of explaining one aspect of nineteenth-century secularization. Writing about Charles Lyell in 1975, Martin Rudwick also deplored distortions produced by Draper and White, arguing that abandoning their outdated historiography would solve puzzles surrounding Lyell's time at King's College, London. Examining nineteenth-century European thought in *History, Man, and Reason* (1971), the philosopher-historian Maurice Mandelbaum rejected what he called "the conventional view of the place of religion in the thought of the nineteenth century," which "holds that science and religion were ranged in open hostility, and that unremitting warfare was conducted between them" (Mandelbaum 1971, 28).

Why did these historians believe that the conflict thesis was sufficiently alive and well to require refutation? For one thing, even those historians who were most significant in undermining the conflict thesis did not reject it entirely. Moreover, they made statements that could be construed as more supportive of the thesis than perhaps they intended. "Conflict with science" was the only subheading under "Religion" in the index to Gillispie's *The Edge of Objectivity* (1960), and it directed the reader to statements that seemed to support the conflict model. What geology in the 1830s "needed to become a science was to retrieve its soul from the grasp of theology" (Gillispie 1960, 299). "There was never a more unnecessary battle than that between science and theology in the nineteenth century" (Gillispie 1960, 347). Even Gillispie's *Genesis and Geology* was criticized by Rudwick in 1975 as only a more sophisticated variety of the "positivist" historiography of Draper and White. Westfall, in a preface to the

1973 paperback edition of his book, wrote: "In 1600, Western civilization found its focus in the Christian religion; by 1700, modern natural science had displaced religion from its central position" (Westfall 1973, ix). Greene introduced the subjects of the four chapters in his *Darwin and the Modern World View* (1961) as four stages in "the modern conflict between science and religion" (Greene 1961, 12). Surely, the most widely known book written by a historian of science, Kuhn's *Structure of Scientific Revolutions* (1962), excluded those philosophical and religious views that Kuhn had earlier (in his *Copernican Revolution* [1957]) labeled "equally fundamental" aspects of astronomy. This exclusion undoubtedly aided the view that a conflict existed, a view that was the ally of internalism. The 1970s were a period in which past scientists' religious statements could still be dismissed as "ornamental or ceremonial flourishes" or as "political gestures." The "orthodoxy" of internalism among historians of science in the 1960s and early 1970s was the target of the fascinating autobiographical account of life as a student and teacher at Cambridge University by Robert Young in his contribution to *Changing Perspectives in the History of Science* (1973). And even Young, whose own pathbreaking nonconflictive articles from around 1970 were later reprinted in *Darwin's Metaphor* (1985), wrote in his 1973 piece that "the famous controversy in the nineteenth century between science and theology was very heated indeed" (Young 1973, 376).

A second factor was the prevailing view among scientists themselves, which influenced historians of science, who either had their own early training in science or maintained regular contact with scientists, or both. In this regard, we might consider the work of the scientist-historian Stephen Jay Gould, one of the most successful popularizers of both science and the history of science. A collection of his popular essays appeared in 1977 as *Ever since Darwin.* Gould stoutly rejected the "simplistic but common view of the relationship between science and religion—they are natural antagonists" (Gould 1977, 141). However, the book's specific instances came preponderantly from the conflict theorist's familiar bag of examples: the church's disagreeing with Galileo; T. H. Huxley's "creaming" Bishop "Soapy Sam" Wilberforce; natural selection's displacing of divine creation; and, as Freud said, man's losing his status as a divinely created rational being at the center of the universe because of the science of Copernicus, Darwin, and Freud himself. Gould's most sympathetic chapter was his discussion of Thomas Burnet's late-seventeenth-century geological explanations of biblical events like Noah's flood. Even here, however, Gould regarded the views of Burnet's opponents as dogmatic and antirationalist, reflecting the same unhappy spirit that, wrote Gould, later possessed Samuel Wilberforce, William Jennings Bryan, and modern creationists. "The Yahoos never rest" (Gould 1977, 146).

Whatever the reasons for the continued survival of the conflict thesis, two other books on the nineteenth century that were published in the 1970s hastened its final demise among historians of science. In 1974, Frank Turner carved out new conceptual territory in *Between Science and Religion*. He studied six later Victorians (including Alfred Russel Wallace, the co-inventor of the theory of evolution by natural selection) who rejected both Christianity and the agnostic "scientific naturalism" of the time. In their various ways, they used different methods, including the empiricism of science (but not the Bible), to support two traditionally religious ideas: the existence of a God and the reality of human immortality. Even more decisive was the penetrating critique "Historians and Historiography" that James Moore placed at the beginning of his *Post Darwinian Controversies* (1979). In what would have been a small book in itself, Moore's analysis adroitly explored the historical origins of Draper and White's "military metaphor" and went on to show how the metaphor promulgated false dichotomies: between science and religion, between scientists and theologians, between scientific and religious institutions. The metaphor simply could not handle, for example, a case of two scientist-clergymen who disagreed about a scientific conclusion partly because of their religious differences. Finally, Moore called for historians to write "non-violent" history, of which the remainder of his book was a prodigious example. Examining Protestant responses to Darwin's ideas, he concluded that it was an "orthodox" version of Protestantism that "came to terms" with Darwin more easily than did either a more liberal or a more conservative version and, in addition, that much anguish would have been spared had this orthodoxy prevailed.

The Complexity Thesis

By the 1980s and 1990s, there had been nearly a complete revolution in historical methodology and interpretation. Setting aside his own views of science and religion, the historian was expected to write non-Whiggish history to avoid what Maurice Mandelbaum called the "retrospective fallacy." This fallacy consisted of holding an asymmetrical view of the past and the future, in which the past was seen as like a solid, with all of its parts irrevocably fixed in place, while the future was viewed as fluid, unformed, and unforeseen. The problem for the historian was to transpose his mind to such an extent that a historical figure's future (which was part of the historian's own past) lost the fixity and inevitability that the historian perceived in it and, instead, took on the uncertainty that it had for the historical figure. The concern for what led to the present and the extent to which it was right or wrong by present standards thus dissipated. A good test for the historian was whether he could write a wholly sympathetic account of a historical figure with whom he totally disagreed or whose ideas he found repugnant. Would the historical figure, if by some magic given the chance to read the historian's reconstruction, say that, indeed, it explained what he thought and his reasons for doing so? To be valid, any broader historical generalization had to be based on specific, non-Whiggish studies that accurately represented past thought.

This radically different methodology yielded a very different overall conclusion about the historical relationship of science and religion. If "conflict" expressed the gist of an earlier view, "complexity" embodied that of the new. The new approach exposed internalism as incomplete and conflict as distortion. Past thought turned out to be terribly complex, manifesting numerous combinations of scientific and religious ideas, which, to be fully understood, often required delineation of their social and political settings.

From this mainstream perspective, moreover, historians could deem other approaches unacceptable. Zeal for the triumph of either science or religion in the present could lure historians into Whiggish history. The works not only of Draper and White, but also of Hooykaas and Jaki fell into that category. Kenneth Thibodeau's review in *Isis* of Jaki's *Science and Creation,* for example, declared it "a lopsided picture of the history of science" that "minimizes" the accomplishments of non-Christian cultures and "exaggerates" those of Christian ones (Thibodeau 1976, 112). In a review in *Archives Internationale d'Histoire des Sciences,* William Wallace found Hooykaas's *Religion and the Rise of Modern Science* to be "a case of special pleading." In their historiographical introduction to the book they edited, *God and Nature* (1986), David Lindberg and Ronald Numbers judged that Hooykaas and Jaki had "sacrificed careful history for scarcely concealed apologetics" (Lindberg and Numbers 1986, 5). Likewise, some historians found Moore's nonviolent history unacceptable: He "sometimes seems to be writing like an apologist for some view of Christianity" (La Vergata 1985, 950), criticized Antonella La Vergata in his contribution to *The Darwinian Heritage* (1985).

Among the multitude of articles and books that argued for a relatively new, non-Whiggish complexity thesis, two exemplars were Lindberg and Numbers's *God and Nature* and John Brooke's *Science and Religion* (1991). Though similar in outlook, they differed in format. The first was a collection of eighteen studies by leading scholars in their own areas of specialty, while the second was a single scholar's synthesis of a staggering amount of scholarship, an appreciable portion of which was his own specialized research.

Turning in their introduction to the contents of their own volume, Lindberg and Numbers rightly observed that "almost every chapter portrays a complex and diverse interaction that defies reduction to simple 'conflict' or 'harmony'" (Lindberg and Numbers 1986, 10). Medieval science, for example, was a "handmaiden" to theology (but not suppressed), while the close interlocking of science and religion that developed by the seventeenth century began to unravel in the eighteenth. To examine briefly the complexity of only one chapter, consider James Moore's (nonapologetic) discussion of "Geologists and Interpreters of Genesis in the Nineteenth Century." Moore focused on British intellectual debates occurring in a variegated context of geographical, social, generational, institutional, and professional differences. Around 1830, professional geologists (i.e., those with specialist expertise) tended to "harmonize" Genesis and geology by using geology to explain the sense in which the natural history of Genesis was true. They were opposed by nonprofessional "Scriptural geologists," who used Genesis to determine geological truths. By the 1860s, a new generation of professional geologists did their geology independently of Genesis. They were in agreement with a new generation of professional biblical scholars in Britain, who believed that Genesis and geology should be understood separately. Meanwhile, the earlier conflicting traditions of harmonization and scriptural geology were kept going by amateurs. Hence, while debate over how to meld Genesis and geology was a social reality in late-Victorian Britain, it did not perturb the elite level of the professionals. Numbers expanded his own chapter in *God and Nature* (1986) into *The Creationists* (1992), an outstanding treatment of such issues at the nonelite level in the twentieth century.

Brooke's volume targeted general readers in a way that Lindberg and Numbers's did not. In his historiographical remarks, Brooke considered the very meanings of the words *science* and *religion*, resisting specific definitions for them. The problem, Brooke explained, was that the words had so many meanings. It could even be misleading to refer to Isaac Newton's *science*, when Newton called what he was doing "natural philosophy," a phrase connoting quite different issues in the seventeenth century than did "science" in the twentieth.

As did Lindberg and Numbers, Brooke found complexity: "The principal aim of this book," he wrote, "has been to reveal something of the complexity of the relationship between science and religion as they have interacted in the past" (Brooke 1991, 321). As for Lindberg and Numbers, so also for Brooke, complexity did not preclude general theses. He concluded, for example, that science went from being "subordinate" to religion in the Middle Ages to a position of relative equality in the seventeenth century, not separate from religion but "differentiated" from it.

Conclusion

This essay, in rejecting presentist histories of science and religion, may itself seem somewhat presentist. Though it tries fairly to present the opposite point of view, it favors the recent historiographical revolution in advocating a contextualist approach, with all its attendant complexities. Though the new point of view has decided advantages over the old, it has the potential of leading historians astray. Pursuit of complexity could produce ever narrower studies that are void of generalization. Moreover, awareness of the great variation of views in different times and places could lead to the mistaken conclusion that those ideas were nothing but reflections of their own "cultures." Instead, in thinking about science and religion, as in most human endeavors, there have always been the relatively few who have done their work better than the rest. Existence of differences among them does not mean that they have not thought through and justified their own positions. In fact, that they have done so is an example of a contextualist generalization—one that is not only in harmony with the evidence of the past but also relevant to present discussions.

Indeed, the whole non-Whiggish enterprise might inform the present in other ways, too, though scholars are understandably wary of drawing very specific lessons from history for the present. Consider, however, a few general points. Study of past ideas on their own terms might provide a kind of practice for working out one's own ideas or for nourishing tolerance for the ideas of others. There have been and, no doubt, always will be disagreements among our strongest thinkers, as well as questions of the relationship between their ideas and those of the population at large. Moreover, things always change, though not predictably or necessarily completely. Indeed, the most influential thinkers seem fated to have followers who disagree with them even while invoking their names. Even the most well-founded, well-argued, and well-intentioned ideas about science and religion are liable to later change or eventual rejection. The same is true for historiographical positions, including, of course, the complexity thesis itself.

BIBLIOGRAPHY

Bowler, Peter. *Reconciling Science and Religion: The Debates in Early Twentieth-Century Britain*. Chicago: University of Chicago Press, 2001.

Brooke, John Hedley. *Science and Religion: Some Historical Perspectives*. Cambridge: Cambridge University Press, 1991.

Brooke, John Hedley, and Geoffrey Cantor. *Reconstructing Nature: The Engagement of Science and Religion*. Oxford: Oxford University Press, 1998.

Brooke, John Hedley, Margaret J. Osler, and Jitse M. van der Meer, eds. *Science in Theistic Contexts: Cognitive Dimensions*. *Osiris* 16 (2001).

Butterfield, Herbert. *The Whig Interpretation of History*. 1931. Reprint. New York: Norton, 1965.

———. *Origins of Modern Science, 1300–1800*. 1949a. New edition. London: G. Bell and Sons, 1962.

———. *Christianity and History*. 1949b. London: G. Bell and Sons, 1950.

Daston, Lorraine. "A Second Look. History of Science in an Elegiac Mode: E. A. Burtt's *Metaphysical Foundations of Modern Physical Science* Revisited." *Isis* 82 (1991): 522–31.

Draper, John William. *History of the Conflict between Religion and Science*. 1874. Reprint. New York: Appleton, 1928.

Fisch, Menachem, and Simon Schaffer, eds. *William Whewell: A Composite Portrait*. Oxford: Clarendon, 1991.

Gillispie, Charles Coulton. *Genesis and Geology: A Study in the Relations of Scientific Thought, Natural Theology, and Social Opinion in Great Britain, 1790–1850*. 1951. Reprint. New York: Harper Torchbooks, 1959.

———. *The Edge of Objectivity: An Essay in the History of Scientific Ideas*. Princeton: Princeton University Press, 1960.

Gould, Stephen Jay. *Ever since Darwin: Reflections in Natural History*. New York: Norton, 1977.

Greene, John C. *The Death of Adam: Evolution and Its Impact on Western Thought*. Ames: Iowa State University Press, 1959.

———. *Darwin and the Modern World View*. 1961. Reprint. New York: New American Library, 1963.

Hooykaas, R. *Natural Law and Divine Miracle: The Principle of Uniformity in Geology, Biology, and Theology*. 1959. 2d imp. Leiden: Brill, 1963.

———. *Religion and the Rise of Modern Science*. 1972. Reprint. Edinburgh: Scottish Academic Press, 1973.

Howell, Kenneth J. *God's Two Books: Copernican Cosmology and Biblical Interpretation in Early Modern Science*. Notre Dame: University of Notre Dame Press, 2001.

Jaki, Stanley L. *The Relevance of Physics*. Chicago: University of Chicago Press, 1966.

———. *Science and Creation: From Eternal Cycles to an Oscillating Universe*. Edinburgh: Scottish Academic Press, 1974.

Koyré, Alexandre. "Galileo and Plato." *Journal of the History of Ideas* 4 (1943): 400–428.

———. *From the Closed World to the Infinite Universe*. Baltimore: Johns Hopkins Press, 1957.

Kuhn, Thomas S. *The Copernican Revolution: Planetary Astronomy in the Development of Western Thought*. New York: Vintage, 1957.

———. *The Structure of Scientific Revolutions*. 1962. 2d ed. Chicago: University of Chicago Press, 1970.

La Vergata, Antonello. "Images of Darwin: A Historiographic Overview." In *The Darwinian Heritage*, ed. David Kohn, with bibliographical assistance from Malcolm J. Kottler. Princeton: Princeton University Press, 1985, 901–72.

Lindberg, David C., and Ronald L. Numbers, eds. *God and Nature: Historical Essays on the Encounter between Christianity and Science*. Berkeley: University of California Press, 1986.

Mandelbaum, Maurice. *History, Man, and Reason: A Study in Nineteenth-Century Thought*. Baltimore: Johns Hopkins Press, 1971.

Merton, Robert K. *Science, Technology, and Society in Seventeenth Century England*. 1938. Reprint. New York: Harper and Row, 1970.

Moore, James R. *The Post-Darwinian Controversies: A Study of the Protestant Struggle to Come to Terms with Darwin in Great Britain and America, 1870–1900*. Cambridge: Cambridge University Press, 1979.

Numbers, Ronald L. *The Creationists: The Evolution of Scientific Creationism*. New York: Knopf, 1992.

Rudwick, Martin. "The Principle of Uniformity." Review of R. Hooykaas, *Natural Law and Divine Miracle*. *History of Science* 1 (1962): 82–86.

———. "Charles Lyell, F.R.S. (1797–1875) and His London Lectures on Geology, 1832–1833." *Notes and Records of the Royal Society of London* 29 (1975): 231–63.

Rupke, Nicolaas A. "A Second Look: C. C. Gillispie's *Genesis and Geology*." *Isis* 85 (1994): 261–70.

Thibodeau, Kenneth F. Review of *Science and Creation*, by Stanley L. Jaki. *Isis* 67 (1976): 112.

Turner, Frank Miller. *Between Science and Religion: The Reaction to Scientific Naturalism in Late Victorian England*. New Haven: Yale University Press, 1974.

Wallace, William A. Review of R. Hooykaas, *Religion and the Rise of Modern Science*, and J. Waardenburg, *Classical Approaches to the Study of Religion*. *Archives Internationale d'Histoire des Sciences* 25 (1975): 154–56.

Westfall, Richard S. *Science and Religion in Seventeenth-Century England*. 1958. Reprint. Ann Arbor: University of Michigan Press, 1973.

Westman, Robert S. "A Second Look: Two Cultures or One? A Second Look at Kuhn's *The Copernican Revolution*." *Isis* 85 (1994): 79–115.

White, Andrew Dickson. *The Warfare of Science*. New York: Appleton, 1876.

———. *A History of the Warfare of Science with Theology in Christendom*. 2 vols. New York: Appleton, 1896.

Wilson, David B. "On the Importance of Eliminating *Science* and *Religion* from the History of Science and Religion: The Cases of Oliver Lodge, J. H. Jeans, and A. S. Eddington." In *Facets of Faith and Science*, ed. Jitse M. van der Meer. Vol. 1: *Historiography and Modes of Interaction*. Lanham, Md.: Pascal Centre for Advanced Studies in Faith and Science / University Press of America, 1996, 27–47.

———. "Galileo's Religion versus the Church's Science? Rethinking the History of Science and Religion." *Physics in Perspective* 1 (1999): 65–84.

Young, Robert M. "The Historiographic and Ideological Contexts of the Nineteenth-Century Debate on Man's Place in Nature." In *Changing Perspectives in the History of Science: Essays in Honour of Joseph Needham*, ed. Mikulas Teich and Robert M. Young. London: Heinemann, 1973, 344–438. Reprinted in Robert M. Young, *Darwin's Metaphor: Nature's Place in Victorian Culture*. Cambridge: Cambridge University Press, 1985, 164–247.

———. *Darwin's Metaphor: Nature's Place in Victorian Culture*. Cambridge: Cambridge University Press, 1985.

Part II : : The Premodern Period

3 Aristotle and Aristotelianism

Edward Grant

Aristotle's works influenced three major civilizations—Greek, Islamic, and Latin—and were a significant factor for approximately two thousand years. Aristotle's greatest impact, however, was on Western Europe from about 1200 to 1650. Contrary to common misperceptions, Europe in this period was a thriving region. It had survived the barbarian invasions from the seventh to the early eleventh centuries and then, under relatively peaceful conditions, made dramatic advances in agriculture, technology, commerce, manufacturing, and political organization. Europe underwent rapid urbanization, and the new cities became a powerful political force. Within this urban setting a new educational system emerged, first in the form of cathedral schools and, by 1300, in the mode of universities. Because of their emphasis on logic and reason and the wide range of topics they covered, Aristotle's newly translated works were viewed as ideal for the course needs of the new universities. The faculty of arts became wholly dependent on Aristotle's works, while the faculties of law, medicine, and theology also used them extensively.

Edward Grant is distinguished professor emeritus of history and philosophy of science and professor emeritus of history at Indiana University, Bloomington. His recent books include *Planets, Stars, and Orbs: The Medieval Cosmos, 1200–1687* (Cambridge: Cambridge University Press, 1994); *The Foundations of Modern Science in the Middle Ages: Their Religious, Institutional, and Intellectual Contexts* (Cambridge: Cambridge University Press, 1996); and *God and Reason in the Middle Ages* (Cambridge: Cambridge University Press, 2001).

NO ONE in the history of civilization has shaped our understanding of science and natural philosophy more than the great Greek philosopher and scientist Aristotle (384–322 B.C.), who exerted a profound and pervasive influence for more than two thousand years, extending from the fourth century B.C. to the end of the seventeenth century A.D. During this long period, Aristotle's numerous Greek treatises were translated into a variety of languages, most notably Arabic and Latin. He was, thus, a dominant intellectual force in at least three great civilizations that ranged over a vast geographical area, embracing sequentially the Byzantine Empire (which succeeded the Roman Empire in the east), the civilization of Islam, and the Latin Christian civilization of western Europe in the late Middle Ages.

Aristotle's dazzling success is not difficult to understand. Early on, and before anyone else, he left treatises on a breathtaking variety of topics that included logic, natural philosophy, metaphysics, biology, ethics, psychology, politics, poetics, rhetoric, and economics (or household management). Because they seemed to embrace almost all knowledge worth having, Aristotle's works could readily serve as guides to an understanding of the structure and operation of the physical universe, as well as to human and animal behavior. Aristotle's collected works bulked so large in history because relatively little survived intact from the works of his predecessors on the subjects about which he wrote. In some of those subjects, logic and biology, for example, there is reason to believe that he may have written the first comprehensive treatises and been the first to define those disciplines. Moreover, until the sixteenth and seventeenth centuries, Aristotle's interpretation of the world had few significant rivals. He was usually regarded as the preeminent guide for understanding the material and immaterial worlds.

Aristotle covered such a wide variety of learning that many subsequent scholars found it convenient to present their own ideas on those subjects by way of commentaries on one or more of his works. Over the centuries, many such commentaries were written, primarily in the Greek, Arabic, and Latin languages. Taken collectively, they form the phenomenon called Aristotelianism, although that phenomenon is much broader because Aristotle's ideas were also injected into other disciplines, most notably medicine and theology.

Although Aristotle's natural philosophy and metaphysics reveal an interest in the divine, they were written without regard for any of the religious concerns that subsequently proved critical to the civilizations of Christianity and Islam, in which Aristotle's philosophy was a major factor. Aristotle's treatises on natural philosophy and metaphysics form the basis of his interpretation of the material and immaterial entities that make up our world. His natural philosophy essentially comprises five treatises: the *Physics,* which treats of motion, matter and form, place, vacuum, time, the infinite, and the Prime Mover; *On the Heavens (De caelo),* which is devoted largely to cosmology; *On Generation and Corruption,* which is concerned with elements, compounds, and material change generally; *Meteorology*, which describes phenomena in the upper atmosphere just below the moon; and *On the Soul (De anima)*, in which Aristotle treats different levels of soul and discusses perception and the senses. Aristotle's metaphysics, or "first philosophy," or "theology," as it was sometimes called, is embodied in his work titled *Metaphysics*. In this basic work, Aristotle analyzes the nature of immaterial being wholly divorced from matter. Despite its problematic nature, Christian theologians found the *Metaphysics* an invaluable resource for confronting difficult problems about God's nature and existence.

Aristotle's Theology

Aristotle's metaphysics and natural philosophy had a profound, and often disquieting, effect on the theologians and religious guardians of the monotheistic religions of Christianity, Islam, and Judaism. Various elements in Aristotle's philosophy were relevant to theology, most notably his conviction that the world is eternal: that it had no beginning and would never have an end. Aristotle could find no convincing argument for supposing that our world could have come into being naturally from any prior state of material existence. For if the world came from a previously existing material thing, say B, we would then have to inquire from whence did B come, and so on through an infinite regression, since it was assumed that the world could not have come from nothing. To avoid this dilemma, Aristotle concluded that the world had no beginning and, therefore, that it could have no end, for if it could end, it could, necessarily, have had a beginning.

Despite his conviction that the world was uncreated, Aristotle did believe in a divine spirit, or God. But the attributes he assigned to his God, whom he called an "Unmoved Mover," would have been strange, and perhaps repugnant, to anyone raised in one of the three traditional monotheistic religions. Obviously, Aristotle's God was not the creator of our world, since it is uncreated. Indeed, he is not even aware of the world's existence and, therefore, does not, and could not, concern himself with anything in our world. Such a deity could not, therefore, be an object of worship. The only activity fit for such a God is pure thought. But the only thoughts worthy of his exalted status are thoughts about himself. Totally remote from the universe, Aristotle's God thinks only about himself.

Despite his total isolation from the world, the God of Aristotle unknowingly exerted a profound influence on it. He was its "Unmoved Mover," causing the orbs and the heavens to move around with eternal circular motions. The celestial orbs move around eternally because of their love for the Unmoved Mover. By virtue of these incessant motions, the celestial orbs cause all other motions in the world. Thus was Aristotle's Unmoved Mover, or God, the final cause of all cosmic motions.

Aristotle's view of the human soul also proved problematic. He regarded the soul as a principle of life that was inseparable from its body. Aristotle distinguished three levels of soul: (1) the nutritive, or vegetative, soul, which is found in both plants and animals and is solely concerned with the nutrition essential for the sustenance of the organism's life; (2) the sensitive soul, which is possessed only by animals and oversees motion, desire, and sense perception; and (3) the rational soul, which is found only in humans and subsumes the two lower levels of soul to form a single, unified soul in each human being. Each human soul contains an active and a passive intellect. Our thoughts are formed from images abstracted by the active intellect and implanted in the passive intellect as concepts. Except for the active intellect, which is immortal, the soul perishes with the body. Whether Aristotle regarded an individual's active intellect as personally immortal or whether he thought it loses its individuality when it rejoins the universal active intellect is unclear. Those who wished to "save" Aristotle and reconcile his view with the Christian conception of the soul opted for the first alternative, even though such an interpretation involved an elastic view of Christian doctrine.

Aristotle's strong sense of what was possible and impossible in natural philosophy posed serious problems for his Christian, Muslim, and Jewish followers. On a number of vital themes, he presented demonstrations to show that nature was necessarily constrained to operate in one particular way rather than another way. The question confronting Aristotle's followers, then, was whether God could have created our world to operate in ways that Aristotle regarded as impossible.

Later Antiquity, Byzantium, and Islam

Such difficulties were of little concern to the earliest commentators, who were pagans like Aristotle himself. Not until members of the great monotheistic religions began to comment on the works of Aristotle did problems arise. Although the names of the earliest commentators are unknown, commentaries on Aristotle's works were probably written in the Hellenistic period (323–30 B.C.), shortly after his death, and they continued on through the duration of the Roman Empire (30 B.C.–A.D. 476). The historical emergence of the Aristotelian commentary tradition took place in the Greek-speaking area that would become the Byzantine Empire in the eastern Mediterranean world. Here, beginning in the third century A.D., a group of commentators writing in Greek began the historical development of Aristotelianism (the tradition of commenting on the works of Aristotle). The most prominent of these were Alexander of Aphrodisias (fl. second or third century), Themistius (317–c. 388), Simplicius (d. 540), and, especially, the Christian Neoplatonic author John Philoponus (d. c. 570), who rejected many of Aristotle's basic concepts about the nature of the world. After the translation of some of their works into Arabic, these Greek commentators exercised a significant influence on Islam.

Because of religious hostility to Aristotle's natural philosophy, the number of Aristotelian commentators in Islam was not large. The most important of them—al-Kindi (800–70), al-Farabi (873–950), Avicenna (Ibn Sina [980–1037]), and Averroës (Ibn Rushd [1126–98])—were translated into Latin and exerted a major influence in the European Middle Ages—in some instances, playing a greater role in Christendom than in Islam. Three of the most important charges against Aristotle's natural philosophy were (1) his advocacy of the eternity of the world, (2) the conviction that his natural philosophy was hostile to the basic Muslim belief in the resurrection of the body, and (3) his concept of secondary causation. In Islamic thought, the term *philosopher* (*faylasuf*) was often reserved for those who assumed, with Aristotle, that natural things were capable of causing effects, as when a magnet attracts iron and causes it to move or when a horse pulls a wagon and is seen as the direct cause of the wagon's motion. On this approach, God was not viewed as the immediate cause of every effect. Philosophers believed, with Aristotle, that natural objects could cause effects in other natural objects because things had natures that enabled them to act on other things and to be acted upon. By contrast, most Muslim theologians believed, on the basis of the Koran, that God caused everything directly and immediately and that natural things were incapable of acting directly on other natural things. Although secondary causation was usually assumed in scientific research, most Muslim theologians opposed it.

The European Middle Ages

Aristotle's ideas were destined to play a monumental role in western Europe. In the twelfth and thirteenth centuries, his works, along with much of Islamic science and natural philosophy, were translated from Arabic into Latin. The Arabic commentaries on Aristotle spoke favorably of the philosopher and, once translated, were often used in Europe as guides to his thought. Indeed, Averroës, who was probably the greatest of all Aristotelian commentators, was known to all simply as "the Commentator."

Until the thirteenth century, Aristotelian commentators were a disparate group scattered in time and place. All of this changed, however, with the emergence of universities around 1200. Aristotle's thought achieved a widespread prominence as his logic, natural philosophy, and metaphysics became the basis of the curriculum leading to the baccalaureate and master of arts degrees. As a result, universities in western Europe—and by 1500 there were approximately sixty of them in Europe, extending as far east as Poland—became the institutional base for Aristotelianism. A relatively large class of professional teachers developed who were specialists in Aristotelian thought, and a much larger class of nonteaching scholars emerged who had studied Aristotelian natural philosophy and metaphysics in depth. Nothing like this had ever been seen before. For the first time in history, natural philosophy, the exact sciences (primarily geometry, arithmetic, astronomy, and optics), and medicine were permanently rooted in an institution, the university, that has endured for approximately eight hundred years and has been established worldwide.

By the end of the thirteenth century, the method of teaching both theology and Aristotle's natural philosophy, which significantly influenced theology, was to proceed by way of a series of questions. Indeed, the very titles of many of the treatises indicate their pedagogical method. For example, in his *Questions on Aristotle's Book on the Heavens,* John Buridan (c. 1295–c. 1358), perhaps the greatest arts master of the Middle Ages, proposed and responded to the following questions: "whether there are several worlds"; "whether the sky is always moved regularly"; "whether the stars are self-moved or moved by the motion of their spheres"; "whether the earth always rests in the middle [or center] of the world"; and many others about the terrestrial and celestial regions. Similar questions were posed in various treatises on Aristotle's other works. Peter Lombard (c. 1100–c. 1160), for example, employed the questions format in his *Sentences.* Composed in the 1140s, *Sentences* was the great theological textbook of the late Middle Ages, an essential work on which all bachelors in theology lectured and commented. Many of the questions fused natural philosophy and theology, utilizing natural philosophy to resolve theological issues, especially in the second book, which considered the Creation. They covered such matters as "whether God could make a better universe"; "whether the empyrean heaven is luminous"; "whether light is a real form"; "whether the heaven is the cause of these inferior things"; "whether every spiritual substance is in a place"; "whether God could make something new"; and "whether God could make an actual infinite."

Logic, natural philosophy, and the exact sciences, along with the scholastic methods for treating these subjects, became permanent features of medieval universities. But the entry of natural philosophy into the curriculum of the University of Paris, the premier university of western Europe during the Middle Ages, differed markedly from its entry into other contemporary universities, such as Oxford and Bologna. While the exact sciences and medicine encountered little opposition, Aristotelian natural philosophy met a different fate. Christianity, during its first six centuries, had adjusted fairly easily to pagan Greek learning and had adopted the attitude that Greek philosophy and natural philosophy should be used as "handmaidens to theology" (i.e., they should be studied for the light they shed on Scripture and theological problems and for any insights they might offer for a better understanding of God's Creation). Nevertheless, some influential theologians in Paris, specifically those at the university, were deeply concerned about the potential dangers that Aristotelian natural philosophy posed for the faith. During the first half of the thirteenth century, the Parisian authorities first banned the works of Aristotle, decreeing in 1210 and 1215 that they were not to be read in public or private. Subsequently, in 1231, they sought to expurgate his works, an intention that was apparently never carried out. By 1255, Aristotle's works had been adopted as the official curriculum at the University of Paris. Efforts to deny entry of Aristotelian natural philosophy into the University of Paris failed utterly. The reason is obvious: For Christians, the value of Aristotle's works and the commentaries thereon far outweighed any potential danger they might pose.

During the second half of the thirteenth century, a number of conservative theologians, who were still concerned about the impact of Aristotelian thought, changed their means of attack. Rather than attempt to ban or expurgate Aristotle's works, they now sought to identify and condemn specific ideas that they believed were dangerous to the faith. When it became apparent that repeated warnings about the perils of secular philosophy were to no avail, the traditional theologians appealed to the bishop of Paris, Etienne Tempier (d. 1279). In 1270, Tempier intervened and condemned 13 articles that were derived from the teachings of Aristotle or were upheld by his great commentator, Averroës. In 1272, the masters of arts at the University of Paris instituted an oath that compelled them to avoid consideration of theological questions. If, for any reason, an arts master found himself unable to avoid a theological issue, he was further sworn to resolve it in favor of the faith. The intensity of the controversy was underscored by Giles of Rome's (c. 1243–1316) *Errors of the Philosophers,* written sometime between 1270 and 1274, in which Giles compiled a list of errors drawn from the works of the non-Christian philosophers Aristotle, Averroës, Avicenna, al-Ghazali (1058–1111), al-Kindi, and Moses Maimonides (1135–1204). When these countermoves failed to resolve the turmoil or abate the controversy, a concerned Pope John XXI instructed the bishop of Paris, still Etienne Tempier, to initiate an investigation. Within three weeks, in March 1277, Tempier, acting on the advice of his theological advisers, issued a massive condemnation of 219 articles. Excommunication was the penalty for holding or defending any one of them. Although the list of condemned articles was drawn up in haste without apparent order and with little concern for consistency or repetition, many, if not most, of the 219 articles reflected issues that were directly associated with Aristotle's natural philosophy and, hence, form part of the history of the reception of Aristotelian learning. Some 27 of the articles—more than 10 percent—condemned the eternity of the world in a variety of guises. Numerous articles were condemned because they set limits on God's absolute power to do things that were deemed impossible in Aristotle's natural philosophy.

Scattered through the works of Aristotle were propositions and conclusions demonstrating the natural impossibility of certain phenomena. For example, Aristotle had shown that it was impossible for a vacuum to occur naturally inside or outside the world, and he had also demonstrated the impossibility that other worlds might exist naturally beyond ours. Theologians came to view these Aristotelian claims of natural impossibility as restrictions on God's absolute power to do as he pleased. Just because Aristotle had declared it impossible, why should an omnipotent God not be able to produce a vacuum inside or outside the world, if he chose to do so? Why could he not create other worlds, if he wished to do so? Why should he not be able to produce an accident without a subject? And why should he not be able to produce new things in the world that he had created long ago? A condemned article was issued for each of these restrictions on God's power. As if to reinforce all of the specific articles, the bishop of Paris and his colleagues included a separate article (147) that condemned the general opinion that God could not do what was judged impossible in natural philosophy.

By appeal to the concept of God's absolute power, medieval natural philosophers introduced subtle and imaginative questions that often generated novel responses. By conceding that God could create other worlds, they inquired about the nature of those worlds. By assuming that God could, if he wished, create vacuums anywhere in the universe, they were stimulated to pose questions about the behavior of bodies in such hypothetical vacuums. They asked, for example, whether bodies would move with finite or infinite speeds in such empty spaces. They posed similar questions about a variety of imaginary, hypothetical physical situations. Although these speculative questions and their responses did not cause the overthrow of the Aristotelian worldview, they did challenge some of its fundamental principles and assumptions. They made many aware that things might be quite otherwise than were dreamt of in Aristotle's philosophy.

Despite the adverse theological reaction to some of Aristotle's ideas and attitudes during the thirteenth century, it would be a serious error to suppose that medieval theologians in general opposed Aristotelian natural philosophy. If the majority of theologians had chosen to oppose Aristotelian learning as dangerous to the faith, it could not have become the focus of studies in the universities. But theologians had no compelling reason to oppose it. Western Christianity had a longstanding tradition of using pagan thought for its own benefit. As supporters of that tradition, medieval theologians treated the new Greco-Arabic learning in the same manner—as a welcome addition that would enhance their understanding of Scripture.

Indeed, we can justifiably characterize medieval theologians as theologian–natural philosophers, since almost all of them were thoroughly trained in natural philosophy, which was a virtual prerequisite for students entering the higher faculty of theology. So enthusiastically did these theologian–natural philosophers incorporate natural philosophy into their theological treatises that the church had to admonish them, from time to time, to refrain from frivolously employing it in the resolution of theological questions. Some of the most significant contributors to science, mathematics, and natural philosophy came from the ranks of theologians, as is obvious from the illustrious names of Albertus Magnus (1193–1280), Robert Grosseteste (c. 1168–1253), John Pecham (d. 1292), Theodoric of Freiberg (d. c. 1310), Thomas Bradwardine (c. 1290–1349), Nicole Oresme (c. 1320–82), and Henry of Langenstein (fl. 1385–93). The positive attitude of medieval theologians toward Aristotelian natural philosophy and their belief that it was a useful tool for the elucidation of theology must be viewed as the end product of a longstanding attitude that was developed and nurtured during the first four or five centuries of Christianity and was maintained thereafter in the Latin West.

The Early Modern Period

Because Aristotle's works formed the basis of the medieval university curriculum, Aristotelianism emerged as the primary, and virtually unchallenged, intellectual system of western Europe during the thirteenth to fifteenth centuries. It not only provided the mechanisms of explanation for natural phenomena but also served as a gigantic filter through which the world was viewed. Whatever opposition theologians may once have offered to it, by the thirteenth century that opposition had long ceased. Aristotelian physics and cosmology were triumphant and dominant. By the sixteenth and seventeenth centuries, however, rival natural philosophies had materialized following a wave of translations of new Greek philosophical texts previously unknown in the West. Opposition to Aristotelianism now became widespread. As a direct consequence of the new science that was emerging in the first half of the seventeenth century, the positive medieval attitude toward science and natural philosophy that prevailed in the late Middle Ages underwent significant change. In the aftermath of the Council of Trent (1545–63), the Roman Catholic Church came to link the defense of the faith with a literal interpretation of those biblical passages that clearly placed an immobile earth at the center of the cosmos. By this move, it aligned itself with the traditional Aristotelian-Ptolemaic geocentric universe in opposition to the Copernican heliocentric system. In 1633, the church condemned Galileo (1564–1642) for upholding the truth of the Copernican heliocentric planetary theory. By condemning Galileo, the church and its theologians came to be viewed as obscurantists who were hostile to science and natural philosophy. Instead of confining that opinion to Aristotelian scholasticism of the seventeenth century, when Aristotelian natural philosophy was under assault and nearing the end of its dominance in European intellectual life, the critics of Aristotle and Aristotelianism indiscriminately included the late Middle Ages in that judgment and viewed it as an equally unenlightened period.

In this way, medieval attitudes toward science and natural philosophy have been seriously distorted. The aftermath of Galileo's condemnation produced hostility and contempt toward the late Middle Ages. Not only was the attitude of theologians toward natural philosophy misrepresented, but the positive role that Aristotle and Aristotelian natural philosophers played in the history of science was ignored, as was the legacy they bequeathed to the seventeenth century. More than anyone else in the history of Western thought, it was Aristotle who molded and shaped the scientific temperament. He was the model for the Middle Ages. Gradually, medieval scholars reshaped and supplemented Aristotle's methods and insights by their own genius and fashioned a more sophisticated body of natural philosophy, which they passed on to the scientists and natural philosophers of the seventeenth century. This legacy included a variety of methodological approaches to nature that had been applied to a large body of important questions and problems about matter, motion, and vacuums. These questions were taken up by nonscholastic natural philosophers in the seventeenth century. Initially, the scientific revolution involved the formulation of successful responses to old questions that had been posed during the Middle Ages.

Embedded in the vast medieval Aristotelian commentary literature was a precious gift to early modern science: an extensive and sophisticated body of terms that formed the basis of scientific discourse. Terms such as *potential, actual, substance, property, accident, cause, analogy, matter, form, essence, genus, species, relation, quantity, quality, place, vacuum,* and *infinite* formed a significant component of scholastic natural philosophy. The language of medieval natural philosophy, however, did not consist solely of translated Aristotelian terms. New concepts, terms, and definitions were added, most notably in the domains of change and motion, in which new definitions were fashioned for concepts like uniform motion, uniformly accelerated motion, and instantaneous motion.

The universities of the Middle Ages, in which natural philosophy and science were largely conducted, also conveyed a remarkable tradition of relatively free, rational inquiry. The medieval philosophical tradition was fashioned in the faculties of arts of medieval universities. Natural philosophy was their domain, and, almost from the outset, masters of arts struggled to establish as much academic freedom as possible. They sought to preserve and expand the study of philosophy. Arts masters regarded themselves as the guardians of natural philosophy, and they strove mightily for the right to apply reason to all problems concerning the physical world. By virtue of their independent status as a faculty with numerous rights and privileges, they achieved a surprising degree of freedom. During the Middle Ages, natural philosophy remained what Aristotle had made it: an essentially secular and rational enterprise. It remained so only because the arts faculty, whose members were the teachers and guardians of natural philosophy, struggled to preserve it. In the process, they transformed natural philosophy into an independent discipline that had as its objective the rational investigation of all problems relevant to the physical world. However, the success of the arts masters was dependent on the theological faculties, which were sympathetic to the development of natural philosophy. Despite the problems of the thirteenth century, medieval theologians were as eager to pursue that discipline as were the arts masters. If that had not been so, medieval natural philosophy would never have attained the heights it has reached, nor would it have been so extensively employed. For not only was natural philosophy imported into theology, especially into theological commentaries on the *Sentences* of Peter Lombard, the textbook of the theological schools for five centuries, but it was also integrated into medicine, both in the standard textbooks of physicians such as Galen (129–c. 210), Avicenna, and Averroës and in the numerous medical commentaries by physicians who were thoroughly acquainted with Aristotle's natural philosophy and recognized its importance for medicine. Even music theorists occasionally found it convenient to introduce concepts from natural philosophy to elucidate musical themes and ideas.

Finally, the seventeenth century also inherited from the late Middle Ages the profound sense that all of these activities were legitimate and important, that discovering the way the world operated was a laudable undertaking. Without the crucial centuries of medieval Aristotelianism to serve as a foundation, the scientific revolution of the seventeenth century would have been long delayed or might still lie in the future.

BIBLIOGRAPHY

Barnes, Jonathan. *The Cambridge Companion to Aristotle.* Cambridge: Cambridge University Press, 1995.

Callus, D. A. "The Introduction of Aristotelian Learning to Oxford." *Proceedings of the British Academy* 29 (1943): 229–81.

Courtenay, W. J. "Theology and Theologians from Ockham to Wyclif." In *The History of the University of Oxford.* Vol. 2. Ed. J. I. Catto and Ralph Evans. Oxford: Clarendon, 1992, 1–34.

Evans, Gillian R. *Philosophy and Theology in the Middle Ages.* London: Routledge, 1993.

Fletcher, John M. "Some Considerations of the Role of the Teaching of Philosophy in the Medieval Universities." *British Journal for the History of Philosophy* 2(1) (1994): 3–18.

Gabriel, Astrik L. "Metaphysics in the Curriculum of Studies in the Mediaeval Universities." In *Die Metaphysik im Mittelalter,* ed. P. Wilpert. Berlin: Walter de Gruyter, 1963, 92–102.

Grant, Edward. "Aristotelianism and the Longevity of the Medieval World View." *History of Science* 16 (1978): 93–106.

———. "The Condemnation of 1277, God's Absolute Power, and Physical Thought in the Late Middle Ages." *Viator* 10 (1979): 211–44.

———. "Science and the Medieval University." In *Rebirth, Reform, and Resilience: Universities in Transition, 1300–1700,* ed. James M. Kittelson and Pamela J. Transue. Columbus: Ohio State University Press, 1984, 68–102.

———. "Science and Theology in the Middle Ages." In *God and Nature: Historical Essays on the Encounter between Christianity and Science,* ed. David C. Lindberg and Ronald L. Numbers. Berkeley: University of California Press, 1986, 49–75.

———. "Ways to Interpret the Terms 'Aristotelian' and 'Aristotelianism' in Medieval and Renaissance Natural Philosophy." *History of Science* 25 (1987): 335–58.

———. "Medieval Departures from Aristotelian Natural Philosophy." In *Studies in Medieval Natural Philosophy,* ed. Stefano Caroti. Biblioteca di Nuncius. Florence: Olschki, 1989, 237–56.

———. *The Foundations of Modern Science in the Middle Ages: Their Religious, Institutional, and Intellectual Contexts.* Cambridge: Cambridge University Press, 1996.

———. "God, Science, and Natural Philosophy in the Late Middle Ages." In *Between Demonstration and Imagination: Essays in the History of Science and Philosophy Presented to John D. North.* Leiden: Brill, 1999, 243–67.

———. *God and Reason in the Middle Ages.* Cambridge: Cambridge University Press, 2001.

Iorio, Dominick A. *The Aristotelians of Renaissance Italy.* Lewiston, N.Y.: Edwin Mellen, 1991.

McKeon, Richard P. "Aristotelianism in Western Christianity." In *Environmental Factors in Christian History,* ed. John Thomas McNeill. Chicago: University of Chicago Press, 1939, 206–31.

Murdoch, John E. "The Analytic Character of Late Medieval Learning: Natural Philosophy without Nature." In *Approaches to Nature in the Middle Ages,* ed. L. D. Roberts. Binghamton, N.Y.: Center for Medieval and Early Renaissance Studies, 1982, 171–213. See also "Comment" by Norman Kretzmann, 214–20.

———. "The Involvement of Logic in Late Medieval Natural Philosophy." In *Studies in Medieval Natural Philosophy,* ed. Stefano Caroti. Biblioteca di Nuncius. Florence: Olschki, 1989, 3–28.

North, J. D. "Natural Philosophy in Late Medieval Oxford." In *The History of the University of Oxford.* Vol. 2. Ed. J. I. Catto and Ralph Evans. Oxford: Clarendon, 1992, 65–102.

Sabra, Abdelhamid I. "Science and Philosophy in Medieval Islamic Theology: The Evidence of the Fourteenth Century." *Zeitschrift für Geschichte der Arabisch-Islamischen Wissenschaften* 9 (1994): 1–42.

Schmitt, Charles B. *Aristotle and the Renaissance.* Cambridge: Published for Oberlin College by Harvard University Press, 1983.

Steenberghen, Fernand Van. *Aristotle in the West: The Origins of Latin Aristotelianism.* Trans. L. Johnston. Louvain: Nauwelaerts, 1955.

———. *The Philosophical Movement in the Thirteenth Century.* Edinburgh: Nelson, 1955.

Sylla, Edith D. "Autonomous and Handmaiden Science: St. Thomas Aquinas and William of Ockham on the Physics of the Eucharist." In *The Cultural Context of Medieval Learning,* ed. John E. Murdoch and Edith D. Sylla. Dordrecht: Reidel, 1975, 349–96.

———. "Physics." In *Dictionary of the Middle Ages.* Vol. 9. Ed. Joseph R. Strayer. New York: Charles Scribner's Sons, 1987, 620–28.

Wallace, William A. "Aristotle in the Middle Ages." In *Dictionary of the Middle Ages.* Vol. 1. Ed. Joseph R. Strayer. New York: Charles Scribner's Sons, 1982, 456–69.

4 Early Christian Attitudes toward Nature

David C. Lindberg

The early Christian Church has often been portrayed as a fountainhead of antirationalistic and antiscientific sentiment, one of the agents responsible for propelling Europe into what are popularly referred to as the dark ages. This portrayal has been achieved in large part through selective quotation and, especially, by the choice of the church father Tertullian (c. 160–c. 220) as representative of early Christian attitudes to science. The historical reality, as this essay attempts to reveal, was both a great deal more complicated and a great deal more interesting.

David C. Lindberg earned the Ph.D. in history and philosophy of science from Indiana University in 1965. He has taught for thirty-four years in the History of Science Department of the University of Wisconsin at Madison, where he is Hilldale Professor Emeritus. His publications include *The Beginnings of Western Science: The European Scientific Tradition in Philosophical, Religious, and Institutional Context, 600 B.C. to A.D. 1450* (Chicago: University of Chicago Press, 1992); *God and Nature: Historical Essays on the Encounter between Christianity and Science*, co-edited with Ronald L. Numbers (Berkeley: University of California Press, 1986); and *Science and the Christian Tradition: Twelve Case Histories*, co-edited with Ronald L. Numbers (Chicago: University of Chicago Press, 2002).

The Christian Intellectual Tradition

When we refer to Christian attitudes toward nature, we are referring to the attitudes of a small, highly educated Christian elite. This elite emerged during the second and third centuries of the Christian era as educated Christians, attempting to come to terms with Greco-Roman intellectual culture and entering into dialogue with pagans on critical philosophical and theological issues. In the course of this dialogue, they took important steps toward the definition, refinement, and defense of the fundamentals of Christian belief and practice. Many who belonged to this Christian intelligentsia had been the recipients of a pagan literary, rhetorical, and philosophical education before their conversion to Christianity, and inevitably they brought with them attitudes and ideals acquired in the Greco-Roman schools. Although they frequently turned against significant portions of the content learned in this prior educational experience, especially where it touched upon theological issues, the broad intellectual values and methodology of this pagan schooling had been absorbed too deeply to be easily abandoned.

The early church has often been portrayed as a haven of anti-intellectualism, and evidence apparently favorable to this opinion is not hard to find. The Apostle Paul (whose influence in shaping Christian attitudes was enormous) warned the Colossians: "Be on your guard; do not let your minds be captured by hollow and delusive speculations, based on traditions of man-made teaching centered on the elements of the natural world and not on Christ" (Col. 2:8 New English Bible, substituting an alternative translation provided by the translators for one phrase). In his first letter to the Corinthians, he admonished: "Make no mistake about this: if there is anyone among you who fancies himself wise . . . he must become a fool to gain true wisdom. For the wisdom of this world is folly in God's sight" (I Cor. 3:18–19 New English Bible). Tertullian (c. 160–c. 220), who frequently expressed similar sentiments, elaborated these thoughts in a celebrated passage:

> What indeed has Athens to do with Jerusalem? What concord is there between the Academy and the Church? What between heretics and Christians? . . . Away with all attempts to produce a mottled Christianity of Stoic, Platonic, and dialectic composition! We want no curious disputation after possessing Christ Jesus, no inquisition after enjoying the gospel! With our faith, we desire no further belief. For once we believe this, there is nothing else that we ought to believe (Tertullian 1986, 246b, with minor revision).

Denunciations of Greek philosophy for its vanity, its contradictions, its occupation with the trivial and disregard for the consequential, and its instigation of heresy became standard, almost formulaic, elements in the works of Tertullian and other early Christian writers.

But to stop here would be to present an incomplete and highly misleading picture. The very writers who denounced Greek philosophy also employed its methodology and incorporated parts of its content into their own systems of thought. In the battle for the minds of the educated, Christian apologists had no alternative but to meet pagan intellectuals on their own ground. From Justin

Martyr (c. 100–165) to St. Augustine (354–430) and beyond, Christian scholars allied themselves with Greek philosophical traditions that they considered congenial to Christian thought. Chief of these traditions was Platonism or Neoplatonism, but borrowing from Stoic, Aristotelian, and Neo-Pythagorean philosophy was also common. Even the denunciations issuing from Christian pens, whether of specific philosophical positions or of philosophy generally, often reflected an impressive command of the philosophical tradition.

The Church Fathers and Natural Philosophy

But where and how did science enter the picture? In the first place, we must understand that there was no activity and no body of knowledge during the patristic period that bore a close resemblance to modern science. However, there *were* beliefs about nature: about the origins and structure of the cosmos, the motions of celestial bodies, the elements, sickness and health, the explanation of dramatic natural phenomena (thunder, lightning, eclipses, and the like), and the relationship between the cosmos and the gods. These are the ingredients of what would develop centuries later into modern science, and, if we are interested in the origins of Western science, they are what we must investigate. The best way of denoting these ingredients is by the expression *natural philosophy*. The term is useful because it calls attention to the relationship between the philosophy of nature and the larger philosophical enterprise (although the expressions *science* and *natural science* will also be used occasionally in the remainder of this essay). As an integral part of philosophy, natural philosophy shared the latter's methods and its fate, and it became a concern of Christians and entered into their sermons, debates, and writings insofar as it impinged on Christian doctrine and Christian worldview, as it frequently did. After all, Christians had as much need of a cosmology as did pagans.

Among Christian writers, we find expressions of hostility toward natural philosophy, just as we do toward philosophy in general. Tertullian, for example, attacked the pagan philosophers for their assignment of divinity to the elements and the sun, moon, other planets, and stars. In the course of his argument, he vented his wrath on the vanity of the ancient philosophers:

> Now pray tell me, what wisdom is there in this hankering after conjectural speculations? What proof is afforded to us . . . by the useless affectation of a scrupulous curiosity, which is tricked out with an artful show of language? It therefore

> served Thales of Miletus quite right, when, stargazing as he walked, . . . he had the mortification of falling into a well. . . . His fall, therefore, is a figurative picture of the philosophers; of those, I mean, who persist in applying their studies to a vain purpose, since they indulge a stupid curiosity on natural objects (Tertullian 1986, 133).

But it *is* an argument that Tertullian presents, and, to a very significant degree, he builds it out of materials and by the use of methods drawn from the Greco-Roman philosophical tradition. He argues, for example, that the precise regularity of the orbital motions of the celestial bodies (a clear reference to the findings of the Greek astronomical tradition) bespeaks a "governing power" that rules over them, and, if they are ruled over, they surely cannot be gods. He also introduces the "enlightened view of Plato" in support of the claim that the universe must have had a beginning and, therefore, cannot partake of divinity. In this and other works, he "triumphantly parades" his learning (as one of his biographers puts it) by naming a long list of other ancient authorities (Barnes 1985, 196).

Basil of Caesarea (c. 330–79), representing a different century and a different region of the Christian world, reveals similar attitudes toward Greek natural philosophy. He sharply attacked philosophers and astronomers who "have wilfully and voluntarily blinded themselves to the knowledge of the truth." These men, he continued, have "discovered everything, except one thing: they have not discovered the fact that God is the creator of the universe." Elsewhere he inquired why we should "torment ourselves by refuting the errors or rather the lies of the Greek philosophers, when it is sufficient to produce and compare their mutually contradictory books." And he attacked belief in the transmigration of souls by admonishing his listeners to "avoid the nonsense of those arrogant philosophers who do not blush to liken their own soul to the soul of a dog" (Amand de Mendieta 1976, 38, 31, 37).

But, while attacking the errors of Greek natural philosophy—and what he didn't find erroneous, he generally found useless—Basil also revealed a solid mastery of its content. He argued against Aristotle's (384–322 B.C.) fifth element, the quintessence; he recounted the Stoic theory of cyclic conflagration and regeneration; he ridiculed theories of the eternity and divinity of the cosmos; he applauded those who employ the laws of geometry to refute the possibility of multiple worlds (a clear reference to Aristotle's argument for the uniqueness of the cosmos); he derided the Pythagorean notion of music of the planetary spheres; he proclaimed the vanity of mathematical astronomy; and he revealed familiarity with various opinions about the shape of the earth and (for those who believed it to be spherical) calculations of its circumference.

Tertullian and Basil have generally been portrayed as outsiders to the philosophical tradition, attempting to discredit and destroy what they regarded as a menace to the Christian faith. Certainly, much of their rhetoric supports such an interpretation, as when they appealed for simple faith as an alternative to philosophical reasoning. But we need to look beyond rhetoric to actual practice: It is one thing to deride natural philosophy or declare it useless, another to abandon it. Despite their derision, Tertullian, Basil, and others like them were continuously engaged in serious philosophical argumentation. It is no distortion of the evidence to see them as insiders, attempting to formulate an alternative natural philosophy based on Christian principles and opposed, not to the enterprise of natural philosophy, but to specific principles of natural philosophy that they considered both erroneous and dangerous.

The most influential of the church fathers and the one who codified Christian attitudes toward nature was St. Augustine, bishop of Hippo in North Africa. Like his predecessors, Augustine had deep reservations about the value of natural philosophy. But his criticism was more muted and qualified by an acknowledgment, in both word and deed, of legitimate uses to which natural knowledge might be put and a recognition that it may even be of religious utility. In short, Augustine certainly did not devote himself to the promotion of natural science, but neither did he fear pagan versions of it to the degree that some of his predecessors had.

Scattered throughout Augustine's voluminous writings are worries about pagan philosophy (including natural philosophy) and admonitions for Christians not to overvalue it. In his *Enchiridion,* he assured his readers that there is no need to be "dismayed if Christians are ignorant about the properties and the number of the basic elements of nature, or about the motion, order, and deviations of the stars, the map of the heavens, the kinds and nature of animals, plants, stones, springs, rivers, and mountains. . . . For the Christian, it is enough to believe that the cause of all created things . . . is . . . the goodness of the Creator" (Augustine 1955, 341–42). In his *On Christian Doctrine,* he commented on the uselessness of astronomical knowledge:

> Although the course of the moon . . . is known to many, there are only a few who know well the rising or setting or other movements of the rest of the stars with-

> out error. Knowledge of this kind in itself, although it is not allied with any superstition, is of very little use in the treatment of the Divine Scriptures and even impedes it through fruitless study; and since it is associated with the most pernicious error of vain [astrological] prediction it is more appropriate and virtuous to condemn it (Augustine 1976, 65–66).

And, in his *Confessions,* he argued that "Because of this disease of curiosity . . . men proceed to investigate the phenomena of nature, . . . though this knowledge is of no value to them: for they wish to know simply for the sake of knowing" (Augustine 1942, 201, slightly edited). Knowledge for the sake of knowing is without value and, therefore, illegitimate.

But, once again, this is not the whole story. Natural philosophy may be without value for its own sake, but from this we are not entitled to conclude that it is entirely without value. Knowledge of natural phenomena acquires value and legitimacy insofar as it serves other, higher purposes. One such purpose is biblical exegesis, since ignorance of mathematics, music (conceived as a mathematical art in Augustine's day), and natural history renders us incapable of grasping the literal sense of Scripture. For example, only if we are familiar with serpents will we grasp the meaning of the biblical admonition to "be as wise as serpents and as innocent as doves" (Matt. 10:16). Augustine also conceded that portions of pagan knowledge, such as history, dialectic, mathematics, the mechanical arts, and "teachings that concern the corporeal senses," contribute to the necessities of life (Augustine 1976, 74).

In his *Literal Meaning of Genesis,* in which he put his own superb grasp of Greek cosmology and natural philosophy to good use, Augustine expressed dismay at the ignorance of some Christians:

> Even a non-Christian knows something about the earth, the heavens, and the other elements of this world, about the motion and orbit of the stars and even their size and relative positions, about the predictable eclipses of the sun and moon, the cycles of the years and the seasons, about the kinds of animals, shrubs, stones, and so forth, and this knowledge he holds to as being certain from reason and experience. Now it is a disgraceful and dangerous thing for an infidel to hear a Christian . . . talking nonsense on these topics; and we should take all means to prevent such an embarrassing situation, in which people show up vast ignorance in a Christian and laugh it to scorn (Augustine 1982, I:42–43).

Insofar as we require knowledge of natural phenomena—and Augustine is certain that we do—we must take it from those who possess it: "If those who are called philosophers, especially the Platonists, have said things which are indeed true and are well accommodated to our faith, they should not be feared; rather, what they have said should be taken from them as from unjust possessors and converted to our use" (Augustine 1976, 75). All truth is ultimately God's truth, even if found in the books of pagan authors; we should seize it and use it without hesitation.

In Augustine's view, then, knowledge of the things of this world is not a legitimate end in itself, but, as a means to other ends, it is indispensable. Natural philosophy must accept a subordinate position as the handmaiden of theology and religion: The temporal must be made to serve the eternal. Natural philosophy is not to be loved, but it may be legitimately used. This attitude toward scientific knowledge was to flourish throughout the Middle Ages and well into the modern period.

But does endowing natural philosophy with handmaiden status constitute a blow against scientific progress? Are the critics of the early church right in viewing it as the opponent of genuine science? We need to make three points here. First, it is certainly true that the early church was no great patron of the natural sciences. These had low priority for the church fathers, for whom the major concerns were establishment of Christian doctrine, defense of the faith, and the edification of believers. Second, low priority was far from no priority. Throughout the Middle Ages and well into the modern period, the handmaiden formula was employed countless times to justify the investigation of nature. Indeed, some of the most celebrated achievements of the Western scientific tradition were made by scholars who justified their labors by appeal to the handmaiden formula. Third, there were no institutions or cultural forces during the patristic period that offered more encouragement for the investigation of nature than did the Christian Church. Contemporary pagan culture was no more favorable to disinterested speculation about the cosmos than was Christian culture. It is at least arguable that the presence of the Christian Church enhanced, rather than damaged, the prospects for the natural sciences.

Three Illustrative Examples

We cannot end this account without touching briefly on a trio of examples that illustrate how Christian attitudes toward natural philosophy worked themselves out in actual practice: First, Augustine on Creation; second, the shape of the earth; third, medicine and the supernatural.

Augustine not only authorized the use of natural philosophy in biblical exegesis, he also practiced what he preached. In his *Literal Meaning of Genesis,* Augustine produced a verse-by-verse exposition of the biblical account of Creation as it appears in the first three chapters of Genesis. In the course of this work of his mature years, he brought to bear all knowledge that would help elucidate the meaning of the biblical text, including the pagan tradition of natural philosophy. In so doing, he transmitted to medieval scholars (before the thirteenth century) one of their richest sources of cosmological, physical, and biological knowledge.

It is almost universally held that Europeans of the Roman and medieval periods believed in a flat earth and that biblical literalism had something to do with this belief. The truth is quite otherwise. The sphericity of the earth was proposed by Pythagorean philosophers no later than the fifth century B.C. The sphericity of the earth was never seriously doubted after Aristotle, and the earth's circumference was satisfactorily calculated by Eratosthenes (c. 275–194 B.C.). But what about Christian opinion? Did the literal interpretation of certain biblical passages compel Christians to deny the earth's sphericity? The shape of the earth was not a source of controversy during the patristic period, and the evidence is, therefore, thin. Scholars have been able to discover only two Christian writers of the patristic period who denied the sphericity of the earth: the Latin church father Lactantius (c. 240–320) and the Byzantine merchant Cosmas Indicopleustes (fl. 540). Evidently, early Christians did *not* reject the powerful arguments of Greek cosmologists for a spherical earth in favor of a literal interpretation of biblical passages that seemed to suggest otherwise.

Finally, can we learn anything by exchanging the purely theoretical subjects of cosmology and mathematical geography for the far more practical realm of medicine? Much has been made of Christian supernaturalism and its incompatibility with aspects of Greco-Roman medical theory and practice. But, in fact, the tension, though not totally absent, was not as serious as alleged. In the first place, religious elements (including miracle cures) were also an important part of Greco-Roman medicine. Second, belief in sickness as a divine visitation did not rule out simultaneous belief in natural causes. When Christians maintained that disease could be both natural and divine, conceiving natural causes as instruments of divine purpose, they were not breaking new ground, for this was a commonplace of the Hippocratic tradition. Third, belief in the existence of supernatural medicine and active pursuit of supernatural cures did not prevent Christians from availing themselves simultaneously of secular, naturalistic medicine—just as many of the sick in our own day participate simultaneously in conventional and nonconventional medical therapies.

BIBLIOGRAPHY

Amand de Mendieta, Emmanuel. "The Official Attitude of Basil of Caesarea as a Christian Bishop towards Greek Philosophy and Science." In *The Orthodox Churches and the West,* ed. Derek Baker. Oxford: Blackwell, 1976, 25–49.

Amundsen, Darrel W. "Medicine and Faith in Early Christianity." *Bulletin of the History of Medicine* 56 (1982):326–50.

Armstrong, A. H., ed. *The Cambridge History of Later Greek and Early Medieval Philosophy.* Cambridge: Cambridge University Press, 1970.

Armstrong, A. H., and R. A. Markus. *Christian Faith and Greek Philosophy.* London: Darton, Longman, and Todd, 1960.

Augustine, St. of Hippo. *Confessions.* Trans. F. J. Sheed. New York: Sheed and Ward, 1942.

———. *Confessions and Enchiridion.* Trans. Albert C. Outler. Philadelphia: Westminster, 1955.

———. *On Christian Doctrine.* Trans. D. W. Robertson Jr. 1958. Reprint. Indianapolis: Bobbs-Merrill, 1976.

———. *The Literal Meaning of Genesis.* Trans. John Hammond Taylor, S. J. In *Ancient Christian Writers: The Works of the Fathers in Translation,* ed. Johannes Quasten, W. J. Burghardt, and T. C. Lawler. Vols. 41–42. New York: Newman, 1982.

Barnes, Timothy David. *Tertullian: A Historical and Literary Study.* Rev. ed. Oxford: Clarendon, 1985.

Cochrane, Charles N. *Christianity and Classical Culture.* Oxford: Clarendon, 1940.

Ferngren, Gary B. "Early Christianity as a Religion of Healing." *Bulletin of the History of Medicine* 66 (1992): 1–15.

———. "Early Christian Views of the Demonic Etiology of Disease." In *From Athens to Jerusalem: Medicine in Hellenized Jewish Lore and Early Christian Literature,* ed. Samuel Kottek, Manfred Horstmanshoff, Gerhard Baader, and Gary B. Ferngren. Rotterdam: Erasmus, 2000, 195–213.

Grant, Robert M. *Miracle and Natural Law in Graeco-Roman and Early Christian Thought.* Amsterdam: North-Holland, 1952.

Lindberg, David C. "Science and the Early Church." In *God and Nature: Historical Essays on the Encounter between Christianity and Science,* ed. David C. Lindberg and Ronald L. Numbers. Berkeley: University of California Press, 1986, 19–48.

———. "Science as Handmaiden: Roger Bacon and the Patristic Tradition." *Isis* 78 (1987): 518–36.

———. *The Beginnings of Western Science: The European Scientific Tradition in Philosophi-*

cal, Religious, and Institutional Context, 600 B.C. to A.D. 1450. Chicago: University of Chicago Press, 1992.

Pelikan, Jaroslav. *Christianity and Classical Culture: The Metamorphosis of Natural Theology in the Christian Encounter with Hellenism.* New Haven: Yale University Press, 1993.

Scott, Alan. *Origen and the Life of the Stars: A History of an Idea.* Oxford: Clarendon, 1991.

Tertullian. *Writings,* in *The Ante-Nicene Fathers.* Ed. Alexander Roberts and James Donaldson, rev. ed. A. Cleveland Coxe, Vol. 3. Reprint. Grand Rapids, Mich.: Eerdmans, 1986.

5 Medieval Science and Religion

David C. Lindberg

The Middle Ages have served as a historical arena within which two schools of thought have done battle—one school accusing the medieval church of actively opposing the advancement of scientific learning, the other praising the medieval church and its theology for laying a foundation that made modern science possible. This essay mediates between these two alternatives, offering a nuanced portrayal of the complicated relationship between medieval religion and medieval science.

David C. Lindberg earned the Ph.D. in history and philosophy of science from Indiana University in 1965. He is Hilldale Professor Emeritus of the history of science at the University of Wisconsin. His publications include *The Beginnings of Western Science: The European Scientific Tradition in Philosophical, Religious, and Institutional Context, 600 B.C. to A.D. 1450* (University of Chicago Press, 1992); *God and Nature: Historical Essays on the Encounter between Christianity and Science*, co-edited with Ronald L. Numbers (Berkeley: University of California Press, 1986); and *Science and the Christian Tradition: Twelve Case Histories*, co-edited with Ronald L. Numbers (Chicago: University of Chicago Press, 2002).

Remarks on Methodology

Discussions of the relationship between science and religion in the Middle Ages have long been dominated by a bitter debate between the defenders of two extreme positions. At one extreme are the nineteenth-century popularizers and polemicists John William Draper (1811–82) and Andrew Dickson White (1832–1918), who formulated what has come to be called the *warfare thesis*, according to which the Christian Church set itself up as the arbiter of truth and the opponent of the natural sciences, thereby retarding the development of genuine science for a thousand years.

The warfare thesis has retained a following throughout the twentieth century, at both a scholarly and a popular level, but it has also elicited strong opposition from scholars (some with a religious agenda) who have attempted to demonstrate that the Christian Church was not the opponent of science but its ally—that Christian theology was not an obstacle to the development of modern science but its necessary condition. Pierre Duhem (1861–1916) pioneered this line of argument early in the twentieth century. Stanley Jaki is its most notable contemporary champion.

It seems quite possible that the debate between defenders of the warfare thesis and their opponents will never entirely disappear, but it has been pushed off the center of the scholarly stage by a determined effort to gain a more dispassionate, balanced, and nuanced understanding. There are several definitions and methodological precepts prerequisite to the success of this venture. First, we must continually remind ourselves that "science," "Christianity," "theology," and "the church" are abstractions rather than really existing things, and it is a serious mistake to reify them. What existed during the Middle Ages were highly educated scholars who held beliefs about both scientific and theological (and, of course, many other) matters. Science and theology cannot interact, but scientists and theologians can. Therefore, when the words *science, theology,* and *the church* are employed in the following pages, the reader should understand that such locutions are shorthand references to beliefs and practices of scientists, theologians, and the people who populated the institutions of organized Christianity.

Second, scholars who made scientific beliefs their business and scholars who made religious or theological beliefs their business were not rigidly separated from one another by disciplinary boundaries. It is true that the nontheologian who encroached on theological territory ran certain risks (which varied radically with time and place), but all medieval scholars were both theologically and scientifically informed, and all understood that theological beliefs necessarily entailed scientific consequences *and conversely.* Indeed, the scientist and the theologian were often the very same person, educated in the full range of medieval disciplines—capable of dealing with both scientific and theological matters and generally eager to find ways of integrating theological and scientific belief.

Third, we need to agree on what is meant when we talk about medieval "science." There was nothing in the medieval period corresponding even approximately to modern science. What we do find in the Middle Ages are the roots, the sources, of modern scientific disciplines and practices—ancestors of many of the pieces of modern science, which bear a family resemblance to their offspring without being identical to them. In short, medieval scholars had ideas about nature, methods for exploring it, and languages for describing it. Many of these ideas, methods, and languages were drawn from the classical tradition, the corpus of philosophical thought that originated in ancient Greece and was transmitted by various complicated processes to medieval Europe, where it became the object of intense scholarly discussion and dispute. Within the classical tradition, thought about nature was not sharply separated from thought about other subjects; all belonged to the general enterprise known as "philosophy," within which there was considerable methodological unity and interlocking content. If one wished to refer specifically to the aspects of philosophy concerned with nature, the expression *natural philosophy* was readily available. In the account that follows, the expressions *natural philosophy, science,* and *natural science* are employed as approximate synonyms, with the context being relied upon to make clear any shades of meaning. Furthermore, because natural philosophy interacted with Christianity not as a distinct enterprise but as one aspect of philosophy more generally, it will frequently prove useful to refer to *philosophy* (without qualification); it is to be understood, in such cases, that philosophy and natural philosophy shared approximately the same fate.

Fourth, medieval natural philosophy as a collection of theories was not uniform or monolithic; as an activity, its pursuit was as varied as the scholars who pursued it. The relations between medieval natural philosophy and medieval Christianity, therefore, varied radically over time, from place to place, from one scholar to another, and with regard to different issues, and it will never be possible to characterize those relations in a catchy slogan, such as that old standby, "the warfare of science and theology." A useful historical account must take the variations seriously, make distinctions, and reveal nuance. In short, the interaction between science and religion in the Middle Ages was not an abstract encounter between bodies of fixed ideas but part of the human quest for understanding. As such, it was characterized by the same vicissitudes and the same rich variety that mark all human endeavor.

The Patristic Period and the Early Middle Ages

The church father who has most often been taken to represent the attitude of the early Christian Church toward Greco-Roman philosophy is Tertullian (c. 160–c. 220), a North African, born of pagan parents, knowledgeable in philosophy, medicine, and law. Though superbly educated, Tertullian has been presented (through selective quotation) as radically anti-intellectual, preferring blind faith to reasoned argument. The truth is that Tertullian had considerable respect for philosophical argument and frequently demonstrated argumentative prowess on behalf of his religious beliefs. But it is also true that he was no great friend of pagan philosophical systems, including systems of natural philosophy.

A wider examination of attitudes within the early church reveals a range of reactions to pagan philosophy, most of them more favorable than that of Tertullian. The first serious encounter between Christianity and Greco-Roman philosophical culture occurred in the second century of the Christian era. Plagued by internal doctrinal disputes and external persecution, the Christian Church turned to Greek philosophy for help. The result was the emergence of a Christian intellectual tradition, which employed Greek philosophy for apologetic purposes, attempting to demonstrate not merely that Christian doctrine was true but also that Christianity measured up to the highest aspirations of the Greek philosophical tradition. Thus, Justin Martyr (c. 100–165), Clement of Alexandria (c. 155–c. 220), and Origen of Alexandria (c. 185–c. 251) adopted an eclectic mixture of Greek philosophies, dominated by Platonism or Neoplatonism, but with an admixture of influences from Stoic, Aristotelian, and Neo-Pythagorean sources.

But if Justin, Clement, and Origen appear to have been generally receptive to influences coming from Greek philosophy, others were less welcoming. Justin's student Tatian (second century A.D.) deeply disapproved of Greek philosophy and issued a strong condemnation of its errors and perversions. And, of course, there was Tertullian, lashing out against philosophical conclusions that ran counter to Christian doctrine. Indeed, Justin, Clement, and Origen themselves understood the problematic character of Greek philosophy and frequently expressed ambivalent feelings toward it. For example, all three had cosmological interests, and all incorporated elements of Greek cosmology into their own cosmologies; at the same time, all perceived the dangers of uncritical acceptance of Greek cosmological doctrine and the difficulties of reconciling portions of it with the teachings of Christian theology.

Similar ambivalence is apparent in the writings of Augustine of Hippo (354–430), the most influential of all of the early church fathers and codifier of early Christian attitudes toward Greek philosophy and science. Augustine made no attempt to conceal his worries about the Greek philosophical tradition, firmly elevating divine wisdom, as revealed in Scripture, over the results of human rational activity. As for nature, Augustine maintained that there is nothing to worry about if Christians are ignorant of its workings; it is sufficient for them to understand that all things issue from the Creator.

But this is only part of the story. In other contexts, Augustine admitted that knowledge of the natural world is mandatory for the assistance it can provide in the task of biblical exegesis. And he rebuked Christians for opening themselves to ridicule by refusing to accept knowledge about nature from the Greek philosophers who possessed it. Insofar as the philosophical tradition contains truth (and there was no question in Augustine's mind that the truth it contained was substantial), and insofar as this truth is of religious or theological importance, it is to be seized upon and put to use by Christians. This is a clear statement of what has come to be known as the *handmaiden formula:* the acknowledgment that science or natural philosophy is not an end in itself but a means to an end. It is to be cultivated by Christians insofar as it contributes to the interpretation of Scripture or other manifestly religious ends. Augustine did not repudiate the natural sciences; he christianized them and subordinated them to theological or other religious purposes. Science became the handmaiden of theology.

It was largely this attitude that motivated pursuit of the natural sciences through the early Middle Ages (c. A.D. 500–1000). During this period of political disintegration, social turmoil, and intellectual decline that came after the barbarian conquest of the western Roman Empire in the fifth century A.D., natural philosophy was an item of low priority. But when it was cultivated, as it sometimes was, it was cultivated by people in positions of religious authority or with a religious purpose, motivated by its perceived religious or theological utility. A sketch of the lives of five leading scholars, one each from the sixth through the tenth centuries and all interested in natural philosophy, may illuminate this claim.

Cassiodorus (c. 485–c. 580), a Christian member of the Roman senatorial class, founded the monastery of Vivarium, to which he retreated after his departure from public life. There he established a scriptorium, where secular Greek authors in substantial numbers were translated into Latin for use by the monastic community. Cassiodorus also wrote a manual of the liberal arts, the *Institutiones*, in which he discussed mathematics, astronomy, and other scientific subjects.

Isidore (c. 560–636), bishop of Seville, was the outstanding scholar of the seventh century. Recipient (probably) of a monastic education, Isidore found time in his busy ecclesiastical career to write books, one of which, the *Ety-*

mologiae (or *Origines*), became extraordinarily influential. In it he surveyed contemporary knowledge in biblical studies, theology, liturgy, history, law, medicine, and natural history.

In the eighth century, the Venerable Bede (c. 673–735), a monk from Northumbria in England, wrote a series of books for his fellow monks, including one on natural philosophy, entitled *On the Nature of Things,* and two on timekeeping and the calendar (which dealt largely with astronomical matters). In the ninth century, John Scotus Erigena (c. 810–c. 877), a product (in all likelihood) of Irish monastic schools, composed a sophisticated synthesis of Christian theology and Neoplatonic philosophy, including a well-articulated natural philosophy.

Finally, in the tenth century, a monk named Gerbert (945–1003), from the monastery at Aurillac in south-central France, crossed the Pyrenees to study the mathematical sciences in Catalonia with Atto, bishop of Vich. Gerbert occupied teaching and administrative posts after his return from Catalonia and ended his career as Pope Sylvester II (999–1003). What is particularly noteworthy is the extent to which Gerbert, throughout a busy career, consistently advanced the cause of the mathematical sciences.

Several striking characteristics are shared by these five scholars. First, all had religious vocations: One was a bishop, another a pope, and a third a monk; one of the remaining two founded a monastery, to which he then belonged; another (Erigena), though apparently associated primarily with the court of Charles the Bald (823–77), emerged as an important theologian. Second, all evidently had monastic educations, except Cassiodorus, who became a monastic educator, and it was out of the educational experience in the monastery, rather than in repudiation of it, that their interest in the natural sciences grew. Third, all wrote treatises that revealed their interest in the natural sciences and helped enlarge the role of the sciences in European culture. And, finally, it can be plausibly argued in each case that the handmaiden formula supplied the motivation for writing about the natural sciences: The natural sciences are worth pursuing because ultimately they are a religious necessity. The Christian scholar cannot fulfill his calling without them.

But there is an important question that we must face: Can such studies, pursued as handmaidens of theology, count as genuine science? First, can science be a handmaiden of anything and remain science? Of course! Who would deny the status of genuine science to research on the atomic bomb during World War II (as handmaiden of the war effort) or to pharmacological research pursued by modern pharmaceutical corporations (as handmaiden of commerce)? Indeed, it is not easy to find scientific research during any period of Western history that was not the handmaiden of some ideology, social program, practical end, or profit-making venture.

Second, it must be understood that an important aspect of every scientific tradition is the preservation and transmission of accumulated scientific knowledge; it is primarily to these functions that the early Middle Ages contributed. The church played a crucial positive role in the process, principally as the patron of European education, and its patronage extended to all aspects of learning, including the natural sciences. There can be no question about where the church placed priority. The natural sciences were not its primary interest, but they were given a small place in the curriculum of the schools and the writings of the leading scholars. Most assuredly, the early medieval church did not mobilize the resources of European society in support of the natural sciences, but no element in the European social or cultural fabric contributed more than did the church to the preservation of scientific knowledge during this intellectually precarious period.

Eleventh- and Twelfth-Century Renewal

Europe saw dramatic political, social, and economic renewal in the eleventh and twelfth centuries. The causes are complex, involving the restoration of centralized monarchies, reduction of the ravages of warfare, and the revival of trade and commerce. They led to rapid urbanization, the multiplication and enlargement of schools, and the growth of intellectual culture. Education shifted from the countryside to the cities, as cathedral and municipal schools replaced the monasteries as the principal educational institutions. Although cathedral schools shared the monastic commitment to education that was exclusively religious in its aims, the curriculum of the cathedral schools reflected a far broader conception of the range of studies that were religiously beneficial and might, therefore, be legitimately taught and learned.

This broader curriculum of the schools had important consequences. In the first place, new emphasis was placed on the Latin classics (including Greek works available in Latin translation), which had long been available but had been little studied during the early Middle Ages. The most important of these for natural philosophy was Plato's (c. 427–347 B.C.) *Timaeus*, which became the principal source for cosmological instruction and speculation in the twelfth-century schools. A major preoccupation of twelfth-century natural philosophers became the task of harmonizing Plato's account of the construction of the cosmos by the Demiurge (Plato's divine craftsman) with the account of Creation in Genesis. An associated development was increasing insistence on the principle that natural phenomena were to be explained exclusively in naturalistic terms. This was not a result of skepticism about the divine origins of the universe, but the product of a growing conviction that investigation of the secondary causes established by the Creator was a legitimate (and perhaps the only legitimate) means of studying natural philosophy.

A second development associated with the schools was a rationalistic turn: the attempt to extend the application of reason or philosophical method to all realms of human activity, including theology. Much of the impetus came from Aristotle's (384–322 B.C.) logic and commentators on it (especially Boethius [c. 475–c. 525]). Illustrative of this movement and its potential dangers was the attempt by Anselm of Bec and Canterbury (1033–1109) to extend rational methodology into the theological realm by proving God's existence without any reliance on biblical authority or the data of revelation. While this does not appear, on the face of it, to be a perilous activity, what would happen if, having made reason our guide, we found that it led us to the wrong answer?

Third, a development that was certainly stimulated by educational developments and that was, in turn, to have momentous consequences for the schools was the translation movement of the eleventh and twelfth centuries. Through an intricate process spread over about two centuries, an enormous body of new learning became available in Latin through translation from Greek and Arabic. The classical tradition of natural philosophy became available virtually in its entirety, including almost the entire Aristotelian corpus, the medical writings of the Galenic tradition, and the mathematical works produced in ancient Greece and the world of medieval Islam. As this new learning was assimilated by Europeans in the thirteenth century, it enormously complicated the relations of science and Christianity.

Finally, as the schools grew in size and sophistication, some of them were transformed into universities, offering a higher level of learning, including graduate education in one or more of three advanced subjects: theology, medicine, and law. From the thirteenth century onward, these universities were the scene of much scientific activity and corresponding tension between scientific and theological doctrine.

The Later Middle Ages

The project that dominated the intellectual life of the thirteenth century was the organization and assimilation of the new learning, Greek and Arabic in origin, made available through the activity of the translators. Tensions between this new learning and the blend of Platonic philosophy and Christian theology that had come to dominate European thought over the previous millennium set an agenda that would challenge many of the best European minds. On the one hand, the new literature was enormously exciting because of its breadth and explanatory power. In almost every area, the new treatises surpassed in scope and sophistication anything the West had hitherto known. Moreover, they exhibited methodological principles that held the promise of future intellectual gains for those who would take the trouble to master them. On the other hand, substantial portions of the new literature impinged on Christian doctrine and not always benignly. If the Platonism of the early Middle Ages proved itself relatively congenial to developing Christian theology, the Aristotelianism that arrived with the translations would prove itself far more troublesome.

Aristotle was known during the early Middle Ages mainly through a group of logical works called the "old logic." Now, as a result of the translations, scholars had access to practically the entire Aristotelian corpus, including the works of important Muslim commentators (especially Avicenna [980–1037] and, by the middle of the thirteenth century, Averroës [1126–98]). There was much in Aristotelian philosophy that could be immediately put to use, and it is this (rather than coercion from some source of authority in medieval culture) that explains the overwhelming popularity and influence of the Aristotelian system. But Aristotelian philosophy also had theological implications that threatened central Christian doctrines and posed a serious challenge for scholars who were unwilling either to abandon the theology to which they were firmly committed or to ignore the enormous promise of the new Aristotle.

What were the major problems? Perhaps the most obvious and one of the most contentious was Aristotle's claim that the world is eternal—that the cosmos had no beginning and will have no end. This belief obviously clashed with the Christian doctrine of Creation. The nature of the soul also posed problems, for it was not easy, within an Aristotelian framework, to view individual souls as separable from the body and eternal. The determinism and naturalism of the Aristotelian system presented further difficulties. The Aristotelian cosmos was a network of natural causes (associated with the natures of things) operating deterministically. Such a cosmos threatened the Christian doctrines of divine omnipotence and providence and, especially, of miracles. Before the end of the century, some philosophers at Paris, inspired by Aristotelian naturalism and determinism, went so far as to explain biblical miracles in naturalistic terms—and, thus, in the opinion of the theological authorities, to explain them away.

Finally, these troublesome Aristotelian doctrines were manifestations of a general outlook that pervaded the Aristotelian corpus, namely the view that Aristotelian demonstration, with its exclusive reliance on sense perception and rational inference, was the only way of achieving truth or of testing truth claims. Aristotelian philosophy thus arrived under the banner of extreme rationalism, which, if taken seriously, excluded biblical revelation and church tradition as sources of truth and made human reason the measure not only of philosophical claims but of theological ones as well.

The trouble was first felt at Paris (the leading theological center in Europe) early in the thirteenth century. In 1210 and again in 1215, the teaching of Aristotle's natural philosophy in the faculty of arts was banned, first by a council of bishops and subsequently by the papal legate. A papal bull (and subsequent letter) issued by Gregory IX (b. 1147, p. 1227–41) in 1231 acknowledged both the value and the dangers of Aristotelian philosophy, mandating that Aristotle's writings on natural philosophy be "purged of all suspected error" (Thorndike 1944, 38) so that, once erroneous matter had been removed, the remainder could be studied by Parisian undergraduates. (There is no evidence that the commission appointed by the pope ever met or that a purged version of Aristotle was ever produced.)

Within a decade of the papal bull of 1231, these early bans had lost their effectiveness. By the late 1230s or early 1240s, lectures on Aristotle's natural philosophy began to make their appearance in the faculty of arts at Paris. One of the first to give such lectures was the English friar Roger Bacon (1213–91). By 1255, all restrictions on the use of Aristotle either had been rescinded or were being ignored, for in that year the faculty of arts passed new statutes mandating lectures on all known Aristotelian works. In a remarkable turning of the tables, Aristotelian philosophy had moved from a position of marginality, if not outright exclusion, to centrality within the arts curriculum.

It is clear why this occurred. The Aristotelian corpus offered a convincing framework and a powerful methodology for thinking and writing about cosmology, meteorology, psychology, matter theory, motion, light, sensation, and biological phenomena of all kinds. The persuasive power of Aristotelian philosophy was so great as to preclude its repudiation. Traditionalists might be terrified by its theological implications, but Aristotelian philosophy was simply too valuable to relinquish. The task confronting those who wrestled with this problem would be the domestication, rather than the eradication, of Aristotle.

How was the domestication of Aristotle to be accomplished? Robert Grosseteste (c. 1168–1253), first chancellor of the University of Oxford, made an early and influential attempt to understand and explain Aristotle's method and some of his physical doctrines, while reconciling them with certain aspects of Plato's philosophy and a variety of other non-Aristotelian teachings. A generation later, Roger Bacon, a great admirer of Grosseteste but equipped with a much fuller knowledge of the newly translated learning, wrote an impassioned plea for papal support of the new learning (not only of Aristotle and his commentators, but also of the mathematical, and what Bacon called the "experimental," sciences). Bacon's case was based on claims of the utility of the new learning: for biblical exegesis, for proving the articles of the Christian faith, for establishing the religious calendar, for prolonging life, for producing devices that would terrorize unbelievers and lead to their conversion, and much more. This was Augustine's handmaiden formula, skillfully applied by Bacon to fresh circumstances in which the quantity and variety of knowledge available to be enlisted as handmaiden was far larger and more problematic.

But perhaps the most influential actors in this drama were a pair of theologians (both with Parisian connections) writing after midcentury. Albert the Great (Albertus Magnus [1193–1280]) and Thomas Aquinas (c. 1225–74), both Dominican friars, undertook to interpret the whole of Aristotelian philosophy, correcting it where necessary, supplementing it from other sources where possible, and, in the process, attempting to define the proper relationship between the new learning and Christian theology. As a single example, both Albert and Thomas took up the question of the eternity of the world. Albert's early opinion was that philosophy can offer no definitive answer to this question, so that one is obliged to accept biblical teaching. Later, he concluded that Aristotle's opinion was *philosophically* absurd and must be rejected. Thomas argued that philosophy is incapable of resolving the question but that there is no philosophical reason why the universe could not be both eternal (having no beginning or end) and created (dependent on God for its existence).

What Albert and Thomas accomplished (assisted, of course, by Grosseteste, Bacon, and many others) was to find a solution to the problem of faith and reason—perhaps not a permanent solution but one that proved satisfactory to many in the Middle Ages and that continues to attract a significant following at the beginning of the twenty-first century. They produced an accommodation between Aristotelian philosophy and Christian theology by christianizing Aristotle (correcting Aristotle where he was theologically unacceptable or had otherwise gone astray) and "Aristotelianizing" Christianity (importing major pieces of Aristotelian metaphysics and natural philosophy into Christian theology).

But not everybody was interested in accommodation. The freedom at the University of Paris and elsewhere that allowed Albert and Thomas to think creatively about the reconciliation of Aristotelian philosophy and Christian theology allowed others to promote Aristotle's philosophical program with little regard for its theological risks. In short, where there were those of liberal outlook, like Albert and Thomas, there would be others with more radical purposes. The radical faction at the University of Paris, led by Siger of Brabant (c. 1240–84), adopted Aristotle's rationalistic and naturalistic agenda, setting aside theological concerns or constraints to engage in the single-minded application of philosophical method to philosophical problems. Moreover, these "radical Aristotelians" were apparently teaching dangerous Aristotelian doctrines, such as the eternity of the world and denial of divine providence, in the faculty of arts. They attempted to protect themselves from anticipated criticism by noting that, although such conclusions are the proper and necessary conclusions of philosophy, truth lay on the side of theology. In short, philosophical and theological inquiry, each properly pursued, may lead in different directions, which is, of course, to free philosophy from servitude as the handmaiden of theology.

It should come as no surprise that scholars and ecclesiastical authorities of more conservative outlook should come to regard Siger and his group as a threat that required decisive action. The decisive action came in 1270, when the bishop of Paris, Etienne Tempier, condemned 13 philosophical propositions allegedly taught by Siger and his fellow radicals in the faculty of arts. The decree was renewed, with an enlarged list of propositions, now numbering 219, in 1277. The Aristotelian claims identified above as dangerous are all represented on the latter list: the eternity of the world, rationalism, naturalism, and determinism. But a miscellaneous collection of other propositions impinging on natural philosophy was also included. Among them were several astrological propositions that were apparently perceived as dangerous because of the risk of astrological determinism. But the most interesting of the propositions condemned in 1277, from the standpoint of science and religion, were several that described what God could not do (e.g., move the universe in a straight line, create multiple universes, or create accidents without subjects), because these things were judged impossibilities within the framework of Aristotelian metaphysics and natural philosophy. Tempier's point in condemning them was to remind Parisian scholars that divine freedom and omnipotence were not to be compromised by the dictates of Aristotelian philosophy—that God can do anything that involves no self-contradiction and, therefore, could have created, had he wished, a world that violates the principles of Aristotelian metaphysics and natural philosophy.

A great deal of ink has been spilled over the significance of the condemnations of 1270 and 1277. Pierre Duhem, who regarded them as the "birth certificate" of modern science, argued that Tempier's attack on entrenched Aristotelianism provided scholars with the freedom and the incentive to explore non-Aristotelian alternatives and that this theologically sanctioned exploration led ultimately to the emergence of modern science. Most historians of science would now judge Duhem's position to be overblown. A more modest assessment of the condemnations might look like this: In the first place, the condemnations were clearly the product of a conservative backlash against liberal attempts to extend the application of philosophy into the theological realm. They reveal the strength of the opposition and must surely be judged a victory—not for modern science but for theological conservatives at the University of Paris. Their purpose and their effect were to impose limits on philosophical freedom. That they achieved their intended effect is nicely illustrated by the extreme caution exercised by the Parisian master of arts Jean Buridan (c. 1295–c. 1358), writing at the University of Paris about the middle of the fourteenth century. Having strayed into theological territory by arguing against the existence of angelic movers of celestial spheres, Buridan adds that he makes these assertions tentatively, seeking "from the theological masters what they might teach me in these matters as to how these things take place" (Clagett 1959, 536).

But, in the second place, historical reality is not orderly. While inhibiting philosophical speculation in some directions, the condemnations encouraged it in others. There can be no question that the condemnations' stress on God's absolute freedom and power to have made any sort of world he wished, including a non-Aristotelian one, sanctioned the exploration of non-Aristotelian cosmological and physical possibilities. Within the framework of Aristotelian natural philosophy, there was no possibility that the universe (conceived as a single mass) could be put in motion. But scholars committed to the proposition that God could have put it in motion, had he so wished, felt compelled to develop new theories of motion consistent with such an imaginary but (in view of God's absolute power) possible state of affairs. Likewise, no good Aristotelian believed in void space, but stress on God's ability to have created void within or outside the universe led to important speculations about what such a world would have been like, and these speculations contributed, in the long run, to belief in the actual existence of void space. A concrete example may again be useful. In 1377, a full century after the condemnations, the Parisian theologian Nicole Oresme (c. 1320–82) defended the opinion that the cosmos is surrounded by void space and that it is not logically impossible for the universe to be moved in a straight line through this void space, reminding his reader that "the contrary is an article condemned at Paris" (Oresme 1968, 369).

Third, and finally, we need to take the long view. The condemnations of 1270 and 1277 represent a victory for conservatives wishing to restrict the range of philosophical speculation, but surely an ephemeral victory. As the thirteenth century yielded to the fourteenth and the fourteenth to the fifteenth, it became clear that the administrators and scholars who staffed the church bureaucracy and the universities had neither the power nor the desire to place philosophy (especially natural philosophy) on a short leash. Compromises were made, working arrangements were developed, and the church found itself in the role of the great patron of natural philosophy through its support of the universities. Certainly, there were theological boundaries that scholars trespassed at great risk, and there would continue to be skirmishes on specific, sensitive issues. But the late-medieval scholar rarely experienced the coercive power of the church and would have regarded himself as free (particularly in the natural sciences) to follow reason and observation wherever they led. There was no warfare between science and the church. The story of science and Christianity in the Middle Ages is not a story of suppression nor one of its polar opposite, support and encouragement. What we find is an interaction exhibiting all of the variety and complexity with which we are familiar in other realms of human endeavor: conflict, compromise, understanding, misunderstanding, accommodation, dialogue, alienation, the making of common cause, and the going of separate ways. Out of this complex interaction (rather than by repudiation of it) emerged the science of the Renaissance and the early modern period.

BIBLIOGRAPHY

Amundsen, Darrel W. "The Medieval Catholic Tradition." In *Caring and Curing: Health and Medicine in the Western Religious Traditions,* ed. Ronald L. Numbers and Darrel W. Amundsen. 1986. Reprint. Baltimore: Johns Hopkins University Press, 1998, 65–107.

Amundsen, Darrel W., and Gary B. Ferngren. "The Early Christian Tradition." In *Caring and Curing: Health and Medicine in the Western Religious Traditions,* ed. Ronald L. Numbers and Darrel W. Amundsen. 1986. Reprint. Baltimore: Johns Hopkins University Press, 1998, 40–64.

Armstrong, A. H., and R. A. Markus. *Christian Faith and Greek Philosophy.* London: Darton, Longman, and Todd, 1960.

Chenu, M.-D. *Nature, Man, and Society in the Twelfth Century: Essays on New Theological Perspectives in the Latin West.* Ed. and trans. Jerome Taylor and Lester K. Little. Chicago: University of Chicago Press, 1968.

Clagett, Marshall. *The Science of Mechanics in the Middle Ages.* Madison: University of Wisconsin Press, 1959.

Cochrane, Charles N. *Christianity and Classical Culture.* Oxford: Clarendon, 1940.

Dales, Richard C. *Medieval Discussions of the Eternity of the World.* Leiden: Brill, 1990.

Draper, John William. *History of the Conflict between Religion and Science.* 1874. 7th ed. London: King, 1876.

Duhem, Pierre. *Medieval Cosmology: Theories of Infinity, Place, Time, Void, and the Plurality of Worlds.* Ed. and trans. Roger Ariew. Chicago: University of Chicago Press, 1985.

Funkenstein, Amos. *Theology and the Scientific Imagination from the Middle Ages to the Seventeenth Century.* Princeton: Princeton University Press, 1986.

Grant, Edward. "Science and Theology in the Middle Ages." In *God and Nature: Historical Essays on the Encounter between Christianity and Science,* ed. David C. Lindberg and Ronald L. Numbers. Berkeley: University of California Press, 1986, 49–75.

———. *Planets, Stars, and Orbs: The Medieval Cosmos, 1200–1687.* Cambridge: Cambridge University Press, 1994.

Hooykaas, Reijer. "Science and Theology in the Middle Ages." *Free University [of Amsterdam] Quarterly* 3 (1954): 77–163.

Jaki, Stanley L. *The Road of Science and the Ways to God.* Chicago: University of Chicago Press, 1978.

Knowles, David. *The Evolution of Medieval Thought.* New York: Vintage, 1962.

Lindberg, David C. "Science and the Early Church." In *God and Nature: Historical Essays on the Encounter between Christianity and Science*, ed. David C. Lindberg and Ronald L. Numbers. Berkeley: University of California Press, 1986, 19–48.

———. "Science as Handmaiden: Roger Bacon and the Patristic Tradition," *Isis* 78 (1987): 518–36.

———. *The Beginnings of Western Science: The European Scientific Tradition in Philosophical, Religious, and Institutional Context, 600 B.C. to A.D. 1450*. Chicago: University of Chicago Press, 1992.

———. "Medieval Science and Its Religious Context." *Osiris* n.s. 10 (1995): 61–79.

———. "The Medieval Church Encounters the Classical Tradition: St. Augustine and Roger Bacon." In *Science and the Christian Tradition: Twelve Case Histories*, ed. David C. Lindberg and Ronald L. Numbers. Chicago: University of Chicago Press, 2002.

———. "Science and the Medieval Church." In *The Cambridge History of Science*. Vol. 2: *Medieval Science*, ed. David C. Lindberg and Michael H. Shank. Cambridge: Cambridge University Press, forthcoming.

Oresme, Nicole. *Le Livre du ciel et du monde*. Ed. and trans. A. D. Menut and A. J. Denomy. Madison: University of Wisconsin Press, 1968.

Southern, Richard W. *Robert Grosseteste: The Growth of an English Mind in Medieval Europe*. Oxford: Clarendon, 1986.

Steneck, Nicholas H. *Science and Creation in the Middle Ages: Henry of Langenstein (d. 1397) on Genesis*. Notre Dame: University of Notre Dame Press, 1976.

Sylla, Edith D. "Autonomous and Handmaiden Science: St. Thomas Aquinas and William of Ockham on the Physics of the Eucharist." In *The Cultural Context of Medieval Learning*, ed. John E. Murdoch and Edith D. Sylla. Dordrecht: Reidel, 1975, 349–96.

Thorndike, Lynn, ed. and trans. *University Records and Life in the Middle Ages*. New York: Columbia University Press, 1944.

Van Steenberghen, Fernand. *Thomas Aquinas and Radical Aristotelianism*. Washington, D.C.: Catholic University of America Press, 1980.

White, Andrew Dickson. *A History of the Warfare of Science with Theology in Christendom*. 2 vols. New York: Appleton, 1896.

Wippel, John F. "The Condemnations of 1270 and 1277 at Paris." *Journal of Medieval and Renaissance Studies* 7 (1977): 169–201.

6 Islam

Alnoor Dhanani

With its close links to the legacy of Greek science and its transmission of scientific texts and concepts to the medieval Latin world, science in Islam provides an opportunity for the comparative study of science and religion in premodern social contexts. The practice of science in medieval Islam raises the same set of questions as it does in any milieu: What was the attitude toward science? Was there a conflict between religion and science? These complex questions require engagement with the dynamic of actual historical and social contexts. There are several aspects to this engagement: an analysis of the social milieu of medieval Islam and of the dynamic of interaction between political, intellectual, and religious elites at different times and in different places; an investigation into the establishment and role of formal institutions, as well as the informal networks between scientists and the enabling material conditions (particularly the availability of cheap paper) that were conducive to the scientific enterprise; and, finally, an examination of the vexed question of the factors that contributed to the decline of science in Islam.

Alnoor Dhanani received his Ph.D. from Harvard University. He is currently an independent scholar. He is the author of *The Physical Theory of Kalam* (Leiden: Brill, 1994), as well as of several articles on science and philosophy and their interaction with religion in Islamic civilization.

PRIMARILY, *Islam* denotes the monotheistic religion established in the seventh century by the Arabian prophet Muhammad. Secondarily, it denotes the world-historical consequences of this religion, namely, the empire founded by Muslim rulers that, in the ninth century, extended from the Atlantic to the borders of China. *Islam* also denotes a distinct world civilization—Islamic civilization—that first emerged within the heartland of this empire but continued within successor states well into the eighteenth and nineteenth centuries. Any discussion of science and religion in Islam must take into account these three interrelated aspects of Islam: as religion, empire, and civilization.

Background: Religion, Empire, and Civilization

The religion of Islam was established by the Prophet Muhammad, born in A.D. 570 in the Arabian pilgrimage center and commercial town of Mecca. Deeply distressed by the decline in social values in this predominantly pagan milieu, Muhammad would retreat to the surrounding mountains to reflect and meditate. During one such retreat, the angel Gabriel appeared to Muhammad with the first of many revelations from Allah, the one true God (which were collected into the Qur'an), declaring to Muhammad that he was the chosen messenger of Allah. Soon thereafter, Muhammad began preaching monotheism—belief in Allah—and declared himself to be his prophet in the tradition of earlier Semitic prophets, warning of the impending day of judgment, when individuals would have to account for their actions and enter paradise or be banished to hell. He was particularly critical of social injustices resulting from the breakdown of traditional values of charity and hospitality and of the accumulation of wealth without regard to the needy. Not surprising, the Meccan elite rejected Muhammad's message; the handful who answered his call were slaves and others of low socioeconomic status. In A.D. 622, deteriorating relations forced Muhammad to flee to the northern oasis town of Yathrib (thereafter renamed Medina, the city of the Prophet). Here he established the first Muslim community and built its first mosque. Muhammad became the spiritual leader of the Muslims as well as ruler of the multifaith community of Medina. Ten years later, after a series of skirmishes, Muhammad triumphantly reentered Mecca. Thereafter, so many Arabian tribes acknowledged Muhammad's political supremacy and religious mission that the Arabian peninsula was almost united by the time of Muhammad's death in A.D. 632.

Muhammad's death was the first of many crises for the nascent community. Who was to succeed Muhammad? What was to be the basis of authority? These questions were critical to the historical development of the political and religious institutions of Islam. At this early juncture, two tendencies were manifested: One favored continuing Muhammad's religious and political authority under the leadership of his cousin and son-in-law 'Ali (it would later crystallize into the Shia interpretation of Islam); the other held that religious authority had ended (it would crystallize into the majority Sunni interpretation of Islam) but favored political leadership under the first caliph, Abu Bakr (ruled 632–34), a close companion to Muhammad. Abu Bakr's immediate task was to reintegrate the Arabian peninsula under the Islamic banner, preparing the way for his successors' conquests of Palestine, Syria, Egypt, and Iran within two decades of Muhammad's death. Internal division caused by conflict over the distribution of wealth and booty, as well as charges of nepotism, characterized the reign of the third caliph, Uthman (ruled 644–56), leading to a civil war during the reign of the fourth caliph, 'Ali (ruled 656–61). Thereupon the "rightly guided Caliphate" of Muhammad's close companions came to an end, and political control passed on to the Ummayad dynasty (661–750).

The Ummayad dynasty established many of the normative features of Muslim polity, including dynastic political succession and conflict between political and religious elites (despite the nominal title of the caliph as "commander of the faithful"). The latter is significant, as the dialectic of frequent opposition and infrequent cooperation between Muslim political rulers and religious scholars forms the backdrop for the relationship of science and religion in Islam. Thus, pious religious scholars declined Ummayad appointment to judicial positions. However, the Ummayads continued the conquests, pressing east to the Indus Valley and west over the Straits of Gibraltar into Spain, incorporating established centers of science and learning within a single empire. This large empire, with its multiethnic, multifaith, and multinational populations and retaining preconquest administrative structures, was becoming increasingly fractious and ungovernable. In A.D. 696, 'Abd al-Malik (ruled 692–705) adopted measures to unify the empire by introducing Arabic coinage and making Arabic the language of administration. But the significance of these measures was greater, for Arabic was to extend beyond being the language of revelation and now the language of administration to become the language of literature, art, and science. However, opposition to the Ummayads did not subside and culminated in the Abbasid revolution in A.D. 750.

The Abbasids had harnessed several strands of anti-Ummayad sentiment arising from the dynasty's alleged impiety, disregard of religious scholars, and blatant Arabism. While the Abbasids successfully draped themselves with the banner of Islam, they quickly adopted policies, once they were in power, that were opposed by many religious scholars. Divisive tendencies also continued to plague the empire so that, by the mid–ninth century, it was no longer unitary but consisted of semi-independent kingdoms that paid nominal allegiance to the caliph but retained tax revenues. Nevertheless, a unitary vision of Islamic civilization—united by language, common political and religious institutions, burgeoning trade and commerce within the market of a vast empire and beyond, shared aesthetic sensibilities, and an emerging sense of "Islamic" values—began to take hold. It is within such a milieu that religion and science, to say nothing of literature and the arts, which had hitherto undergone modest development, were to blossom and reach the remarkable level that was to become the hallmark of Islamic civilization.

The Appropriation and Naturalization of Science

Within the milieu of Islamic civilization, science and philosophy came to be denoted by several Arabic terms, including *'ulum al-aw'ail* or *'ulum al-qudama'* (the sciences of the ancients), *al-'ulum al-qadima* (ancient sciences), *al-'ulum al-nazariyya* (rational sciences), *al-'ulum al-'aqliyya* (intellectual sciences), and *al-'ulum al-falsafiyya* (philosophical sciences). These terms emphasized the pre-Islamic origins of science and philosophy, their rational character, and their universality, as was well known even in the fourteenth century to the famous historian and judge Ibn Khaldun (1332–1406), who, in his *Introduction to History* (1377), states: "The intellectual sciences are natural to man, in as much as he is a thinking being. They are not restricted to any particular religious group. They are studied by the people of all religious groups who are equally qualified to learn them and to do research in them. They have existed (and been known) to the human species since civilization had its beginning in the world." On the other hand, terms like *'ulum al-'arab* (the sciences of the Arabs), *al-'ulum al-naqliyya* (transmitted science), and *'ulum al-din* (religious sciences) were used for linguistic and religious disciplines, such as grammar, lexicography, religious law, Qur'anic commentary, and philosophical theology. These sciences were considered to be particular insofar as they were practiced only by Arabs or Muslims.

The stage for the process of the appropriation and naturalization of these intellectual sciences was set as a result of the Muslim conquests of the seventh century, when several pre-Islamic centers of science and learning were incorporated into the nascent Islamic Empire. Initially, intellectual activity at these centers continued with little disruption despite the change of rulers. As before, they continued to provide skilled practitioners to the court and to wealthy patrons. Ummayad rulers availed themselves of physicians and astrologers, and the Ummayad prince Khalid Ibn al-Walid (d. c. 704), who was interested in alchemy, sponsored the translation of some alchemical texts into Arabic. Such mostly utilitarian Ummayad interest in science pales when compared to the interest of early Abbasid rulers, particularly Harun al-Rashid (ruled 786–809) and al-Ma'mun (ruled 813–33), who established and generously funded the institution of the House of Wisdom (*Bayt al-Hikma*). The primary function of this institution was to appropriate past wisdom and learning and to enhance it. Apart from its director, the House of Wisdom included translators, copyists, and binders, as well as scientists. It was the royal institution for the translation of Greek, Syriac, Pahlavi, and even Sanskrit scientific and philosophical texts into Arabic. Al-Ma'mun's keen interest is evident in the report that he would attend the weekly salons at the House of Wisdom. Astronomy was of particular interest: Ptolemy's (second century A.D.) *Almagest* was translated into Arabic, as were Sanskrit astronomical texts. In addition, programs of solar observation and terrestrial measurement were conducted at observatories in Baghdad and Damascus, some of whose personnel were also affiliated with the House of Wisdom, followed by the publication of revised astronomical tables. Significantly, al-Ma'mun was also a keen supporter of the nascent discipline of Islamic philosophical theology (*kalam*).

The House of Wisdom, albeit a royal institution, was one among several sponsors of scientific research and translation. The attitude of these patrons toward earlier civilizations is reflected in the statement of the scientist and philosopher Abu Ya'qub al-Kindi (800–870): "We ought not to be ashamed of appreciating truth and of acquiring it wherever it comes from, even if it comes from races distant and nations different from us." Significantly, al-Kindi's remark demonstrates a conscious commitment to appropriating learning and knowledge. We may surmise that this commitment was shared by al-Kindi's royal patrons, al-Ma'mun and al-Mu'tasim (ruled 833–42). True to his dictum, al-Kindi sponsored an early Arabic translation of Aristotle's (384–322 B.C.) *Metaphysics*. The active sponsorship of translation led to Arabic versions of the available Greek scientific and philosophical corpus, as well as some Syriac, Pahlavi, and Sanskrit texts.

Constant efforts to improve translations of texts laid the basis for Arabic "naturalization" of the pre-Islamic scientific and philosophical heritage. Knowledge of Greek or other languages was no longer a requisite for scientific activity. Rather, from Spain to the borders of China, scientific activity was conducted in Arabic, utilizing a vocabulary coined by the translators but naturalized in subsequent scientific works. The ready availability of cheap paper since the end of the seventh century allowed booksellers, even in smaller provincial towns, to stock their shelves with translated, as well as a growing corpus of original Arabic, scientific texts. Moreover, royal and provincial courts and wealthy patrons vied to sponsor scientific work and establish public and private libraries. The education of Abu 'Ali Ibn Sina (980–1037), known in Latin as Avicenna, is illustrative in this regard. Born in a village in Turkistan, at a very young age he moved with his family to the provincial capital of Bukhara, where his formal learning started with the Qur'an and literature. His father then sent him to a vegetable seller to learn arithmetic. Later, when someone who claimed to be a philosopher came to Bukhara, his father had him stay in their house and tutor the young Ibn Sina, who soon surpassed his teacher. He began to study on his own, reading Euclid's (fl. c. 300 B.C.) *Elements*, Ptolemy's *Almagest*, and Aristotle's logical, physical, and metaphysical works and their commentators. He then taught himself medicine and became so proficient that distinguished physicians began to read medicine with him. Aristotle's *Metaphysics* raised difficulties. Rereading this work forty times, he despaired until he was persuaded in the booksellers' quarter to purchase al-Farabi's (873–950) *On the Purposes of Metaphysics* for the cheap price of three dirhams. With the aid of this text, he overcame his difficulties. At this time, the ruler of Bukhara was ill, and the young Ibn Sina was summoned to participate in his treatment. Here Ibn Sina entered the royal library, "a building with many rooms . . . in one room were books on the Arabic language and poetry, in another jurisprudence, and so on in each room a separate science. I looked through the catalogue of books by the ancients and requested those which I needed."

Ibn Sina's account illustrates the fact that scientific education was not imparted through formal institutions but, rather, via informal, personal contacts. Notable exceptions include hospitals, where medicine was taught in a master-apprentice setting, and possibly astronomical observatories and academies like the House of Wisdom of Baghdad or the House of Knowledge founded by the Fatimid caliphs of Cairo in the tenth and eleventh centuries, as well as Christian theological schools where Aristotelian texts were studied within a centuries-old curriculum. (That these Christian theological schools had survived into the eleventh century is evident in Ibn Sina's disparaging and acerbic critique of their dogmatic and rigid philosophical views.) The lack of formal educational institutions does not, however, entail the absence of a curriculum. There was an established order of study and of texts. Moreover, scientific texts were available both in private and public libraries and for purchase in thriving booksellers' markets. Courts patronized scientists in their practical roles as engineers, astronomers, astrologers, and physicians, thereby providing them with the wherewithal to pursue their research interests. Hence, despite the lack of formal educational institutions for the study of science, the environment of Islamic civilization was conducive to the pursuit of science and philosophy, as is obvious from its scientific and philosophical legacy. Individuals from diverse religious communities, spread over a vast geographical span, participated in this enterprise. They were critical of the scientific theories of predecessors and contemporaries and made substantial advances in their chosen fields.

In addition to critical evaluation of previous theories and continuing research programs, several peculiar problems occupied scientists in an enterprise that has been termed "science in the service of Islam." These problems, which arose from the requirements of religious practice, included the determination of inheritance shares, the determination of the direction of Mecca from any locality so that the faithful might know in which direction to face for prayer, and the determination of times of prayer, some of which were formulated in terms of shadow lengths. In his foundational work on algebra, the mathematician and astronomer al-Khwarizmi (d. c. 847) shows how arithmetic and algebra can be applied to solve problems of Muslim inheritance law. The other two problems of determining the direction of prayer and its times occupied many of the mathematicians of Islamic civilization, who, using novel trigonometric approaches, proposed several solutions.

However, despite the engagement of scientists in such uniquely Muslim problems, the close links between science and Hellenistic philosophy established in Greek antiquity continued to be maintained in the Islamic milieu, as, indeed, in western Europe until the scientific revolution. Thus, the scientific worldview was the predominant Neoplatonized version of Aristotelianism formulated in late antiquity. While God was the ultimate cause and Creator of the cosmos, he played no direct role in its activity. Rather, creation was an eternal process of emanation, from God to celestial intellects, celestial souls, and thence to entities lower in the chain of being. The eternal cosmos consisted of the incorruptible celestial and the corruptible terrestrial realms. In the celestial realm, planets, which were intelligent living beings, revolved around the earth, their motion caused by their souls. Planets influenced events on the earth, though there were different views about whether this influence was predictable, as claimed by astrology. In the terrestrial realm, combinations of the four elements of air, water, earth, and fire gave rise to inanimate minerals and animate plants, animals, and humans. Animation was the result of vegetative, animal, and rational souls, which were susceptible to the influence of planetary souls. There were differing views on the nature of soul, whether it was immaterial or whether it was produced by an equilibrium of the combination of elements. The philosophical worldview also incorporated a modified version of Platonic political philosophy, deriving from the *Republic.* The phenomena of prophecy and religion were understood within the political and psychological perspectives of this system. There was great emphasis on the role of celestial planetary beings in producing and sustaining these phenomena. Such a worldview was far removed from the literal text of revelation. Hence, scientists and philosophers were, to varying degrees, proponents of an allegorical and interpretive reading of the Qur'an.

Not surprisingly, religious groups and scholars sympathetic to an allegorical interpretation of revelation were quick to adopt this worldview, with some modification to bring it into conformity with their specific doctrines. One such group were the Isma'ilis, a branch of the Shia, who founded the Fatimid state in North Africa in the tenth century. Later, in their new capital of Cairo, they established institutions like the House of Science (*Dar al-'Ilm*) and the university-mosque of al-Azhar and patronized the activities of several scientists. An anonymous group sympathetic to the Isma'ilis was responsible for the encyclopedic *Treatises of the Brothers of Sincerity,* written in the tenth century. These treatises present a popular account of the Neoplatonic-Aristotelian worldview within a pietistic, Muslim framework. The popularity of these treatises, which were read even by Ibn Sina and the religious scholar al-Ghazali (1058–1111), among others, attests to the widespread naturalization of this worldview within the milieu of medieval Islamic civilization among nonscientists, particularly, belletrists, poets, and others.

Attitudes toward the "Foreign" Sciences

The relationship of religion to science in Islamic civilization is complex. The roots of such complexity lie in the long-drawn-out and sustained dynamic among local context, political power, religious authority, patronage, and competition among elites, as well as individual epistemological commitments. Attitudes expressed by proponents and opponents of the "foreign" sciences do not reveal this larger setting. Rather, the participants in this discourse formulated their positions within distinctive analytical frameworks and categories. From our perspective, this discourse can be framed by asking certain questions. How, from a religious point of view, should science be evaluated? By which religious disciplines? What is to be the analytic framework of evaluation? What was the reaction of the proponents to their opponents' views?

The disciplines of religious law (*shar'ia*) and philosophical theology (*kalam*) are paramount to understanding the attitudes of religious scholars regarding the relationship of religion to science. The status and roles of these two disciplines varied over the course of Islamic history and were quite different from those in the medieval West. While the origins of these disciplines lie in the seventh century, their mature formulation and formalization were achieved later. Their formative period was contemporaneous with the translation movement to appropriate the scientific and philosophical heritage of earlier civilizations. Interpretive differences, made evident in sectarianism but also arising from differences in local contexts, played a substantial role in the formation of competing schools of religious law and philosophical theology. In the case of the former, pressure to normatize was manifested in the quest to establish a methodology that would restrict the derivation of religious law to "valid" sources only. According to the ninth-century formalization by the Sunni religious scholar al-Shafi'i (d. 820), these sources must consist of only prescriptions of the Qur'an, the tradition of the prophet Muhammad, and the consensus of religious scholars. Earlier Sunni authorities had allowed reasoning by analogy and judicial personal opinion as legitimate sources, thereby incorporating local customary practices into their formulation of law. On the other hand, Shia scholars maintained that the continuity of the Prophet's religious authority in the *imam* meant that the *imam* was the primary source of religious law and Qur'anic interpretation. Despite some convergence with the passage of time, such differences over the sources of law and over legal prescriptions manifested themselves in opposing schools of law and in competition between scholars even within the same school. (Only four Sunni schools, Hanafi, Maliki, Shafi'i, and Hanbali, survive today; several Shia schools also survive.) Proponents of religious law insisted that it should govern all spheres of life: personal conduct, the activities of state, religious affairs, and commercial transactions, as well as intellectual pursuits.

Diversity notwithstanding, scholars of religious law had developed a system of categories whereby any activity was either required, recommended, neutral, abhorred, or prohibited. Within this scheme, some early religious scholars had advocated the pursuit of only those sciences that had practical utility, whether religious or social. A mature exposition of this is found in the section on knowledge in the *Revival of the Religious Sciences* by the influential Sunni religious thinker Abu Hamid al-Ghazali, who belonged to the Shafi'i school of religious law. After extolling the virtues of knowledge on the basis of Qur'anic references and prophetic traditions, al-Ghazali classifies the sciences as sacred or profane. Sacred sciences are acquired through prophets, not (like arithmetic) through reason, or (like medicine) through experience, or (like language) through "hearing" (i.e., social discourse). Profane sciences, whether acquired through reason, experience, or social discourse, may, from the point of view of religious law, be praiseworthy, blameworthy, or tolerated. Praiseworthy profane sciences are those upon which everyday worldly activities depend, such as arithmetic and medicine. Blameworthy profane sciences include magic, talismanic sciences, and trickery. After discussing the sacred sciences, al-Ghazali has an imaginary interlocutor state: "In your classification, you have mentioned neither philosophy nor philosophical theology, nor have you clarified whether they are praiseworthy or blameworthy." In response, al-Ghazali states that, as for philosophical theology, whatever is of utility in it is already covered in the Qur'an and the prophetic tradition. As for philosophy, it consists of four types of sciences: mathematical sciences, logic, metaphysics, and physical sciences. Mathematical sciences are permissible as long as the mathematician is not led to the blameworthy sciences (al-Ghazali believed that most mathematicians eventually are enticed). Logic studies proofs and definitions, subjects also covered in philosophical theology. Metaphysics studies God and his attributes (also covered by philosophical theology), but here philosophers have branched into schools, some of which are characterized by outright unbelief, while others are heretical. As such, metaphysics is clearly blameworthy. As for the physical sciences, some of them are opposed to religion and truth, while others study bodies, their properties, and their changes. Unlike medicine, which studies human bodies, the latter do not have any utility. Al-Ghazali thus endorses a utilitarian position, which is approving of "practical" sciences such as arithmetic and medicine, skeptical of "theoretical" sciences, and disapproving of metaphysics and some physical sciences.

A proponent's attitude toward the foreign scientists is provided by the famous mathematician and scientist al-Biruni (d. 1050). He frames his attitude within a critique of religious scholars who, lacking the nuanced position of al-Ghazali, considered the pursuit of all science and philosophy to be a blameworthy activity. Al-Biruni was patronized by Sultan Mahmud of Ghazna (ruled 998–1030), the ruler of the eastern edge of the Islamic Empire and the conqueror of India. Al-Biruni characterizes these religious scholars as

> the people of our times . . . who oppose the virtuous and attack those who bear the mark of knowledge. . . . they have settled on the worst of morals. . . . You always see them unashamedly begging with outstretched hands . . . seizing opportunities to grab more and more leading them to reject the sciences and despise its practitioners. The extremist among them relates sciences to the path which leads astray thus making like-minded ignoramuses detest the sciences. He labels sciences as heretical so as to allow himself to destroy their practitioners.

Al-Biruni goes on to state that such religious scholars feign wisdom by questioning the utility of the sciences, ignoring the fact that the virtue of humans over other animals lies in pursuing knowledge and that good can arise, or evil be avoided, only by means of knowledge. Neither in spiritual nor in worldly affairs can we be sure without knowledge. Al-Biruni reiterates the scientists' view that a believer seeking real truth (which, to al-Biruni, is obviously not found in competing religious dogmas) must study creation in order to know the Creator and his attributes. Significantly, the context for these remarks is al-Biruni's work on *The Determination of Coordinates of . . . Cities,* in which he presents methods for determining latitude and longitude of localities, as well as methods to determine the direction of Mecca, toward which Muslims must turn for their daily prayers. As he tells us, "What we have discussed regarding the correction of longitudes and latitudes of towns is beneficial to the majority of Muslims in determining the direction of prayer and hence its performance free of the errors of the unfounded legal determination [by religious scholars]." In his *Exhaustive Treatise on Shadows,* al-Biruni defends the study of astronomy: "The learned in religion who are deeply versed in science know that religious law does not forbid what astronomers practice except [the visibility] of the lunar crescent." (Here al-Biruni accepts the prevalent social attitude based on religious law that the beginning of a lunar month is to be established by the actual observation of the lunar crescent by qualified witnesses rather than by astronomical calculation. The importance of this to religious ritual is evident, for example, in determining the beginning and end of Ramadan, the month of fasting.) In this interesting work, al-Biruni presents several views about Muslim times of prayer, demonstrating his mastery of the religious sciences, and then shows how trigonometrical methods may be used to determine times of prayer.

The discipline of philosophical theology, like religious law, also played a substantial role in the relationship between religion and science in Islam. It had emerged in the early Islamic milieu of intense debate with proponents of other religious traditions and cosmologies, especially in natural philosophy. Characterized by rationalism (insofar as it subscribes to a rationalist epistemology rather than the faith-based epistemology of some Christian theology), it sought to explicate Islamic beliefs regarding God, Creation, prophecy, and human religious obligation. By the mid–ninth century, the Mu'tazili school of philosophical theology reached consensus on several aspects of its concerns: The eternal God was the direct Creator and Sustainer of a created world whose ultimate constituents were atoms and the inherent qualities (like color and taste). They combine to form bodies that we perceive. A particular combination of atoms and qualities constituted in a specific manner, namely the shape of a human being, allowed for the further inherence of qualities of life, will, knowledge, and the capacity for action. Human beings were obligated to God, who had provided them with material sustenance and natural intellect, as well as religious guidance mediated through revelation via prophets, to live in a manner consistent with God's law—a manner that was for the benefit of mankind. Conforming to these obligations would enable man to achieve the felicity of paradise; disregarding them would lead to harm in hell. Causal agency was restricted to direct and indirect action by living agents, namely God and human beings; the concept of natural causation as a result of "natural properties" was unintelligible. Planets were not living beings but inanimate stones with no causal influence whatsoever. This brief summary demonstrates the radically different, non-Aristotelian character of the worldview of philosophical theology compared to the philosophical worldview.

However, philosophical theology was not, as is evident in al-Ghazali's account in the *Revival of the Religious Sciences,* without its detractors from within, who, critical of its rationalist methods, insisted on a conformist adherence to religious tenets as propounded by orthodox schools of religious law. These schools defined themselves as "followers of the pious ancestors," rejecting any deviation from their views as a heretical innovation and, therefore, against religious law. The attempt by the Abbasid caliph al-Ma'mun to impose Mu'tazili philosophical theology on religious scholars was met with stiff resistance by opponents of this view, forcing his successors to abandon the enterprise. Subsequently, a school of philosophical theology that was more in alignment with the traditionalists but maintained the rationalist stance characteristic of this discipline was founded by al-Ash'ari (d. 935). It became the predominant school of theological philosophy after the eleventh century. Significantly, al-Ash'ari's school restricted causal agency further, holding that God was the only and direct cause of all events, even of human actions.

Despite al-Ghazali's later misgivings about philosophical theology, he had, in his earlier years, embraced al-Ash'ari's critique of the Neoplatonic-Aristotelian worldview of the scientists and philosophers. His thorough acquaintance with the philosophers' worldview, in particular as explicated by Ibn Sina, is reflected in his work *On the Aims of the Philosophers,* which is a summary of the doctrines of the Islamic philosophers. (In the Latin West, al-Ghazali was erroneously considered to be one of the principal Islamic philosophers primarily through the translation of this work.) Al-Ghazali followed this preliminary study with a masterful critique, *The Incoherence of the Philosophers,* which argued against key elements of the philosophical worldview that were contrary to the literal meaning of revelation and to philosophical theology, such as the doctrines of the eternity of the world, God's lack of knowledge of particular events in the world, and the impossibility of physical resurrection. On this basis, he accused the philosophers of unbelief. Moreover, he rejected the philosophers' theory of natural causation, which was fundamental to their physical doctrines. In conformity with al-Ash'ari's philosophical theology, he denied any connection between causes and effects. Rather, God is the only cause of all events, and, as a free agent, he can alter habitual patterns of presumed effects following their presumed causes (e.g., burning when fire and cotton are brought together) because their connection is arbitrary. Such observed patterns are entirely within God's causal activity, which is wholly volitional, for God can choose to alter such patterns at any juncture.

A direct answer to al-Ghazali's critique was made in the westernmost part of the Islamic Empire by the famous Spanish Muslim philosopher and judge Ibn Rushd (1126–98), known to the Latin West as Averroës. A biographical account tells how Ibn Rushd was introduced by the philosopher and royal physician Ibn Tufayl (d. 1186) to the Almohad ruler Abu Ya'qub Ibn Yusuf (ruled 1163–84). Prompted by the ruler to discuss the controversial question of the eternity of the world, Ibn Rushd hesitated. But the ruler displayed his familiarity with philosophy by discussing the question with Ibn Tufayl, easing the young Ibn Rushd's anxieties. Abu Ya'qub then commissioned Ibn Rushd to write the Aristotelian commentaries that earned him the epithet of the Commentator in the Latin West. Ibn Rushd was also appointed judge of Seville and later chief judge of Cordoba, continuing the family tradition of engagement with religious law. Within this context, Ibn Rushd was also the author of a major Maliki treatise on religious law. Ibn Rushd adopted a legal framework in his work *The Decisive Treatise on the Harmony between Religion and Philosophy,* in which he reexamined al-Ghazali's question of the validity of the pursuit of science and philosophy. Rather than discuss this question in terms of whether the science is blameworthy, praiseworthy, or tolerated, as al-Ghazali had, Ibn Rushd examined whether, from the perspective of religious law, philosophical inquiry and logic are permissible, prohibited, or required. Ibn Rushd argued that philosophy studies creation and reflects upon the Creator. Since revelation commands believers to recognize the Creator, he concluded that philosophy provides the best method for this, for it is a demonstrative science. However, not everyone is suited to pursue demonstrative science, and, for the masses of believers, Islam allows rhetorical and dialectical knowledge provided by the religious sciences, including theological philosophy. Philosophy does not conflict with revelation. Rather, the true intent of revelation is accessible only to the philosopher, who, grounded in demonstrative science, is best suited to undertake the interpretive task. Such interpretation is necessary, for, in its apparent form, revelation must be couched in symbols to be accessible to the masses so that they may be persuaded to believe, perform religious acts, and maintain social order. Conflicts arise only when unqualified masses are provided access to metaphorical meanings of revelation and the results of demonstrative science by religious scholars like al-Ghazali.

Ibn Rushd also responded to al-Ghazali's critique of philosophical doctrines in his *The Incoherence of the Incoherence of the Philosophers* with a point-by-point rebuttal of the doctrines of Ash'ari philosophical theology that al-Ghazali had championed. Ibn Rushd and other scientist-philosophers were put in a precarious position when, in 1190, they lost the support of the Almohad ruler Abu Yusuf (ruled 1184–99), who, under the pressure of a growing Reconquista threat by the Portuguese, acquiesced in the antagonism of Maliki legal scholars toward scientists and philosophers. Ibn Rushd was rehabilitated a few months before his death, no doubt because of improvement in Abu Yusuf's political position.

Ibn Rushd's case illustrates the complexity of the relationship between religion and science in Islam, but it would be inappropriate to draw general conclusions about this relationship on the basis of such cases. This is because diversity is a characteristic of Islamic religious doctrines and institutions, as is evident from differing views of religious law. As such, the case of Ibn Rushd is not as representative as it is sometimes made out to be. This is not surprising. Islamic doctrines and institutions developed across a vast geographical area and over a long temporal period, sometimes in the context of opposition to political authority and almost always within localized milieus that had retained pre-Islamic concepts and institutions and even naturalized them within "Islamic" terms. Moreover, like other world religions, Islam has also been engaged in a continual dialectic with competing worldviews (e.g., Hellenism in the past and secularism in the present), which has given rise to movements that have sought to accommodate, adapt, or reject such worldviews. Such movements and their doctrines, influence, and effects on the relationship of religion and science need to be placed within the dynamic of their historical context.

In the case of Ibn Rushd, the local Andalusian milieu (where predominantly Maliki religious scholars opposed the theological and legal views of their Almohad rulers), the anti-Ash'arism (both of Almohad theology and Andalusian Malikism), Almohad literalist legal theory, and the pressure of the Reconquista movement are all factors that need to be considered to understand the environment that, in good times, sustained a substantial number of scientists and philosophers but, in bad times, led to their persecution. Even al-Ghazali's opposition to science and philosophy needs to be placed in the context of the ongoing pro-Shia Fatimid military and intellectual challenge to the Abbasids; the growing strength of the Sunni Seljuqs, who were the principal backers of the Abbasids; and the establishment by the Seljuq vizier Nizam al-Mulk (1018–92) of a state-sponsored system of colleges of religious law (*madrasas*), in which Shafi'i religious law and Ash'ari philosophical theology were primarily taught (al-Ghazali was a professor in such a college in Baghdad). Moreover, due consideration needs to be given to al-Ghazali's positive attitude to logic, resulting in his wholesale appropriation of it into religious law. For this reason, al-Ghazali was viewed with suspicion by some religious scholars and even condemned by others.

Clearly, the lack of a central hierarchical religious institution in Islam analogous to the church inhibited uniformity of doctrine and religious law. Rather, uniformity was possible only through the widespread consensus of religious scholars, which was hard to achieve. When it was achieved, it was largely the result of the decline of opposing views or of the domination of one group through its alliance with political authority. As a result, political power and religious authority were neither allied nor concentrated in the same manner as in the medieval West but relied much more on local factors, particularly the political ambitions of local elites, which so infuriated al-Biruni. However, when faced with a united and powerful opposition, which in normal circumstances tended to be fractious, political authority was under great pressure to conform, as was the case in Ibn Rushd's Andalusia. The fact that Ibn Rushd himself was a legal scholar and that biographical accounts mention other examples of legal scholars who were well versed in science reveals the complexity of the actual situation and the difficulty of maintaining the view of a simple dichotomy between scientists on one side and religious scholars on the other.

The Decline of Science in Islam

What was the influence of the opposition to scientific activity described above? Some scholars maintain that science and philosophy were "marginal" to Islamic civilization, while others have suggested that such opposition caused the decline of scientific activity in Islamic civilization and explains the absence of a scientific revolution. The marginality thesis suffers from an obvious flaw. It underrates the geographical breadth and the temporal duration of the scientific enterprise, the large number of people who were engaged in it despite the lack of formal educational institutions, the patronage of the court and other wealthy citizens, and, finally, its achievements.

That scientific activity declined is indisputable. The scientific activity of later centuries was not qualitatively at the same level as that of the earlier period. But when it declined, where it declined, why it declined, and whether it declined as a result of opposition to science are questions requiring detailed study. The thesis that al-Ghazali's critique was responsible for the decline of science in the twelfth and later centuries can no longer be maintained, for substantial progress was made in astronomy and mathematics in the thirteenth and fourteenth centuries as a result of the research program of the Maragha observatory, initially sponsored by the Mongols. Moreover, there were also important developments in medicine and optics during this period.

Even in the case of philosophy, al-Ghazali's influence has been exaggerated. The case of Nasir al-din al-Tusi (d. 1274), the famous mathematician, astronomer, director of the Maragha observatory, and Ibn Sina commentator, is illustrative. Al-Tusi is also the author of several religious texts, including a very influential Twelver Shia text on religious philosophy in which he presented

Twelver philosophical theology within the framework of Ibn Sina's philosophy (earlier Twelver works on philosophical theology were under the influence of Mu'tazilism). Within Twelver intellectual circles, the pursuit of philosophy continued, reaching its peak in the School of Isfahan. The representatives of this school, including Mir Damad (d. 1630) and Mulla Sadra (d. 1640), were part of a broad cultural and intellectual renaissance initiated by the founding of the Safavid state in Iran in 1501. In Sunni circles, too, the impact of al-Ghazali's critique of philosophy and philosophical theology was not definitive. The pursuit of theological philosophy continued in the centuries after his death, producing some of its classic texts. However, Sunni theological philosophy had also undergone transformation, for, while remaining true to al-Ash'ari's central tenet of an absolutely free Creator, it appropriated many elements of philosophy. (Many features of late Ash'ari philosophical theology find resonance with the ideas of seventeenth-century European philosophy. They include a determined anti-Aristotelianism and the concept of an omnipotent and absolutely free God, features that are usually regarded as being fundamental to the seventeenth-century reorientation of science and philosophy.) Hence, without substantive direct evidence, it is difficult to maintain that al-Ghazali's critique of science and philosophy entailed the decline of scientific activity in the world of Islam.

However, historical factors of the disintegrating Islamic Empire of the twelfth and thirteenth centuries (such as the rise of regional "nationalisms" [particularly Iranian]; the gradual displacement of Arabic, which was the language of science, by "vernacular"languages like Persian and Turkish; and the economic and social havoc wreaked by the Mongol conquest) bear further examination. So do changes in patronage patterns that began in the late eleventh century when the Seljuq vizier Nizam al-Mulk diverted substantial public resources into a state system of religious colleges. Such a change in patronage is even found in the case of private donors who now patronized religious colleges as well as Sufi centers. Funds that previously may have been used to support scientific activity were now diverted into these kinds of activities. Nevertheless, several accounts suggest that sciences continued to be taught privately despite being excluded from the official curriculum of most religious colleges. Clearly, where hospitals were allied to religious colleges, allowances must have been made for the study of natural sciences as a prerequisite for those pursuing medical careers.

Regional factors must also be explored. In Syria and Egypt, the position of mosque timekeeper was established in the fourteenth century. His function was to determine times of prayer and times of religious festivals. Most appointees to this position were very skilled astronomers, such as Ibn al-Shatir (fl. c. 1360), who completed the project of reforming astronomy, or al-Khalili (fl. c. 1360), who compiled an extremely accurate set of astronomical tables based on the astronomical work of the Maragha school. In Safavid Iran, there was renewed interest in philosophy, evident in the sixteenth-century School of Isfahan. Evidence from post-fifteenth-century Ottoman religious colleges shows the incorporation of the sciences into the curriculum. Only such continued study of science can explain the proliferation of commentaries on established scientific texts and, indeed, the existence of manuscript copies of the texts themselves from later periods when science is thought to have declined, to say nothing of the continued, albeit less prolific, production of new texts.

The decline of scientific activity is thus a complex phenomenon, but it is usually seen by historians through the lens of the seventeenth-century scientific revolution in Europe. The question has been raised whether religious opposition to science inhibited such a scientific revolution in Islam, given the otherwise high level of scientific achievement. But the premise of this question—that successful scientific activity must lead to a scientific revolution like that of the seventeenth century in Europe—is invalid and introduces misleading hypothetical conjectures.

The relationship of religion and science in the Islamic milieu is complex. That the scientific enterprise in Islam reached the remarkable level of achievement it did shows that science was a valued endeavor. On the other hand, it is also clear that there was opposition to science and that science did, indeed, subsequently decline. But these features, in their more detailed manifestations of the appropriation of science, its subsequent naturalization, its efflorescence, the opposition to it, and its decline, are historical phenomena, and, as subjects for historical investigation, they require detailed study of the local social, political, and epistemic contexts in which they occurred. The need to reiterate such a basic methodological norm of historical study reflects the continued domination of general universalistic explanations grounded in essentialist and ahistorical notions—explanations that have rightfully been abandoned elsewhere—in studies investigating the relationship of religion to science in the Islamic milieu.

BIBLIOGRAPHY

al-Biruni, Abu al-Rayhan. *The Determination of the Coordinates of Positions for the Correction of Distances between Cities.* Trans. Jamil Ali. Beirut: American University of Beirut, 1967.

———. *The Exhaustive Treatise on Shadows.* Trans. E. S. Kennedy. Allepo: Institute for the History of Arabic Science, 1976.

al-Farabi. *Al Farabi on the Perfect State.* Trans. R. Walzer. Oxford: Clarendon, 1985.

al-Ghazali, Abu Hamid. "Autobiography." In *Freedom and Fulfillment: An Annotated Translation of al-Ghazali's Muqidh min al-Dalal and Other Relevant Works of al-Ghazali.* Trans. R. McCarthy. Boston: Twayne, 1980, 61–143.

———. "The Clear Criterion for Distinguishing between Islam and Heresy." In *Freedom and Fulfillment: An Annotated Translation of al-Ghazali's Muqidh min al-Dalal and Other Relevant Works of al-Ghazali.* Trans. R. McCarthy. Boston: Twayne, 1980, 145–74.

———. *The Book of Knowledge, Being a Translation with Notes of the Kitab al-'ilm of al-Ghazali's Ihya 'ulum al-din.* Trans. N. Faris. Delhi: International Islamic Publishers, 1988.

———. *The Incoherence of the Philosophers.* Trans. M. Marmura. Provo: Brigham Young University Press, 1997.

al-Kindi, Abu Ya'qub. *Al-Kindi's Metaphysics.* Trans. A. Ivry. Albany: State University of New York Press, 1974.

Averroës. *On the Harmony of Religion and Philosophy.* Trans. George Hourani. London: Luzac, 1967.

———. *The Incoherence of the Incoherence of the Philosophers.* Trans. Simon van den Bergh. London: Luzac, 1978.

Avicenna. *The Life of Ibn Sina.* Trans. William Gohlman. New York: State University of New York Press, 1974.

Dhanani, Alnoor. *The Physical Theory of Kalam.* Leiden: Brill, 1994.

Gimaret, D. *La Doctrine d'al-Ash'ari.* Paris: Les éditions du Cerf, 1990.

Goldziher, Ignaz. "The Attitude of Orthodox Islam towards the 'Ancient Sciences.'" In *Studies in Islam.* Trans. M. Schwartz. New York: Oxford University Press, 1981, 185–215.

Grunebaum, Gustav von. "Muslim World View and Muslim Science." In *Islam: Essays in the Nature and Growth of a Cultural Tradition.* London: Routledge and Kegan Paul, 1964, 111–26.

Gutas, Dimitri. *Avicenna and the Aristotelian Tradition.* Leiden: Brill, 1988.

———. *Greek Thought, Arabic Culture.* London: Routledge, 1998.

Halm, Heinz. *The Fatimids and Their Traditions of Learning.* London and New York: I. B. Tauris, in conjunction with the Institute of Ismaili Studies, 1997.

Huff, Toby. *The Rise of Early Modern Science: Islam, China, and the West.* Cambridge: Cambridge University Press, 1993.

Ibn Khaldun. *The Muqadimmah: An Introduction to History.* Trans. F. Rosenthal. Princeton: Princeton University Press, 1958.

Khadduri, Majid. *Islamic Jurisprudence: Shafi'i's Risala.* Baltimore: Johns Hopkins P0.ress, 1961.

King, David. *Astronomy in the Service of Islam.* Aldershot, England: Variorum, 1993.

Marmura, Michael. "Al-Ghazali's Attitude towards the Secular Sciences and Logic." In *Essays on Islamic Philosophy and Science,* ed. G. Hourani. Albany: State University of New York Press, 1975, 100–11.

Puig, J. "Materials on Averroës Circle." *Journal of Near Eastern Studies* 51 (1992): 241–61.

Rahman, Fazlur. *Prophecy in Islam: Philosophy and Orthodoxy.* London: Allen and Unwin, 1958.

Rashed, Rushdi. *Encyclopaedia of the History of Arabic Science.* London: Routledge, 1996.

Sabra, A. I. "The Appropriation and Subsequent Naturalization of Greek Science in Medieval Islam: A Preliminary Statement." *History of Science* 25 (1987): 223–43.

———. "Philosophy and Science in Medieval Islamic Theology: The Evidence of the Fourteenth Century." *Zeitschrift für Geschichte der arabisch-islamischen Wissenschaften* 9 (1994): 1–42.

———. "Situating Arabic Science: Locality versus Essence." *Isis* 87 (1996): 654–70.

Saliba, George. *A History of Arabic Astronomy.* New York: New York University Press, 1994.

Sayili, Aydin. *The Observatory in Islam.* Ankara: Türk Tarih Kurumu Basimevi, 1960.

Urvoy, Dominique. *Ibn Rushd (Averroës).* Trans. O. Stewart. London: Routledge, 1991.

Part III : : The Scientific Revolution

7 The Copernican Revolution

Owen Gingerich

The intellectual ferment that swept over Europe in the wake of the invention of printing by movable type and the discovery of a New World to the West included, besides profound religious changes, a fresh approach to nature and to the heavens above. In retrospect, it was perhaps inevitable that these radically evolving viewpoints would become mutually entangled. This essay considers a pivotal episode in the transition from a geocentric to a heliocentric universe, with special attention to the religious implications and interactions.

Owen Gingerich is research professor of astronomy and history of science at the Harvard-Smithsonian Center for Astrophysics. His most recent book is *An Annotated Census of Copernicus'* De Revolutionibus *(Nuremberg, 1543 and Basel, 1566)* (Leiden: Brill, 2001).

IN 1543, the year of his death, Nicholas Copernicus (1473–1543) saw his life work, *De revolutionibus orbium coelestium* (*On the Revolutions of the Heavenly Bodies*), finally printed. A four-hundred-page technical treatise, it laid out a heliocentric framework for the planetary system, thereby providing the essential basis for the Newtonian synthesis that was to follow a century and a half later. During this same interval, the gradual overthrow of the long-accepted geocentric worldview created an upheaval in the sacred geography of the cosmos. These changes, both in technical astronomy and mechanics and in humankind's vision of its physical place in the universe, constitute the Copernican revolution.

Copernicus was born in Torun, Poland, in 1473. His father died when he was ten years old, and his maternal uncle, Lucas Watzenrode, took over responsibility for the young man's education. Watzenrode was in a successful career of ecclesiastical politics, becoming in 1489 bishop of the northernmost Roman Catholic diocese in Poland, and here he provided a position for Copernicus as canon in the Frombork (Frauenburg) Cathedral. Copernicus was never ordained as a priest, but he took minor orders and, after appropriate graduate study in Italy, served as personal physician to his uncle and as the principal legal officer of the cathedral chapter.

Precisely when and where Copernicus caught the vision of a heliocentric system we do not know. He was interested in astronomy even while an undergraduate at Cracow, and he continued to develop his understanding as he studied canon law in Bologna from 1496 to 1500. By 1514, he had written out a short precis of the heliocentric astronomy, the so-called *Commentariolus*, which was, however, not printed until its rediscovery in the 1880s. The Latin edition of Ptolemy's *Almagest* in 1515 showed Copernicus the required scope of any treatise that would challenge Ptolemy's authority, and he began work in earnest on his *De revolutionibus*. He quickly realized that he would need a baseline of nearly twenty years of observations to establish the modern parameters for the planets, so he bided his time with a variety of duties for the cathedral as he slowly collected the fresh data. Only toward the end of his life did he finally pull together the various parts of his extensive and highly mathematical account.

The opening chapters of *De revolutionibus* lay the philosophical foundations for a moving earth and a fixed central sun, leading to the glorious chapter I,10, a powerful rhetorical defense of the heliocentric cosmology, pointing to the sun "as if on a royal throne governing the planets that wheel around it. For in no other way can we find such a wonderful commensurability and sure harmonious connection between the motions of the spheres and their sizes." The chapters that follow include a section on basic trigonometry, a catalog of fixed stars, the theory of the sun (i.e., of the earth's annual orbital motion, as well as a heliocentric explanation of the precession of the equinoxes), the theory of the moon, the theory of planetary longitudes, and, finally, the theory of planetary latitudes.

Copernicus never fully explained his reasons for considering a heliocentric arrangement, and a number of hypotheses have been subsequently proposed, many unconvincing if not outright erroneous. For example, a standard account found in numerous secondary works describes the increasing disparity between actual observations and the planetary predictions based on Ptolemy's theory and the continued addition of more and more epicycles to account for these discrepancies. Eventually, this mythological account runs, the system was ready to collapse under its own weight.

In fact, there is no historical evidence for the addition of epicycles upon epicycles to increase the accuracy of the Ptolemaic system. Furthermore, the ingenious and intricately dovetailed tables provided by Ptolemy and used by all of the medieval astronomers could not be readily modified to accommodate additional epicycles. Finally, because Copernicus used the ancient Ptolemaic observations as his fundamental base, his own predictive system was not substantially more accurate than Ptolemy's, and, if accuracy of prediction were the criterion, then Copernicus's work must be deemed a massive failure. Besides, the accuracy of prediction could have been considerably improved without moving to the heliocentric arrangement. Because of the basic geometric equivalence between the two systems, not only would the predictions not be improved merely by moving to a heliocentric arrangement, but, equally important, no simple observational test could differentiate the two arrangements prior to Galileo's (1564–1642) telescopic observation of the phases of Venus in 1610.

While observational evidence could not have entered directly into Copernicus's enthusiasm for the heliocentric layout, undoubtedly aesthetic considerations played a powerful role. Copernicus describes the pleasure of a theory "pleasing to the mind." When the planets were linked together in the sun-centered arrangement, Mercury, the fastest planet, automatically fell into the innermost position, and Saturn, the slowest, fell farthest from the sun, with a gradation in between. As cited above, Copernicus commended this arrangement "that can be found in no other way." He also noticed that, in his system, the so-called retrograde motion of the superior planets was required to occur when the planet was opposite the sun in the sky, thereby giving a natural explanation to what was just an arbitrary observation in the Ptolemaic scheme.

It is quite possible that Copernicus would never have published his hypotheses except for the persuasive intervention of a young Lutheran astronomer from Wittenberg, Georg Joachim Rheticus (1514–74). Rheticus's initial account of Copernicus's ideas, *Narratio prima (First Narrative* [1540]), did not create the opposition Copernicus feared, so the Polish astronomer gave him permission to take a manuscript of *De revolutionibus* to Nuremberg for publication. The printer arranged for a local Lutheran clergyman, Andreas Osiander (1498–1552), not only to take charge of the proofreading but also finally to add at the very beginning an anonymous warning to readers concerning the hypotheses in the work. In highly abridged form, here is the gist of Osiander's *Ad lectorem* (*To the reader*): "It is the duty of an astronomer to make careful observations, and then to make hypotheses so that the positions of the planets can be predicted. This the author has done very well. But these hypotheses need not be true nor even probable. Perhaps a philosopher will seek after truth, but an astronomer will take whatever is simplest, but neither will learn anything certain unless it has been divinely revealed to him."

When Copernicus, on his deathbed, finally received the front matter of his book (the last part to be printed), he was greatly agitated, but whether this was in disagreement with what Osiander had written or perhaps merely the excitement of having his work completed is unknown. Did Copernicus believe in the physical truth of his heliocentric arrangement? Certainly, some parts of the work well reflect Osiander's instrumentalist stance (e.g., when Copernicus gave three different arrangements of the small circles for the solar theory, remarking with consummate illogic that "it must be one of these since they all yield the same result"). On the other hand, at the end of the cosmological chapter I,10, Copernicus declared, "So vast, without any doubt, is the handiwork of the Almighty Creator." This pious passage was later censored by the Inquisition, apparently because it implied that this was the way God had actually created the cosmos, but its enthusiasm suggests that Copernicus really believed that the Creator had placed the planets heliocentrically.

The Initial Reception of the Copernican Hypothesis

Whether or not Copernicus considered his work simply a mathematical hypothesis, not to be taken as a literal description of the physical world, the astronomers and theologians at the University of Wittenberg (where the book received its first detailed study) were convinced that astronomers used fictional circles in their modeling of the cosmos and that these were not to be confused with the actual physical reality sought by the professors of philosophy. Erasmus Reinhold (1511–52), Wittenberg's beloved and authoritative professor of astronomy, devoured the technical details of *De revolutionibus,* reveling in Copernicus's strict adherence to uniform circular motion (which corrected Ptolemy's heuristic digressions with respect to these aesthetic standards), but he essentially skipped the heliocentric cosmology. His attitude aptly illustrates what historian Robert Westman has called "the Wittenberg interpretation" of Copernicus.

Martin Luther (1483–1546), who heard of Copernicus's cosmology through his Wittenberg astronomers before its publication, made an offhand remark that was recorded in his "table talk," to the effect that "whoever wants to be clever has to do his own thing. This is what that fool does who wants to turn astronomy upside down." His remark has gained publicity out of proportion to its significance; more important is the fact that Wittenberg became the intellectual center for teaching and publishing about Copernicus. Luther's right-hand man, Philip Melanchthon (1497–1560), referred indirectly to Copernicus in the first edition of his *Initia doctrinae physicae* (*Elements of the Knowledge of Natural Science* [1549]), saying: "The joke is not new. . . . The young should know it is not decent to defend such absurd positions publicly," but he promptly watered down his opinion in subsequent editions.

In the initial stages of the reception, any response on the Roman Catholic side was muted. Only later, in the wake of the Galileo affair in the early seventeenth century, was it discovered that a Florentine Dominican, Giovanni Maria Tolosani, had quickly written against Copernicus, but his patron died before the manuscript was printed, and his blast languished on an archival shelf. Because of the vehemence of the later Catholic response, some nineteenth-century commentators, such as Andrew Dickson White (1832–1918), hoped to give the Protestants equal time, and various anti-Copernican sentiments were attributed to John Calvin (1509–64), but careful research has been unable to substantiate any of them.

A principal point of tension in the religious community centered on various scriptural proof texts that seemed to demand a fixed earth or a moving sun. Psalm 104, "The Lord God laid the foundations of the earth that it should not be moved forever," was an often-cited verse, as was Joshua's command for the sun, and not the earth, to stand still to prolong the battle at Gibeon (Josh. 10:12–14). Rheticus supposedly wrote a "Second Narrative" defending the Copernican doctrine, but it remained lost until it was serendipitously recovered by the Dutch historian Reijer Hooykaas in 1973. Rheticus's account had been printed anonymously in the seventeenth century, in a little book now known in only two copies. Rheticus addressed the Scriptures concerning the stability of the earth by saying: "For, although on earth there occur corruptions, generations, and all kinds of alterations, yet the earth itself remains in its wholeness as it was created." He went on to argue that Scripture should be understood to mean that each object (e.g., the earth or the moon) had been founded on its own stability. As for the apparent motion of the sun, he stated: "Common speech, however, mostly follows the judgment of the senses. . . . We must distinguish in our minds between appearance and reality."

The Later Protestant Reception of Copernicanism

De revolutionibus was immediately recognized as an important and magisterial book and was widely quoted in various technical contexts even in its first two decades, though rarely with respect to its cosmology. An interesting exception is Robert Recorde's (c. 1510–58) *Castle of Knowledge* (1556), in which in the dialogue the Scholar protests, "Nay syr in good faith, I desire not to heare such vaine phantasies," to which the Master rejoins, "You are to yonge to be a good judge in so great a matter." An interesting comment was given by the Louvain astronomer Reiner Gemma Frisius, who pointed out in 1555 that Copernicus had provided a reasoned explanation for the retrograde motion at opposition and that it was no longer merely a "fact in itself," as it had been for Ptolemy.

One of the first committed sixteenth-century Copernicans was the English astronomer Thomas Digges (d. 1595), who published an English translation of the cosmological chapter of *De revolutionibus* in 1576, "to the ende such noble English minds might not be altogether defrauded of so noble a part of Philosophy." He proposed that the stars extended infinitely upward and that, therefore, the sun was immovable in this frame. He also provided the first step in revising the sacred cosmology of heaven by locating "the habitacle for the elect, devoid of greefe" among the stars, "garnished with perpetual glorious shining lights innumerable."

Tycho Brahe (1546–1601), the Danish observer, remarked that "Copernicus nowhere offends the principles of mathematics, but he throws the earth, a lazy, sluggish body unfit for motion, into a speed as fast as the ethereal torches." Tycho's name is not closely associated with religion, but he had a pew in the Lutheran church on his fiefdom of Hven, and, in evaluating the Copernican system, he repeatedly said that it offended physics and Holy Scripture, always in that order. Eventually, he proposed his own geoheliocentric version of the Copernican layout, in which the sun revolved around a fixed earth, but the moving sun carried the planets in orbit around itself. The arrangement saved some of the compelling Copernican linkages but destroyed part of the beauty of the system to preserve a fixed, central earth consistent with physics and the Bible.

His contemporary, Michael Maestlin (1550–1631), a Lutheran clergyman before taking up his astronomy professorship and a virulent critic of the new "popish" Gregorian calendar reform, is probably best known for teaching Johannes Kepler (1571–1630) about Copernicus at the University of Tübingen. Maestlin concluded from his own study of the comet of 1577 that the comet's motion was best understood as being seen from a moving earth. Maestlin clearly hoped to find other circumstantial evidence for the Copernican arrangement by looking for parallel changes in the eccentricities of the planetary orbits that could be attributed simply to the change in eccentricity of the earth's orbit, but the ancient observations proved too insensitive for the test. His stance toward the reality of the Copernican system was ambiguous, and he remains an enigmatic but important transitional figure, especially because of his encouragement to Kepler.

Kepler was in the final year of the Lutheran theological program at Tübingen when he was sent as a high-school teacher to Graz in Austria. He had already become a devoted Copernican, believing that the sun-centered cosmos was an image of the Holy Trinity, with God represented by the sun, Christ by the shell of fixed stars, and the Holy Spirit by the intervening space. While in Graz he stumbled upon an imaginative explanation for the Copernican spacing of the planets, a scheme involving the five regular polyhedra. Maestlin helped him publish his book, *Mysterium cosmographicum* (*Cosmographic Mystery* [1596]), but his theological introduction was suppressed when the university senate objected. Kepler simply saved his theological defense of the Copernican system for his greatest book, *Astronomia nova* (*New Astronomy* [1609]). There he explained (as Rheticus had done earlier in his as yet unpublished *Narratio secunda* [*Second Narrative*]) that Scripture is written in common language for universal understanding and is not to be taken as a textbook of science. He wrote especially concerning Psalm 104:

> I implore my reader not to forget the divine goodness conferred on mankind, and which the Psalmist urges him especially to consider. . . . Let him not only extol the bounty of God in the preservation of living creatures of all kinds by the strength and stability of the earth, but let him acknowledge the wisdom of the Creator in its motion, so abstruse, so admirable.
>
> Whoever is so weak that he cannot believe Copernicus without offending his piety, and who damns whatever philosophical opinions he pleases, I advise him to mind his own business and to stay at home and fertilize his own garden, and when he turns his eyes toward the visible heavens (the only way he sees them), let him pour forth praise and gratitude to God the Creator. Let him assure himself that he is serving God no less than the astronomer to whom God has granted the privilege of seeing more clearly with the eyes of the mind (Kepler 1983, 321–22).

The Later Catholic Reception of Copernicanism

Among the Roman Catholics who wrote on Copernican matters was the Spanish theologian Diego de Zuñiga (1536–97), who argued that certain passages in Job could actually be read with a Copernican interpretation. The eclectic philosopher Giordano Bruno (1548–1600), who had only a very faulty technical understanding of the Copernican theory, espoused it as part of his arguments for the plurality of inhabited worlds. While the reasons for his condemnation as a heretic were many and complex, his dalliance with the Copernican doctrine gave pause to many Catholics when he was burned at the stake in 1600.

Although Galileo had written to Kepler in 1597 that he was secretly a Copernican, he kept silent on the subject until his remarkable telescopic discoveries of 1609–10. Then he became increasingly open in his suggestions about the efficacy of heliocentrism. When the question arose at the Florentine court about scriptural objections to Copernicus, his protégé Benedetto Castelli (1578–1643) announced that Galileo could no doubt answer them. Galileo was probably taken by surprise, but he promptly began a review of the relevant materials by the church fathers and produced an essay on scriptural interpretation. The similarity of some of his arguments to those Kepler had used suggests that Galileo knew of the introduction to *Astronomia nova*, but it would have been folly for a Catholic astronomer to quote a Lutheran in such a delicate matter. Galileo's most memorable line, borrowed from the cardinal director of the Vatican Library, was that "the Bible teaches how to go to heaven, not how the heavens go."

Matters came to a head in 1616 when Galileo went to Rome in an effort to keep the Catholic authorities from banning the Copernican system. Galileo was silenced, and *De revolutionibus* was declared erroneous (but not heretical) and placed on the Index of Prohibited Books "until corrected" (along with Zuñiga's book and a few others). For the first and only time for any prohibited book, the Inquisition actually specified the corrections; in 1620 Inquisitors announced ten changes to make Copernicus's book appear more hypothetical. A recent study has shown that about 60 percent of the copies of Copernicus's book in Italy were censored, but essentially none in the other Catholic countries.

Earlier, in 1581, Christopher Clavius (1537–1612), the leading Jesuit astronomer, had written that what the Copernican system showed was that

Ptolemy's arrangement was not the only possibility. Nevertheless, he held firmly to the Ptolemaic cosmology, and he was unenthusiastic when Tycho proposed his alternative geoheliocentric arrangement. After Clavius's death in 1612, and especially after Copernicus's book was placed on the Index, the Jesuits espoused the Tychonic system in their teaching. This had a curious effect on the Jesuit mission to China, which had started out teaching the Copernican system as a demonstration of the advanced state of Western science but, after 1620, rapidly backpedaled to the Tychonic arrangement, leaving Chinese students in great confusion.

At the University of Louvain, the maverick astronomer Libert Froidmond argued in 1631 that the Copernican system should be considered heretical. In France, however, Marin Mersenne (1588–1648), in a careful analysis in 1623, had concluded that the heliocentric cosmology was merely erroneous but not heretical. Earlier, in 1616, internal Vatican examiners had decided that the proposition that the earth moved was erroneous, whereas the belief that the sun was fixed was actually heretical. However, their hastily prepared memorandum was not publicized. After the publication in 1632 of Galileo's *Dialogo,* a brilliant polemical defense of Copernicanism, and after his trial that followed for "a vehement suspicion of heresy," the Copernican doctrine became de facto heretical, and Copernicus's book remained on the Index well after the matter was all but settled in scientific circles.

In 1757, action by Pope Benedict XIV (b. 1675, p. 1740–58) essentially made the heliocentric doctrine acceptable in Catholic schools; nevertheless, the original decree stood, and *De revolutionibus* still appeared in an Index published in Rome in 1819. A pivotal moment arrived when a Catholic astronomer, canon Guiseppe Settele, was refused an imprimatur for his astronomy textbook in 1820 because his book treated the Copernican system as a thesis instead of as a hypothesis. It eventually required a papal command to overrule an obstinate censor, and in 1835 a new edition of the Roman Index finally appeared without a listing for Copernicus, although it had actually been removed from the Index in 1820.

In the mid-twentieth century, Catholic physicists, still embarrassed by the Galileo affair, urged the papacy to "do something" about it. John Paul II (b. 1920, p. 1978–), a pope from Copernicus's homeland, announced to the Pontifical Academy at the time of the Einstein centennial (in 1979) that the case would be reexamined. Thirteen years later, in 1992, with little consultation with the Roman Catholic historians of science who had been commissioned to look into the matter, he made a final statement. Since Galileo had not been found guilty of heresy (as he denied believing in the truth of the Copernican doctrine) but rather of disobedience (for teaching it), Pope John Paul II's options were limited. He said that Galileo had suffered much but that times were different then. He repeated the aphorism "the Bible tells how to go to heaven, not how the heavens go" and declared that Galileo had been a better theologian than those opposing him.

BIBLIOGRAPHY

Fantoli, Annibale. "The 'Galileo Affair' from the Trial's End until Today." In *Galileo: For Copernicanism and for the Church.* Trans. George V. Coyne. Vatican City: Vatican Observatory Press, 1996, 487–532.

Gingerich, Owen. *The Eye of Heaven: Ptolemy, Copernicus, Kepler.* New York: American Institute of Physics, 1993.

Heilbron, John L. *The Sun in the Church: Cathedrals as Solar Observatories.* Cambridge: Harvard University Press, 1999.

Hooykaas, Reijer. *G. J. Rheticus's Treatise on Holy Scripture and the Motion of the Earth.* Amsterdam: North-Holland, 1984.

Kaiser, Christopher B. "Calvin, Copernicus, and Castellio." *Calvin Theological Journal* 21 (1986): 5–31.

Kepler, Johannes. *Astronomia Nova.* Trans. Owen Gingerich and William Donahue. In *The Great Ideas Today, 1983,* ed. Mortimer J. Adler. Chicago: Encyclopedia Britannica, 1983, 306–41.

———. *New Astronomy.* Trans. William H. Donahue. Cambridge: Cambridge University Press, 1992.

Maffei, Paolo. *Giuseppe Settele: His Diary and the Question of Galileo.* Università degli Studi di Perugia Osservatorio Astronomico, *Pubblicazioni: English Supplement to Vol. 1.* Perugia: Edizioni dell' Arquata, 1987.

Randles, W. G. L. *The Unmaking of the Medieval Christian Cosmos, 1500–1760.* Aldershot, England: Ashgate, 1999.

Russell, John L. "Catholic Astronomers and the Copernican System after the Condemnation of Galileo." *Annals of Science* 46 (1989): 365–86.

Westman, Robert S. "The Melanchthon Circle, Rheticus, and the Wittenberg Interpretation of the Copernican Theory." *Isis* 66 (1975): 165–93.

———. "The Copernicans and the Churches." In *God and Nature: Historical Essays on the Encounter between Christianity and Science,* ed. David C. Lindberg and Ronald L. Numbers. Berkeley: University of California Press, 1986, 76–113.

8 Galileo Galilei

Richard J. Blackwell

The "Galileo affair" is the most frequently discussed case of a conflict between science and religion. An oversimplified and false view of it is that Galileo became a martyr of science because of the Roman Catholic Church's opposition to science, but it is now commonly agreed that the facts are quite otherwise. The church had understandable reasons for refusing to reinterpret the Bible in Galileo's favor: Galileo was never able to produce the proof he needed, and the waters were muddied by Galileo's academic enemies and by several misunderstandings, basic mistakes, missed opportunities, and complex theological debates that were rooted in the Protestant Reformation. This essay provides a concise factual account of what actually transpired.

Richard J. Blackwell received his Ph.D. in philosophy from Saint Louis University, where he is professor emeritus of philosophy. He has published many essays on seventeenth-century philosophy and history of science. He is the author of *Galileo, Bellarmine, and the Bible* (Notre Dame: University of Notre Dame Press, 1991) and the translator of Thomas Campanella's *A Defense of Galileo* (Notre Dame: University of Notre Dame Press, 1994).

THE CLASSIC CASE of conflict between Western science and religion is the confrontation between Galileo (1564–1642) and the Roman Catholic Church in the early decades of the seventeenth century. This episode centered on four central issues: (1) the state of the scientific debate at that time over the comparative merits of the older earth-centered astronomy of Claudius Ptolemy (second century A.D.) and the more recent but conflicting sun-centered theory of Nicholas Copernicus (1473–1543); (2) the question of what are the proper exegetical standards to be used in understanding the meaning and the truth of the Bible; (3) the historical events that led the Catholic Church in 1616 to condemn Copernicanism as false and their rationale; and (4) the charges, the legal ground, and the course of events in Galileo's trial and condemnation in 1633.

The Historical Background

In Galileo's day Western European culture was undergoing some basic and disruptive changes. Modern science as we now know it was just beginning to come into existence, partly as a result of Galileo's own scientific work, and its basic epistemological characteristics were not yet fully understood. At the same time Western Christianity was in the middle of a historical split and fundamental transformation. To understand fully the four issues listed above, which constitute what has come to be called the Galileo affair, one needs to be aware of some details about Galileo's early life and work and about the theological debates then under way.

For most of his life, Galileo lived in Florence or in Pisa, where he was born, educated, and later taught at the local university. What passed for science at that time was Aristotelian natural philosophy, whose method of inquiry had deteriorated to the point of view that scientific questions were to be answered by searching the writings of Aristotle, a procedure that even Aristotle himself would not have condoned. In contrast, Galileo adopted the method of observing and conducting experiments directly on actual objects in the natural world. This change of methodology, along with the results he obtained from it, brought Galileo into such strong conflict and academic rivalry with the natural philosophers that he relocated for eighteen years (1592–1610) to the University of Padua, which enjoyed greater freedom of thought, since it was part of the Republic of Venice.

During his years in Pisa, Galileo made several major discoveries that were basic to the birth of classical physics. One was the law governing the free fall of bodies. Aristotle had written that bodies fall because of their weight (*gravitas*), from which he erroneously inferred that their rate of fall is proportional to their weight. Thus, a ten-pound cannonball should fall ten times faster than a one-pound cannonball. Legend has it that Galileo conducted experiments by dropping objects from the leaning edge of the Tower of Pisa to disprove Aristotle's view and to prove that the distance fallen is proportional to the square of the time, independently of the weight of the falling body.

Another early discovery was the law of the pendulum, which states that, given a pendulum of fixed length, the period, or time, of its oscillation is constant, no matter how widely or narrowly the pendulum swings. This law later became of central importance in the construction of accurate clocks. Legend claims that Galileo made this discovery while observing the swinging of a chandelier in the Cathedral of Pisa (perhaps during a dull sermon). Near the end of his life, he published these and other findings in his *Discourses on Two New Sciences* (1638), which became the cornerstone of classical physics.

In 1610 Galileo's work and life changed dramatically. He learned that a Dutch lens grinder, Hans Lipperhey, had developed an optical instrument that made distant objects appear much closer to an observer. Galileo immediately saw the potential uses of the new invention (e.g., the navy could see approaching ships sooner, and, more importantly, new observations could be made of the night sky), and he proceeded to construct increasingly better telescopes. This led him to a series of important astronomical discoveries: that the surface of the moon contains many craters and mountains (contrary to Aristotle's notion that the moon's surface is a perfectly smooth sphere); that Jupiter has four moons that are invisible to the naked eye; that the surface of the sun displays continually changing dark spots that drift from left to right, which indicate that the sun is changeable and that it rotates on its own axis; and that Venus undergoes changing phases, like the moon, which proves that it revolves around the sun and not the earth (again contrary to Aristotle).

Galileo's publication of these and other observations brought him instant fame and controversy. He accepted the well-paid position of ducal philosopher and mathematician back home in Tuscany, where he arrived in triumph. But the academic natural philosophers, whose views were increasingly being refuted and who feared the loss of their favor at court, became more hostile to Galileo. And in Florence he was much closer to possible censorship from Rome. In December 1613 the religious orthodoxy of his pro-Copernican views first came under question at a social event at the ducal palace. In his reply in a short essay, *Letter to Castelli,* which was later denounced before the Holy Office, Galileo was anxious to explain his views on how the findings of science should be related to the Bible.

To understand why this was such a serious concern to Galileo, one needs to recall that the Protestant Reformation and the ensuing Counter Reformation had occurred during the previous one hundred years. Echoes of the Great Schism of 1054, when the Eastern Orthodox Churches split off from Western Christianity, frightened the church authorities in Rome as they witnessed much of northern Europe also breaking away from their control. As a result the Catholic Church in the sixteenth century had become unusually conservative and defensive, especially in regard to theological and scriptural matters. This attitude still predominated at the time of the Galileo affair.

The Protestant Reformation, usually seen as beginning in 1517 with Luther's objections to the sale of indulgences, was originally focused on abuses within the administration of the church and on the moral corruption of the clergy. However, it soon expanded into differences of opinion on the nature of religious revelation and on the interpretation of Scripture. The Catholic Church's response was based primarily on the decisions made at the Council of Trent (1545–63). Its main effect on the Galileo case was its declaration that no individual Christian (including Galileo) should interpret the Scriptures contrary to the common agreement of the early fathers of the church or contrary to the views of the pope and the bishops, who alone have the power to interpret the Bible. Ecclesiastical authorities, of course, interpreted the occasional descriptions of celestial motions in the Bible from the common-sense (and thus Aristotelian) point of view of geocentrism. (Our weather bureaus today still speak of "sunrise" and "sunset.") Given the church's defensive frame of mind, it was in no mood to adopt a new and revolutionary model of the heavens.

It also happened that the Counter Reformation's most prominent opponent of the new wave of Protestantism was Cardinal Bellarmine, who later became the church's spokesman in the Galileo affair. On scriptural interpretation he was not excessively rigid. He made extensive use of St. Augustine's (354–430) principles of exegesis, including the need to distinguish between the literal and the figurative use of language in the Bible. But because of the Reformation he was not inclined to challenge the biblical interpretations of the fathers of the church, including the question of heliocentrism, especially because (as he correctly pointed out) it had not been proven.

If Copernicus's book had been published either one hundred years earlier or one hundred years later, the Galileo affair would probably not have happened. But, in fact, it was published in 1543, when the Reformation was in full bloom and the Counter Reformation was just beginning. Hence it was that by 1616 all of the actors and cultural forces were in place for the drama of the Galileo affair to begin.

The Scientific Dispute

For nearly two thousand years before the Galileo affair, the almost universally accepted view of the heavens in Western culture was the geocentric theory initially proposed by Aristotle (384–322 B.C.) and later considerably refined mathematically by Ptolemy. This universally accepted view, which came to permeate the medieval scientific and religious tradition, looked upon the earth as spherical, motionless, and fixed in the center of the entire universe. The moon, sun, five visible planets, and all of the fixed stars were conceived as rotating daily from east to west around the earth in complex patterns, which the early astronomers had succeeded in reducing to combinations of simple circular motions. All of the then known observational evidence concerning the heavens was consistent with this astronomical model, especially when it was interpreted in the light of Aristotelian natural philosophy.

In 1543 Copernicus published his *Revolutions of the Celestial Spheres,* which modified the earlier view in a major way by locating the sun at the center of the universe and the earth and its moon in motion around the sun above Mercury and Venus. Copernicus had no new evidence to justify his theory; rather, his primary motivation was that he thought that his view had more internal coherence and greater explanatory power than Ptolemy's.

As the generations passed, some new evidence slowly accumulated that tended to make the new cosmic theory more likely to be true. In Galileo's day, however, conclusive proof of Copernicanism still had not been found, despite his own lifelong efforts to establish such a proof. To understand the Galileo affair properly, it is essential to keep in mind that no one, including Galileo himself, was yet able to settle the scientific debate conclusively, although the accumulating evidence, much of it discovered by Galileo himself, spoke more and more in its favor.

Although Galileo had become personally persuaded sometime before 1597 that Copernicanism was true, he did not publicly enter into this debate until early in 1610 when, in a book entitled *The Starry Messenger,* he published his first observations, mostly of the moon, with his newly improved version of the telescope. In the next three years, he published still further telescopic observations, along with his interpretations of them, culminating in two statements in his *Letters on Sunspots* that explicitly endorsed the Copernican theory. In his observations he had discovered numerous new features on the surface of the moon, four of the moons of Jupiter, what we now call the rings of Saturn, sunspots, and the fact that Venus undergoes a regular series of phases, as does the earth's moon. This latter fact was particularly important, since it proved that the original version of the Ptolemaic theory was false, although it did not prove Copernicanism, since still other models of the heavens were not only possible but also actively under consideration at the time. At any rate, by 1613

Galileo was explicitly defending the Copernican theory of the heavens in his published writings.

The Biblical Dispute

At this point Galileo began to come under attack by various opponents, some of whom were motivated by scientific rivalries, while others were moved by overly zealous concerns for religious orthodoxy. In many places the Bible rather explicitly says, contrary to Copernicanism, that the sun moves and the earth is at rest. The most frequently quoted passages were Genesis 1, Ecclesiastes 1:4–6, Joshua 10:12, and Psalm 19:4–6.

In reply to such biblical objections, Galileo wrote a lengthy letter in 1613 addressed to his friend and scientific colleague, the Benedictine priest Benedetto Castelli (1578–1643), which outlined his views on the relations between science and the Bible. This letter was widely circulated privately. Unknown to Galileo, his *Letter to Castelli,* in an adulterated version that was intended to compromise him, was denounced a year later to the Holy Office (the Roman Inquisition) as religiously unorthodox, although these charges were dismissed shortly thereafter. In 1615 Galileo considerably expanded his views into the much longer *Letter to the Grand Duchess Christina.* But by then the topic had become so controversial that this *Letter* was withheld from publication until 1636, three years after his trial.

In writing these letters Galileo was partially motivated by a desire to protect his patronage position at the Tuscan court in Florence, where he had been appointed in 1610 as the ducal philosopher and mathematician. Nevertheless, he must also have seen the dangers in publicly entering into discussions about the interpretation of the Bible. He was trained neither as a theologian nor as a scholar of Scripture and, hence, was vulnerable to charges of not being competent to judge in this field. More seriously, the Council of Trent, session IV (April 8, 1546), had explicitly limited the interpretation of the Bible to the bishops and the councils of the church. No matter how justifiable Galileo's readings of the Bible may have been, as merely a lay member of the church he ran the risk of standing in violation of this restriction, which was an important part of the Catholic reaction to Protestantism during the Counter Reformation. Despite these complications, Galileo's views in the *Letter to the Grand Duchess Christina* have since become commonplace in biblical exegesis and were accepted by the Catholic Church in 1983.

Galileo's views on the relations between science and the Bible can be summarized as follows. God is the common and always truthful author of both the book of revelation (the Bible) and the book of nature (the natural world). As a result it is not possible in principle for the truths of science and the truths of the Bible to be in conflict, provided that we have correctly understood the language and the meaning of both. This proviso is an especially difficult condition to meet in the case of the Bible because God has chosen to reveal the highest spiritual truths in words accommodated to the understanding of the common, uneducated person. Hence, the Bible contains a wide use of metaphorical and figurative expressions as they occur in the common-sense idiom of the era of its human authors. One should thus be very careful not to attribute literal meanings to figurative expressions. The notion of the daily motion of the sun and the heavens from east to west is precisely such a case in point, according to Galileo. Further, the Bible's central purpose is religious and moral, and it is not primarily intended to serve as a source of knowledge about the natural world. Galileo invoked the now famous remark by Cardinal Cesare Baronius (1538–1607) that the Bible tells us how to go to heaven, not how the heavens go.

Cardinal Robert Bellarmine (1542–1621), who was the church's main theological respondent to Galileo, would have agreed with most of these views of Galileo, but he insisted on a critically important modification. Since God is its author, every statement in the Bible must be true, when properly understood, but the loyal Christian believer is also required to accept it as true as a matter of religious faith. Hence, all factual and historical knowledge about the natural world contained in the Bible falls within the scope of religious faith and is governed by the authority of the church.

The biblical dispute between Galileo and Bellarmine came down to the following points. If the scientist could conclusively prove that the heliocentric theory is true (which Galileo persistently attempted but was never able to do), then both sides would agree that contrary remarks in the Bible would need to be reinterpreted as figurative, not literal, in order to maintain the coherent unity of truth. Bellarmine explicitly granted that point. But what about a case (such as Copernicanism at that time) in which a new scientific view was not yet conclusively proven but might become so in the future?

Bellarmine's answer was to retain the traditional view in such a case, since it was supported by the higher truth standards of the Bible and by the common agreement of the fathers of the church, through whom the content of the faith had been handed down through the centuries. But for Galileo, Copernicanism, proven or not, was ultimately not a matter of religious faith, and it was an objectionable procedure to bring biblical passages to bear on the question.

To put the matter in the terms most frequently used by Bellarmine, Copernicanism as an astronomical theory could without ecclesiastical objection be adopted hypothetically but not realistically. In other words, it could be arbitrarily assumed for purposes of making calculations about the motions and positions of the heavenly bodies, but it could not be adopted as an actually true account of the structure of the physical world, since this stronger claim conflicts with the truth of the Bible. This dispute over the use of the Bible in relation to the main scientific debate of the day remained unresolved throughout the Galileo affair.

The Condemnation of Copernicanism

The entire issue reached its climax in the early months of 1616. Galileo had gone to Rome, apparently with the hope of persuading the church to take no action on the matter. The actual results were precisely the opposite.

In February, Pope Paul V requested the opinion of a group of his theologians on the orthodoxy of heliocentrism. They advised him unanimously that Copernicanism was not only false but also formally heretical, since it explicitly contradicts the Scriptures in many places. The pope decided to accept this theological opinion. This decision was then publicly announced to the whole church in a decree issued by the Congregation of the Index, dated March 5, 1616, in which Copernicanism was condemned as "false and completely contrary to the Divine Scriptures." Copernicus's *Revolutions of the Celestial Spheres* was prohibited "until corrected." These corrections, issued four years later, were not extensive. They included the removal of one chapter (book I, chap. 8) and of a few sentences throughout the book that assert that the earth moves. This condemnation of Copernicanism, which closed debate in the Catholic Church on the scientific and biblical issues involved, was a disaster, both at the time and subsequently, for the relationships between science and religion. It is the centerpiece of the Galileo affair.

Neither Galileo nor any of his writings were explicitly mentioned in the decree. A decision had been made to deal with him privately on the matter. The pope ordered Bellarmine to meet with Galileo, to explain to him the decision against Copernicanism, and to ask for his acceptance of the decision, with the threat of imprisonment to be held in reserve in case he refused. This meeting took place on February 26.

Precisely what happened at that meeting is not known, since two inconsistent accounts of it have come down to us. First, in a letter that Galileo requested three months later, Bellarmine says that the only thing that happened was that Galileo was informed of the impending Decree of March 5 to the effect that Copernicanism "cannot be defended or held." He does not report that Galileo resisted this decision.

A second, much stronger account of the meeting, contained in the files of the Holy Office, says that Galileo was served an injunction "not to hold, teach, or defend it [i.e., Copernicanism] in any way whatever, verbally or in writing." Some modern scholars have argued that this file memorandum is a forgery perpetrated in either 1616 or 1632 to entrap Galileo. Although this interpretation has now been abandoned, there is still no general agreement regarding what occurred at the Galileo-Bellarmine meeting, which is quite unfortunate, since Galileo's later trial turned largely on the status of the supposed injunction.

The condemnation of Copernicanism in 1616 and Galileo's understanding with Bellarmine, whatever that may have been, brought the first half of the Galileo affair to an end. The die had now been cast, with no hope of recasting it. The issue that remained was how the decree and the injunction were to be observed in the years that immediately followed.

Galileo's Trial

For the next seven years, Galileo carefully avoided any dealings with the issue of Copernicanism. But in 1623 he was delighted to learn that Matteo Barberini (1568–1644), an admiring personal acquaintance, a fellow Tuscan, and a man of letters in his own right, had been elected pope under the name of Urban VIII. The next year Galileo went to Rome with high hopes that the censure of Copernicanism might be lifted. In a series of six conversations with the new pope (of which, unfortunately, no direct records have come down to us), Urban apparently told Galileo that he could write again about Copernicanism, provided that he kept the discussion at the hypothetical level.

As a result Galileo began to plan the writing of a long fictional dialogue, in the Platonic literary tradition, which would review and evaluate all of the evidence and arguments on both sides of the Copernican question. One spokesman, Salviati, would vigorously present the new ideas; another, Simplicio, would argue doggedly and in detail for the old tradition; and the third, Sagredo, would be an open-minded inquirer who would critically assess the issues from a neutral standpoint. The dialogue format itself would, Galileo hoped, help to place the entire discussion at the required hypothetical level. The result was the *Dialogue Concerning the Two Chief World Systems,* Galileo's scientific manifesto and an Italian literary masterpiece in its own right. The appearance of the book in 1632 created sensational reactions, particularly over the issue of whether it violated the Decree of 1616 against Copernicanism. Most readers, still now as well as then, have concluded that Salviati clearly won the debate, thereby making the *Dialogue* a plea for heliocentrism.

Pope Urban VIII appointed a special commission to investigate the entire matter. In this process the commission uncovered in the files of the Holy Office the above-mentioned document, previously unknown even to Urban VIII, which states that Galileo had been personally served, at his ill-fated meeting with Cardinal Bellarmine in 1616, with an injunction against future writings on Copernicanism. The commission's judgment was that Galileo had also exceeded in the *Dialogue* the instructions to treat the matter only hypothetically.

At this point the pope's friendship toward Galileo turned permanently into anger. A trial on the above charges became inevitable, and it took place in the spring of 1633. The two main legal questions at the trial were (1) whether Galileo had acted improperly in the three years before the book appeared in not gaining the required approvals for publication from lower-level church officials and (2) more importantly, whether Galileo had violated the injunction supposedly served on him in 1616.

The most dramatic moment of the trial occurred when Galileo produced Bellarmine's letter of 1616, previously unknown to the prosecution, which made no mention of any injunction and contained the weaker restriction on "defending and holding" Copernicanism. Not surprisingly, the court gave preference to the Holy Office's document, which Galileo, of course, had never seen before. The substantive questions of the scientific truth of Copernicanism and the proper use of the Bible in relation to science did not arise at the trial itself, although these were the root issues behind it.

The final result, approved by the pope, was the judgment that Galileo was "vehemently suspected of heresy." On June 22 he was forced to read an oath, prepared by the court, in which he denounced his own teachings about Copernicanism. Later stories that he was subjected to torture are quite false. But the old man was disgraced and his spirit broken. Galileo spent most of the remainder of his days under what we would now call house arrest at his villa at Arcetri near Florence. Although the Copernican question was now totally closed to him, he continued his work and writing in theoretical mechanics, resulting in the *Discourse on Two New Sciences*, which is his major scientific contribution to physics.

During the centuries since then, the Galileo affair has cast a long and disturbing shadow over the relations between science and religion. It continues to dominate discussions of the issue, and the distrust it has caused on both sides is often not far beneath the surface.

BIBLIOGRAPHY

Biagoli, Mario. *Galileo Courtier.* Chicago: University of Chicago Press, 1993.

Blackwell, Richard J. *Galileo, Bellarmine, and the Bible.* Notre Dame: University of Notre Dame Press, 1991.

Campanella, Thomas, O.P. *A Defense of Galileo.* Trans. Richard J. Blackwell. Notre Dame: University of Notre Dame Press, 1994.

Coyne, G. V., Michael Heller, and Joseph Zycinski, eds. *The Galileo Affair, a Meeting of Faith and Science: Proceedings of the Cracow Conference, May 24–27, 1984.* Vatican City: Specola Vaticana, 1985.

Drake, Stillman. *Galileo Studies: Personality, Tradition, and Revolution.* Ann Arbor: University of Michigan Press, 1970.

———. *Galileo at Work: His Scientific Biography.* Chicago: University of Chicago Press, 1978.

———. *Galileo: Pioneer Scientist.* Toronto: University of Toronto Press, 1990.

Fantoli, Annibale. *Galileo: For Copernicanism and for the Church.* Trans. George V. Coyne, S.J. 2d English ed. Vatican City: Vatican Observatory, 1996.

Feldhay, Rivka. *Galileo and the Church: Political Inquisition or Critical Dialogue?* Cambridge: Cambridge University Press, 1995.

Finocchiaro, Maurice A., ed. *The Galileo Affair: A Documentary History.* Berkeley: University of California Press, 1989.

Geymonat, Ludovico. *Galileo Galilei.* Trans. Stillman Drake. New York: McGraw-Hill, 1965.

Gingerich, Owen. "The Galileo Affair." *Scientific American* 247 (1982): 132–43.

———. *The Great Copernican Chase and Other Adventures in Astronomical History.* Cambridge, Mass.: Sky, 1992.

Koyré, Alexandre. *Galileo Studies.* Trans. John Mepham. Atlantic Highlands, N.J.: Humanities, 1978.

Langford, Jerome J. *Galileo, Science, and the Church.* Rev. ed. Ann Arbor: University of Michigan Press, 1998.

Lattis, James M. *Between Copernicus and Galileo.* Chicago: University of Chicago Press, 1994.

McMullin, Ernan, ed. *Galileo: Man of Science.* New York: Basic Books, 1967.

———. "Galileo on Science and Scripture." In *The Cambridge Companion to Galileo.* Ed. Peter Machamer. Cambridge: Cambridge University Press, 1998, 271–347.

Pedersen, Olaf. *Galileo and the Council of Trent.* Vatican City: Vatican Observatory, 1991.

Redondi, Pietro. *Galileo Heretic.* Trans. R. Rosenthal. Princeton: Princeton University Press, 1987.

Reston, James, Jr. *Galileo: A Life.* New York: HarperCollins, 1994.

Segre, Michael. *In the Wake of Galileo.* New Brunswick: Rutgers University Press, 1991.

Shea, William R. *Galileo's Intellectual Revolution: Middle Period, 1610–1632.* New York: Science History, 1972.

Sobel, Dava. *Galileo's Daughter.* New York: Walker and Co., 1999.

Wallace, William A., ed. *Reinterpreting Galileo.* Princeton: Princeton University Press, 1984.

Westfall, Richard S. *Essays on the Trial of Galileo.* Notre Dame: University of Notre Dame Press, 1989.

9 Early Modern Protestantism

Edward B. Davis and Michael P. Winship

As the late Richard S. Westfall realized more than forty years ago, many of the most important metaphysical and religious issues related to modern science first rose to prominence during the scientific revolution of the sixteenth and seventeenth centuries, a time when religion and the issues of the Protestant Reformation, not science itself, dominated European thinking and living. Where scholars once saw science as having seized cultural authority from religion in a pitched battle that commenced during this period, we now realize that the interplay of ideas and practices during the Reformation was rich and multifaceted. Scholars continue to debate the ways in which Protestantism and early modern science shaped each other.

Edward B. Davis received his Ph.D. from Indiana University. He is professor of the history of science at Messiah College, Grantham, Pennsylvania. He is editor (with Michael Hunter) of *The Works of Robert Boyle* (14 vols., London: Pickering and Chatto, 1999–2000) and author of numerous articles on the historical relationship of Christianity and science. His current research focuses on the religious beliefs of modern American scientists.

Michael P. Winship holds a Ph.D. from Cornell University. He is associate professor of history at the University of Georgia. He is the author of *Seers of God: Puritan Providentialism in the Restoration and Early Enlightenment* (Baltimore: Johns Hopkins University Press, 1996) and *Making Heretics: Militant Protestantism and Free Grace in Massachusetts, 1636–1641* (Princeton: Princeton University Press, 2002).

THE RELATION between Protestantism and science in the first two centuries after the Reformation involved a creative tension, with important insights in theology coming from the new science and important elements of the new science being shaped by theological assumptions. The salient features of the new science—a new world picture, a new worldview, and new knowledge coupled with a new view of knowledge and its sources—interacted with Christian beliefs in a variety of ways.

The Protestant Reformation

The Protestant Reformation, a wholesale change in the European religious landscape precipitated by Martin Luther's (1483–1546) challenge to the sale of indulgences in 1517, actually involved several reformations by diverse groups of people with different goals and beliefs. They included the Reformed churches that followed John Calvin (1509–64), the Anglicans, the Anabaptists and other radical reformers, and even the Roman Catholics themselves, who sought a renewed spirituality within their own church. However, Luther and Calvin were the two leading architects of the Reformation, and we focus on them here. In response to various Roman Catholic practices stressing that humans must cooperate with God to be saved, Luther and Calvin began with the proposition that salvation depends wholly on the sovereignty of God, who elects to save sinners based solely on his own mercy, not upon any intrinsic merits that sinners might have or any good works they might perform, not even upon a human standard of justice. The just, Luther taught, are saved by faith alone (*sola fide*), where even faith itself is understood as originating in God rather than in ourselves. This particular view of God as utterly sovereign and radically free helped determine how the new science was interpreted and received.

This is not to say that the Protestant mainstream failed to value good works. Quite the contrary; though works themselves could not lead to salvation, Christians were, nevertheless, expected to evidence the presence of saving faith and thus to glorify God by their piety and by the righteousness of their lives. Many Protestants stressed the dignity of labor, and some even saw material success as a sign of God's blessing. Whether beliefs like these encouraged the pursuit of science in Protestant countries has been hotly debated.

Most Protestants also shared a commitment to the primacy of the Bible (*sola scriptura*) as a source of truth over tradition, reason, and experience. The Roman Catholic Church, in their opinion, had developed an erroneous theology by straying too far from the plain words of Scripture. This view was related to another Protestant belief, that individual believers have direct access to God through prayer and the reading of Scripture, apart from the clergy. Hence, believers can read and interpret the Bible profitably, under the guidance of the Holy Spirit, in some cases drawing conclusions contrary to those reached by Roman Catholic clergy. This cluster of beliefs about the Bible affected how Protestants responded to the new world picture.

The Scientific Revolution

The period from 1543 until about 1750, during which early modern science replaced medieval and Renaissance versions of ancient Greek science, is still commonly (though hardly universally) called the *scientific revolution*. This phrase is derived from the philosophes of the eighteenth century, for whom the recent upheavals in science were loaded with ideological overtones, representing not only the triumph of reason over nature, but also the triumph of secular rationalism over the essentially religious (and, therefore, false) worldview of the Middle Ages. Although that interpretation is no longer tenable, the label *scientific revolution* is still appropriate for the period as a whole, not because the changes were rapid but because they were fundamental.

The most famous change was the eventual acceptance of the new heliocentric world picture of Nicholas Copernicus (1473–1543). A series of discoveries and new ideas led most astronomers by 1700 to reject celestial perfection and the circular motion it implied in favor of the heliocentric system, with the sun's powerful gravitational attraction for the planets as the cause of celestial motion. Isaac Newton's (1642–1727) elegant mathematical theory of motion replaced Aristotelian physics in the process. Some Copernicans also accepted an essentially infinite universe, which was often linked with a belief in the existence of other solar systems populated by intelligent beings.

But a change in worldview was actually more fundamental. In the late Renaissance, as the works of Lucretius (c. 99–55 B.C.) and other ancient atomists freshly rediscovered began to be read widely in Europe, the conception of nature as a great concourse of particles moving through an infinite void was revived, leading many early modern natural philosophers to give mechanistic explanations to natural phenomena. Medievals had accepted the organic worldview of Aristotle, according to which motion was to be understood as a process of change from potentiality to actuality, governed by a functional teleology immanent within nature. Mechanical philosophers, by contrast, described motion in mathematical terms without reference to any principles of intelligence or purpose in bodies themselves. Most followed René Descartes (1596–1650) in dividing all things into two kinds of substance: mind (or soul) and matter (or body). Only humans, angels, and God had intelligence and will; animals were just complicated engines like the automatons of master clockmakers and hydraulic engineers, and the world itself was a vast vortex of particles in motion. Although this approach has often been criticized for "dehumanizing" nature, it was also a bold attempt to preserve the transcendence of God, the dignity of humans, and the autonomy of values by placing them all beyond the scope of mechanical explanation.

Just when Europeans were revising their notion of nature, they were also discovering an astonishing variety of new facts about nature. Many came from the numerous voyages of discovery undertaken since the fifteenth century, resulting in a complete reevaluation of the reliability of traditional natural histories and a veritable explosion of botanical and zoological knowledge. Equally important were two new optical instruments, the telescope and the microscope, which opened up wholly new worlds far away and close at hand. Other technical advances, such as the air pump, made possible new types of experiments that led many to question the veracity of older ways of understanding nature.

Rapidly advancing knowledge spurred a recognition that facts themselves must take priority over tradition, leading to a new view of the sources of knowledge. Where medieval scholars had tended to see scientific truth as something to be sought in human books, the older the better, early modern thinkers looked to the book of nature. This renewed emphasis on empiricism (the importance of making systematic observations and experiments) is one of the outstanding features of the scientific revolution. In England, it became the defining characteristic of the new program of learning advocated by Francis Bacon (1561–1626), a lawyer and statesman who served as a prophet of scientific progress, advocating the use of science to alleviate the consequences of the Fall and to improve the human condition. Sometimes empiricism was allied with Hermeticism, a mystical philosophy based on writings attributed to Hermes Trismegistus, a legendary Egyptian sage once believed to have been a contemporary of Moses. It attracted many, including Newton, with its promise of holding the alchemical key to the deep secrets God had hidden in the creation.

Two varieties of rationalism also competed to replace scholasticism during the scientific revolution. One sought scientific truth in mathematical demonstration. Galileo Galilei (1564–1642) held that the kind of knowledge thus obtained was absolutely certain, for it had been arrived at deductively from an analysis of pure forms and was, therefore, superior to any knowledge we might gain from experience alone. The other followed Descartes in seeking certainty from self-evident metaphysical principles rather than from geometrical axioms.

Protestant Beliefs and the Substance of Science

Protestants interacted with the new science in a variety of ways that show both scientific influences on religious beliefs and religious influences on scientific beliefs. The reception of heliocentrism provides one of the clearest examples. Even before Copernicus had published his famous treatise, Luther was quoted by one his students as saying (around the dinner table one evening) that the new astronomy contradicted the tenth chapter of Joshua by placing the earth in motion rather than the sun, but this informal remark has probably been given more attention than it deserves. (Calvin is often said to have made an anti-Copernican statement of his own, but this report has no basis in fact.) Far more significant was the influence of Luther's leading associate, Philip Melanchthon (1497–1560), who viewed Copernicus as a moderate reformer (like himself) because he had sought to purify astronomy by replacing equants with uniform circular motions. Although Melanchthon never accepted the hypothesis of the earth's motion, he positively encouraged the teaching of mathematics (and its subdiscipline, astronomy) at Lutheran universities in Germany.

Three Lutheran astronomers were crucial to the spread of Copernican views. Georg Joachim Rheticus (1514–74), a mathematician from Wittenburg, visited the elderly Copernicus a few years before his death, urged him to publish the details of his cosmology, and received his permission to publish a digest of the new theory under his (Rheticus's) own name; a few years later, the full theory was published at Nuremburg with Rheticus's assistance. He returned to Germany a convinced Copernican. Another Copernican, Michael Mästlin (1550–1631), taught mathematics at Tübingen, where one of his pupils was Johannes Kepler (1571–1630). It was Kepler who showed his fellow Protestants how to reconcile Copernicanism with the Bible. In the preface to his most important book, *The New Astronomy* (1609), Kepler used the Augustinian principle of accommodation to justify the figurative interpretation of biblical references to the motion of the sun. The Bible, he noted, speaks about ordinary matters in a way that can be understood, using common speech to make understandable loftier theological truths. Thus, the literal sense of texts making reference to nature should not be mistaken for accurate scientific statements. Galileo made an identical argument just a few years later in his *Letter to the Grand Duchess Christina,* written privately in Italian in 1615 but later published in Latin and English. Clearly, Galileo behaved rather like a Protestant (as a layman interpreting the Bible for himself in ways contrary to tradition), and the church soon ruled that heliocentrism was heretical. Unfortunately for Catholic scientists, in 1616 Copernicus's book was placed on the Index of Prohibited Books "until corrected," where it remained until 1820. It is not clear, however, how much influence this ban really had, especially outside Italy. Since there was no similar ruling that was binding on Protestants, Protestant scientists generally accepted the arguments of Kepler and Galileo. By the end of the seventeenth century, many Protestant scientists were Copernicans, and many Protestant theologians seemed indifferent to the issue. Indeed, the principle of accommodation, which had made heliocentrism theologically acceptable, henceforth was widely used by theologians and scientists alike for understanding scriptural passages about nature, and it helped immensely to clarify the real purpose of biblical revelation.

The challenge of mechanical philosophies was not so easy to meet. Throughout the scientific revolution, one sees a growing tension with theology over the reality of special providence and the possibility of miracles in a mechanistic universe. In general, scientists and philosophers became increasingly skeptical about reports of miraculous events, including those in the Bible. In part, this skepticism was encouraged by Protestant attacks on Roman Catholic claims that saints had worked miracles. Protestants usually maintained that genuine miracles had ceased with the close of the apostolic age. Skepticism also resulted from repeated failures to demonstrate unambiguously the existence of a supernatural realm in cases of witchcraft and other events thought to involve occult powers. Furthermore, the dualism of soul and body (which ultimately owes more to pagan Greek views than to the Bible), which was commonly invoked by mechanical philosophers, was beset with difficulties, particularly for theologians who were obligated to tackle thorny problems about the origin, nature, and immortality of the human soul.

From these issues alone, one might tend to conclude that theology and mechanistic science have been engaged in a hard-fought battle for cultural supremacy since the seventeenth century. But closer examination reveals a far more complex relationship. In at least two very significant ways, theological assumptions affected the content of the new science. The Reformation emphasis on the saving activity of God alone and the total passivity of sinners was mirrored in the way in which several mechanical philosophers understood matter as utterly passive, possessing no powers or forces of its own, and under the direct manipulation of an ever-active God. This is certainly the way in which Newton understood nature; he actually disowned the clockwork metaphor with which he is so often, and so wrongly, associated. (Some Roman Catholic scientists, such as Nicolas de Malebranche [1638–1715], held similar views of God's relationship to passive matter. Reformation theology was essentially Augustinian and, thus, not the exclusive property of Protestants.)

The emphasis on divine sovereignty had an even more important consequence for the new science. Some of the leading mechanical philosophers believed that, whether or not matter was passive mechanically, it was passive ontologically (i.e., its properties and powers were imposed on it by a free creative act of God, beyond the power of human reason to penetrate). Although Reformation theology similarly stressed the inscrutability of God's will in the central matter of election, in that God's reasons for saving some and not others could not be discovered, seventeenth-century discussions of divine action are linked no less strongly with pre-Reformation theological debates about the relative amount of emphasis to place upon God's will vis-à-vis God's reason. Theological rationalists emphasized divine reason and often viewed human reason as the image of the divine. They tended to have great confidence in our ability to understand the works of God with our reason alone, unaided by experience. Theological voluntarists, on the other hand, emphasized the freedom of the divine will, unfettered by divine or human reason, to do whatever God wished—not only in the original creation of the world but also in its ongoing operations. Thus, reason alone was not sufficient to understand the freely created world; significant data from experience were needed to show us what God has actually done rather than what we think God's reason compelled him to do. And because the world was, at every moment, under the sovereignty of a radically free Creator, the laws of nature were not wholly binding on God's activity, so miracles could not be ruled out.

This dialogue of divine will and reason actually shaped conceptions of science, including notions of proper scientific method, in the seventeenth century. Galileo, for example, held to a rationalist theology and, with it, a rationalist philosophy of science. This is not to say that he did not perform experiments; we know that he did, and they were some of the most clever ever performed. But, in his heart of hearts, he believed that the word *science*, or knowledge, was properly applied only to knowledge that was absolutely certain and could not be otherwise, the kind that only mathematics and logic could provide. This is precisely why he communicated his results in the form of Platonic dialogues and why he repeatedly emphasized the power of mathematics to persuade.

Robert Boyle (1627–91), by contrast, viewed the laws of nature as free creations of an omnipotent God, who could just as easily have made a world of a different kind from that which God actually did create. Consequently, unaided human reason was incapable of telling us anything true about the created order; it was capable merely of comprehending to a limited degree the order revealed to us by our senses. The world, for Boyle, was full of "data" and "facts," things given and things made by a power outside ourselves and, therefore, unknowable by our minds alone. This is why he placed so much emphasis on the experimental life: It was the only way to understand a freely created universe. Newton's view of the inadequacy of pure reason in both science and theology was essentially the same.

Their voluntarist theology, therefore, made it possible for many early modern natural philosophers to baptize mechanical explanations, which surely aided reception of the new ideas. It was easy for them to see how an omnipotent Creator might, by an act of sheer will, endow created matter with any desired properties and powers, which the human investigator then had to discover from the phenomena produced by those properties and powers. Seeing nature in this way encouraged both theologians and scientists to find within nature abundant evidence of God's wisdom, power, and benevolence. As Newton stated in the second edition of his book on *Optiks* (1717), "the main business of natural philosophy" was to arrive at convincing arguments for the existence of God. Many leading scientists of the seventeenth century were convinced that discoveries in science made philosophical atheism literally incredible. What Henry More (1614–87) called "practical atheism" (living a licentious life) was less difficult to understand; yet many scientists took pains to attack it repeatedly and to enlist science as an ally in the religious controversies of the day. Boyle's enthusiasm for the argument from design, especially as seen in the organic world, derived from his conviction that it was the best argument available for producing in people a profound sense of God's existence, the kind of feeling that would move them to repentance. In his will, Boyle endowed a perpetual lectureship to prove the truth of Christianity against "notorious infidels, *viz.*, atheists, theists [i.e., deists], pagans, Jews, and Mahometans," though he stopped short of entering into "any controversies that are among Christians themselves." The first Boyle lecturer, the Anglican cleric Richard Bentley (1662–1742), corresponded with Newton about the details of his physics and, clearly with Newton's approval, proceeded to use the motions of the planets about the sun to argue for the necessity of divine wisdom in making them move as they do.

In time, both scientists and theologians would come increasingly to rely on this kind of natural theology rather than upon the revealed theology of the Bible for propagating and defending the Christian faith. This tendency contributed in the eighteenth century to the popularity of deism, which accepted the doctrine of creation as evident from nature but rejected the doctrine of redemption. Deists saw God as a distant Creator who had made the world with wisdom but was no longer concerned with its day-to-day operation. They had grave doubts about miracles and rejected the Christian message of sin and salvation. Ironically, deists such as Voltaire (François Marie Arouet [1694–1778]) canonized the deceased Newton as their patron saint, yet Newton's own view of God's relation to the world was irreconcilably different from theirs.

The "Merton Thesis"

While some scholars have found a correlation between voluntaristic theology and science, others have sought links between science and a more broadly conceived Protestant religiosity. In 1938, Robert Merton introduced the *Merton thesis*, the best-known example of this approach. Merton asserted that the Reformed Protestant movement known as puritanism shaped and encouraged English science. It did this through what Merton identified as its underlying "sentiments": diligence and industry, worldly vocation, "empiricorationalism," a valuing of education, and the glorification of God through good works of a utilitarian sort and through studying nature. As a pioneering effort to move away from an internalist study of science to a sociological one and to explore the relationship of religion to science in a positive fashion, Merton's thesis was important. As a contribution to understanding seventeenth-century religion and science, however, it was seriously flawed.

The first flaw lay in Merton's definition of puritanism. In sixteenth- and seventeenth-century England, the term was employed primarily as an insult in intra-Protestant religious quarrels. Scholars in the late twentieth century tend to use the term to refer to a zealous style of experiential Calvinist piety that also aimed to purge the Church of England of its remains of Catholicism and the English social order of its sins, this being the sort of posture that was most likely to attract the label *puritanism* from hostile contemporaries. The heuristic value of the term *puritanism* drops off sharply by the middle of the century, although it can be used, with ever-increasing care and ever-diminishing returns, into the beginning of the eighteenth century. Merton sometimes showed awareness that puritanism was a factional impulse within the Church of England, but far more often he made it roughly analogous with a generically conceived, historically flattened-out Calvinism that, he asserted, underlay the religiosity of almost everyone in seventeenth-century England. Merton's use of the term bore little relationship to contemporary realities or common scholarly usage.

Besides its vastly overgeneralized and ahistorical definition of puritanism, Merton's thesis had other flaws. It was sloppy in its use of critical theological concepts, introducing, for example incautious claims about the positive role of good works in salvation that would have appalled seventeenth-century Protestants who accepted the Reformation doctrine of justification by faith alone (*sola fide*). It took specific forms of late-seventeenth-century religious apologetics that stressed the value of reason and scientific endeavors to be representative of Reformed Protestantism in general, and it justified this chronological casualness by relying on dubious teleological assumptions about gradually emerging inherent tendencies in Protestantism. It sometimes made puritanism crucial to the emergence of modern science (which is how the thesis has usually been read), while at other times it made it only one of many factors of an indeterminable importance. This slipperiness allowed fudging on critical comparative questions, such as why, if Protestantism had an inherent bias toward the production of science that Roman Catholicism lacked, for more than one hundred years after the beginning of the Reformation, Roman Catholicism was generating scientific work that was as good if not better, and why, if Calvinism offered such stimulus to science, Calvin's Geneva produced so little.

Merton's thesis in itself was a blind alley; where it was not simply incorrect, it was too amorphous definitionally, chronologically, and causally to have much explanatory value. Later historians made far more historically informed efforts to link puritanism to science, but they have been criticized (like Merton) for using the term *puritan* arbitrarily and with excessive freedom or for ascribing specifically to puritans tendencies that they shared with broader English religious streams.

Despite its many problems, Merton's thesis gave a spur to continuing research into the links between science and religion in seventeenth-century England. Recently, that research has tended to focus on what was in Merton's thesis an acknowledged but unresolved paradox: Science flourished only after the Restoration of Charles II (ruled 1660–85) in 1660 and the decline of puritanism as a political and cultural force. The two decades of Puritan rule preceding the Restoration produced wild sectarian experimentation, religious "enthusiasm," and challenges to traditional hierarchies of authority. Those experiments and challenges claimed ancestry in the Puritan movement's strains of illuminist theology and radical ecclesiology, and they shocked conservatives committed both to a traditional social order and to the idea of a unified national church. In response, many Anglicans attempted to deny legitimacy to the uncontrolled private religious interpretations that had shattered the national church and to recreate consensus in a deeply torn society. Their means, besides retreating from Calvinism, often included emphasizing reasonableness, freedom from dogmatism, and a probabilistic approach to truth. Those emphases could support a wide range of specific political stances, but they would not have had too much purchase with anyone recognizably a Puritan in the first half of the century. People still committed to puritanism found themselves, often reluctantly, in the role of Dissenters, outside the national church altogether, although Dissent slowly assimilated much of the Anglican attack on its values.

Restoration scientists, no less than Anglican apologists, proclaimed the value of reasonableness, freedom from dogmatism, and a probabilistic approach to truth. A number of historians have recently argued from various perspectives that Restoration scientists self-consciously constructed their conceptual frameworks and research protocols out of a desire to avoid the instability of the previous period. Other historians have vigorously disputed these sociologically driven interpretations of Restoration science. But even historians inclined to stress the ideological neutrality of science acknowledge that experimental practices like alchemy and astrology were looked on as potentially subversive and "enthusiastic" in Restoration England. They clashed with the dominant culture's standards of reasonableness and clarity, which attempted to restrict uncontrolled private interpretation in science no less than in religion and to which people attempting to engage in "normative" science adhered. The newly stable, orderly, and benevolent providential world order increasingly evoked by late-seventeenth-century scientific apologists had a great deal in common with that evoked by Anglican apologists. It is hard to imagine that the coeval births of Restoration science and this specific form of early modern Protestantism were coincidental and that they did not mutually reinforce each other and rest on similar "sentiments."

While interactions between Protestantism and early modern science were both complex and uneasy, they hardly warrant a description in terms of conflict. Their complexity arose from the subtlety of both science and Protestantism and their uneasiness from the different goals and methods of two enterprises that both claimed the right to define the world. Above all, because early modern thinkers rarely separated their science sharply from their religiosity, in spite of statements to the contrary, the interactions were as extensive as they were rich and varied.

BIBLIOGRAPHY

Bono, James J. *The Word of God and the Languages of Man: Interpreting Nature in Early Modern Science and Medicine.* Madison: University of Wisconsin Press, 1995.

Brooke, John H. *Science and Religion: Some Historical Perspectives.* Cambridge: Cambridge University Press, 1991.

Cohen, H. Floris. *The Scientific Revolution: A Historiographical Inquiry.* Chicago: University of Chicago Press, 1994.

Davis, Edward B. "Newton's Rejection of the 'Newtonian World View': The Role of Divine Will in Newton's Natural Philosophy." *Science and Christian Belief* 3 (1991): 103–17.

———. "Rationalism, Voluntarism, and Seventeenth-Century Science." In *Facets of Faith and Science.* Vol. 3: *The Role of Beliefs in the Natural Sciences,* ed. Jitse M. van der Meer. Lanham, Md.: Pascal Centre for Advanced Studies in Faith and Science / University Press of America, 1996, 135–54.

Debus, Allen G. *Man and Nature in the Renaissance.* Cambridge: Cambridge University Press, 1978.

Henry, John. "The Matter of Souls: Medical Theory and Theology in Seventeenth-Century England." In *The Medical Revolution of the Seventeenth Century,* ed. Roger French and Andrew Wear. Cambridge: Cambridge University Press, 1989, 87–113.

Hooykaas, R. *Religion and the Rise of Modern Science.* Grand Rapids, Mich.: Eerdmans, 1972.

Jacob, James R. *Robert Boyle and the English Revolution: A Study in Social and Intellectual Change.* New York: Burt Franklin, 1977.

Jacob, James R., and Margaret C. Jacob. "The Anglican Origins of Modern Science: The Metaphysical Foundations of the Whig Constitution." *Isis* 71 (1980): 251–67.

Lindberg, David C., and Ronald L. Numbers. "Beyond War and Peace: A Reappraisal of the Encounter between Christianity and Science." *Church History* 55 (1986): 338–54.

Merton, Robert K. "Science, Technology, and Society in Seventeenth-Century England."

Osiris 4 (1938): 360–632. Reprint with new introduction. New York: Harper and Row, 1970.

Mulligan, Lotte. "Puritans and English Science: A Critique of Webster." *Isis* 71 (1980): 456–69.

Osler, Margaret J., ed. *Rethinking the Scientific Revolution.* Cambridge: Cambridge University Press, 2000.

Popkin, Richard. *A History of Scepticism, from Erasmus to Spinoza.* Rev. ed. Berkeley: University of California Press, 1979.

Webster, Charles, ed. *The Intellectual Revolution of the Seventeenth Century.* London: Routledge, 1974.

———. *The Great Instauration: Science, Medicine, and Reform, 1626–1660.* London: Duckworth, 1975.

Westfall, Richard S. *Science and Religion in Seventeenth-Century England.* 1958. Reprint. New Haven: Yale University Press, 1973.

Westman, Robert S. "The Copernicans and the Churches." In *God and Nature: Historical Essays on the Encounter between Christianity and Science,* ed. David C. Lindberg and Ronald L. Numbers. Berkeley: University of California Press, 1986, 76–113.

Winship, Michael P. *Seers of God: Puritan Providentialism in the Restoration and Early Enlightenment.* Baltimore: Johns Hopkins University Press, 1996.

10 Causation

John Henry

In theology there has never been any doubt that God can cause things to happen, but there has been a great deal of controversy about the precise nature of God's causative activity in nature. God is undeniably the First or Primary Cause, but to what extent has he established secondary causes in nature as intermediaries? Underlying differences of theological opinion are differences in the nature of providentialism. Voluntarist theologians who emphasize the omnipotence of God's arbitrary will insist that he can accomplish everything by secondary causes; to say that he cannot is to circumscribe his omnipotence. Intellectualist or Necessitarian theologians, who believe that God is led by his goodness to create the world in accordance with absolute concepts of goodness, justice, and the like, more frequently conclude that there are some things that even God cannot accomplish by secondary causes but must perform directly. It should be clear, therefore, that ideas about causation are a prime site for understanding the relationship between science and religion.

John Henry received his Ph.D. from the Open University in England. He is currently senior lecturer in the history of science at the University of Edinburgh, Scotland. Recent publications include *Moving Heaven and Earth: Copernicus and the Solar System* (Cambridge: Icon Books, 2001) and *The Scientific Revolution and the Origins of Modern Science*, 2d ed. (Basingstoke: Palgrave, 2002).

THE NATURE OF CAUSATION, how one event or process might be said to produce and so explain another, has been recognized as a site of major philosophical interest in which interconnected difficulties have been discerned. Are all events caused? Can all causes be expressed in the form of general laws? Is there a necessary connection between cause and effect, or is the supposed connection merely the result of inductive inference? Does the concept of cause depend upon a notion of power? Must causes always precede their effects? Fascinating as these and other associated questions are, we do not pursue them here. The aim of this essay is simply to consider theological theories about the role of God in causation and the way these ideas impinged upon and interacted with naturalistic theories of causation. As the theories of causation themselves are not pursued here, we do not even consider the fortunes of teleological accounts of the natural world and the notion of what are called *final causes* (the purposive reasons why particular outcomes are brought about), even though there is a case for saying, as did Sir Thomas Browne (1605–82), author of *Religio Medici* (1642), that God's providence hangs upon the existence of final causes.

Essentially, the theological account is easily told. God was regarded as the first or primary cause, the sine qua non, of the universe and everything in it. On this much everyone in the Judeo-Christian tradition was agreed. There were, however, two principal sources of disagreement. First, opinions were divided about the extent of God's direct involvement in the workings of the universe, some (although this was always a minority view) regarding God as the sole active agent at work in the universe, others recognizing a hierarchy of secondary (or natural) causes descending below God. Second, those who acknowledged that God chose to operate not directly but by delegating various causal powers to the world's creatures disagreed about the fine detail that this picture involved. The resulting disputes seem to be about the level of God's supervision of and involvement in the secondary causes, some thinkers insisting that God's omnipotence is best illustrated by assuming that he delegates all things to secondary causes, others preferring to suppose that he leaves some room for his own direct intervention. These questions were frequently bound up with considerations of the nature of providence—that is, with differing opinions about what it meant to say that God was omnipotent. All were essentially agreed that God could do anything that did not involve a contradiction, but just what was contradictory and what was not was fiercely disputed. For some, God could not create a substance without accidents (roughly speaking, an object without any properties) or create matter that could think. For others, however, such things easily came within God's ability, but it made no sense to say that he could break the law of the excluded middle (according to which, a particular state of affairs either is or is not; there is no third alternative) or create a weight so heavy that his omnipotence could not lift it. In what follows, we try to confine ourselves strictly to the subject of causation, without straying into discussions of providentialism, but it should be borne in mind that this is a somewhat artificial distinction.

Determinism versus Occasionalism

Although the religion of the ancient Greeks was polytheistic, it has been recognized that, among the naturalist philosophers, there was a marked tendency toward monotheism. Believing in a supreme intelligence capable of ordering and creating the world, they argued that a true god need not struggle with other gods to exert his will (as in the various polytheistic myths), and they developed a notion of god that was far removed from human limitations. Deriving principally from their wish to explain all natural phenomena in terms of physical causes, the one god of the philosophers essentially represented the principle of universal and immutable order and was not only physically transcendent but also morally so. The unified divinity of Greek philosophers was unconcerned by the plight of mortals even though, in some interpretations of the "Unmoved Mover" of Aristotle's cosmology, it might have been indirectly responsible, through the chain of cause and effect, for their plight. In Greek natural philosophy, therefore, it followed that accounts of natural processes did not refer back to the divinity (except in the case of creation myths like that presented in Plato's *Timaeus*), it being assumed that nature was entirely and unalterably regular in its operations.

The earliest suggestions that such ideas made an impression upon Christian theology can be seen in discussions among the early church fathers of God's omnipotence, in which it is affirmed that, although God can enact anything through his power, he chooses to enact things in a fitting way, according to what is "just" or correct. This can be seen as a response to criticisms of pagan thinkers like Galen (A.D. 129–c. 210), the great medical authority, who ridiculed the Christians for believing that God could make a horse or a bull out of ashes in contrast to the pagan view that God would not attempt such a natural impossibility but would choose "the best out of the possibilities of becoming" (*On the Usefulness of the Parts of the Body* 11.14). The implication seems to have been that there was some natural necessity that, for a horse really to be a horse, it had to be made of flesh and bone. Something like this idea even seems to have surfaced in Christian popular culture. In a work attributed to the Venerable Bede (c. 673–735), we learn of "a country saying" that "God has the power to make a calf out of a block of wood. Did he ever do it?" Comments like this seem to suggest a recognition among early Christians that nature is best understood in terms of its regular operations and appearances.

In the Muslim tradition, however, the response to Greek philosophy took a somewhat different turn, giving rise to a major examination of theories of causation. As a result of the influence of Greek ideas, beginning in the eighth century and proceeding through the ninth, Muslim theologians were led to emphasize the supreme omnipotence of God. Among the followers of the theologian al-Ash'arī (d. 935), belief in this omnipotence culminated in the rejection of the natural efficacy of "secondary" agents. Based at first on an interpretation of the Koran, which says, for example, that God "created you and your deeds" (37:94), it was later given philosophical underpinning in a critical analysis of causality written by al-Ghazali (1058–1111). Rejecting the determinism of Aristotelian philosophy, which did not allow for any supernatural intervention, al-Ghazali insisted that there is no necessary correlation between what is taken as the cause and what is taken as the effect. The supposed necessary connection between contingencies in the natural world is based on nothing more than psychological habit. Logical necessity is a coherent notion, al-Ghazali declared, but causal necessity is inadmissible, being based on the fallacious assumption that, because an effect occurs with a cause, it must occur through the cause.

Al-Ghazali's rejection of causation went hand in hand with the so-called occasionalist metaphysics of the Muslim Mutakallims, which had been established since the middle of the ninth century. Seeking to prove that God was the sole power, the sole active agent at work in the universe, the Mutakallims had embraced a form of atomism. Believing, not unreasonably, that the existence of indivisibles of time (since, if time were continuous, two indivisible particles might pass each other and be frozen in time at a point when they were halfway past each other, which, of course, is impossible if they are indivisibles), the Mutakallims argued that God must recreate the world from one moment to the next. Just as God created the atomic particles, so he creates the indivisible moments of time one after another. In re-creating the world this way, God re-creates everything as he did before, though with numerous changes. What seems like a continuous pageant of changes in accordance with natural laws of cause and effect is, therefore, merely the result of God's way of re-creating the world in self-imposed accordance with strict patterns and rules. When a natural entity is seen to act, it does not act by its own operation; it is, rather, God who acts through it. There is no other meaning to the notion of cause and effect.

This occasionalism (so-called because it was held that an event did not cause an effect but merely signaled the occasion at which God acts) was vigorously opposed by the Muslim Aristotelian philosopher Averroës (1126–98) and by the Jewish Aristotelian philosopher Moses Maimonides (1135–1204). Both critics insisted that the reality of causal operations could be inferred from sensory experience and argued that knowledge itself depended upon causality, since the distinction between what is knowable and what is not depends upon whether or not causes can be assigned to the thing in question. This last point depends upon acceptance of the important Aristotelian concept of form. A body, according to Aristotle (384–322 B.C.), is made up of matter and form, and it is the form that gives the body its identity, which includes not only its principle of existence but also its principle of activity. Additionally, Averroës objected to the suggestion that activity is legitimately attributable only to an agent having will and consciousness. The distinction between natural and voluntary activity must be maintained, Averroës insisted, because natural agents always act in a uniform way (fire cannot fail to heat), while voluntary agents act in different ways at different times. Besides, by emphasizing God's voluntary action, the Mutakallims were anthropomorphizing God, seeing him as a capricious and despotic ruler of the creation. According to Averroës, voluntary action cannot be attributed to God because it implies that he has appetites and desires that move his will.

The Averroistic position led, however, to an extreme determinism that seemed to circumscribe the power of God. This, among other Averroistic doctrines, enjoyed a certain success with early scholastic natural philosophers in the revival of learning in the Latin West and was included in the condemnation of 219 philosophical propositions issued by the bishop of Paris, Etienne Tempier, in 1277. But what was to become one of the main alternatives in the orthodox Christian view had by then already been worked out by Thomas Aquinas (c. 1225–74). Aquinas wished to maintain the notion of divine providence, effectively rejected in Averroism, while combining it with a recognition of the usefulness of the Aristotelian notion of natural efficacy. The difficulty here, however, is that there seems to be a duplication of effort. If the divine power suffices to produce any given effect, there is no need of a secondary natural cause. Similarly, if an event can be explained in terms of natural causes, there is no need for a divinity. Drawing upon Neoplatonic traditions, Aquinas suggested an emanationist hierarchy of secondary causes in which inferior causes depend ultimately upon the primary cause because they are held to emanate from it in the way that radiance emanates from a light source. This was supposedly in keeping with God's goodness, since it was a case of God's communicating his "likeness" to things, not merely by giving them existence, but by giving them the ability to cause other things.

This was to become the dominant view of causation in Christian orthodoxy. Before pursuing that subject, however, it is worth noting that occasionalism reappeared during the seventeenth century in the Christian West. It emerged from the mechanical philosophy of René Descartes (1596–1650). Seeking to eliminate all unexplained or occult conceptions from his natural philosophy, Descartes tried to explain all physical phenomena in terms of the interactions of invisibly small particles of matter in motion. Apart from the force of impact resulting from a body's motion, he rejected all explanations in terms of forces or powers, regarding them as occultist notions. To characterize the different ways in which moving bodies behaved, Descartes introduced his three laws of nature and, following on from his third law (in which he gives a broad characterization of force of impact), seven rules of impact. By applying Descartes's rules, it is possible in principle to understand or predict how colliding bodies will interact with one another (although, in fact, Descartes's rules incorporate a number of false assumptions).

In keeping with his wish to eliminate occult concepts from his philosophy, however, Descartes was anxious to clarify what he meant by the force of a body's motion: "It must be carefully observed what it is that constitutes the force of a body to act on another body," he wrote in his *Principia philosophiae* (1644). "It is simply the tendency of everything to persist in its present state so far as it can (according to the first law)." But Descartes had already made clear in his discussion of his first law that the tendency of everything to persist in its present state is not the property of a body itself but the result of the immutability of God. Because of his immutability, God preserves motion "in the precise form in which it occurs at the moment that he preserves it, without regard to what it was a little while before." This accounts for the continued motion of projectiles after leaving contact with the projector and for the tangential motion of a body released from a sling. If the precise motions of bodies depend upon an attribute of God (his immutability), it follows that the motions must be directly caused by God. As Descartes wrote in *Le Monde:* "It must be said, then, that God alone is the author of all the movements in the universe."

Although Descartes presented a picture in which God is directly responsible for the motions of every particle in the universe and seems to operate in a discontinuist way, re-creating motions from moment to moment as did the God of the Mutakallims, he was rather coy about drawing attention to the theological implications. A number of his followers took up these ideas, however, with varying degrees of explicitness. The fullest system of occasionalism was developed by Nicolas de Malebranche (1638–1715), who was driven by his own religious commitments to push Cartesianism in a theocentric direction. But there were also several philosophical difficulties with the nature of causation that were avoided by taking an occasionalist line. It was by no means clear to Descartes's contemporaries, for example, that motion could be transferred in a collision from one particle of matter to another, particularly if, as Descartes insisted, the matter was completely passive and inert. As Henry More (1614–87), the Cambridge theologian, wrote in response to the Cartesian account of collision in 1655: "For Descartes himself scarcely dares to assert that the motion in one body passes into the other. . . . [I]t is manifest that one arouses the other from sleep as it were, and in this way aroused bodies transfer themselves from place to place by their own force." Clearly, More's account is too occultist to be acceptable to a Cartesian, but it nicely raises the philosophical issue of causality that confronted Descartes's philosophy. It also illustrates for the modern reader the Humean point that our assumptions about cause and effect are habits of thinking. No modern reader would have any hesitation in accepting the suggestion that motion is transferred from one body to another in a collision, but for thinkers in an earlier age, with different habits of mind, such a view was as absurd as expecting color to be transferred from one object to another in a collision.

Although occasionalism could extricate Descartes from his philosophical difficulties with causation, it brought along with it a number of theological difficulties. For Isaac Barrow (1630–77), the Cartesian system reduced God to a "carpenter or mechanic repeating and displaying *ad nauseam* his one marionettish feat." But worse, as Henry More pointed out, was that God seemed to be directly responsible for all of the evil of the world, and, hence, human free will was made nonsense.

The Absolute and the Ordained Powers of God

In spite of powerful support in both of its historical manifestations, occasionalism never succeeded in becoming part of the philosophical or theological mainstream. The alternative view, that God invests his creatures with causative principles of their own, was certainly the dominant view in the Christian tradition. When Aquinas struck his middle way between the antiprovidential determinism of Averroism and the theistic excesses of occasionalism, he was drawing upon an already established approach, in which the Greek notion of natural efficacy was accommodated to the Christian view of an omnipotent deity. The chief means of making this accommodation was through the distinction between the absolute power of God and his so-called ordained power (*potentia dei absoluta et ordinata*). Although this distinction is not made fully explicit until 1235, when it was used by Alexander of Hales (c. 1170–1245), he was clearly not the first to have thought of it. In about 1260, when it appears in the *Summa Theologiae* of Albertus Magnus (1193–1280), we are told that the distinction is customary. It has been suggested, with some plausibility, that it derives from the earlier distinction, made by Origen (c. A.D. 185–c. 251) and others, between what God can do and what he deems fitting or "just," which was brought to the fore by Peter Damian's (1007–72) *De divina omnipotentia,* an attack on the excessive reliance on logical argument in theology.

By his absolute power, it was held, God could do anything. But, having decided upon the complete plan of Creation, God holds his absolute power in abeyance and uses his ordained power to maintain the preordained order that he chose to effect. Although God is entirely able to use his absolute power to change things, it is safe to assume that all will proceed in accordance with his ordained power. Furthermore, it was generally assumed that, by his ordained power, God had invested his creatures with their own natural powers. Accepting the Aristotelian idea that the natural powers of a particular body were part of its identity, it was believed that, if a body was devoid of any activity of its own, its existence would be pointless. God's ordained power was not, therefore, used to carry out all changes from moment to moment, as in occasionalism. It was a creative power that established the system of the world, delegating causal powers to things, and, subsequently, its role was to uphold the system. To thwart atheistic suggestions that the system was capable of operating without God, it was usually held that the *potentia ordinata* was required to keep the whole system in being. But, given that its existence was maintained, the system functioned by itself in accordance with the laws of nature that God had imposed upon it. Indeed, the so-called laws of nature were recognized to be a shorthand way of referring to the sum total of causal powers possessed by bodies. Inanimate objects were incapable of obeying laws, but natural powers always operated in specific and uniform ways so that bodies might appear to be operating according to law.

Within this broad tradition of causation, however, there were nuances. William of Ockham (c. 1280–c. 1349), accepting the condemnation of 219 Aristotelian propositions of 1277, developed a radical empiricism based on an emphasis on God's absolute power. All that exists are contingencies created by the arbitrary will of God. There are no necessary connections between things: Whatever might be performed by secondary causes might be performed directly by God. So, in a particular case of combustion, an assumption that it was caused by fire might be ill founded if God had directly intervened. Causal relations could be established, therefore, only by experience, not by reason, and even our experiences might be mistaken. Ockham's empiricism proved influential, especially among theological voluntarists, who wished to emphasize the role of God's arbitrary will in Creation, even though it was usually tempered by a perceived need to accept the real and reliable action of secondary causes. The emphasis on experimentalism in the scientific method of Robert Boyle (1627–91) and other leading members of the Royal Society in the late seventeenth century, for example, can be seen to be based on the same kind of theological concern with the unconstrained freedom of God's will, although in other respects Boyle and his colleagues were entirely at ease with the notion of secondary causes and their uniform mode of action.

The famous dispute between Samuel Clarke (1675–1729), speaking for Isaac Newton (1642–1727), and Gottfried Wilhelm Leibniz (1646–1716) included differences over the nature of causation. Ultimately, these differences can be traced back to their opposed positions on the nature of providence, Newton being a voluntarist (who emphasized God's arbitrary will and held him free to make any kind of world he chose) and Leibniz a necessitarian or intellectualist (who held God's reason to be his primary attribute and who believed, therefore, that God was constrained by co-eternal rational and moral principles to create only this world—which must be the best of all possible worlds). Even so, both thinkers subscribed to the general belief that God, the primary cause, had delegated causal efficacy to secondary causes. Leibniz famously suggested that Newton's God was a poor workman, continually obliged to set his work right "by an extraordinary concourse." But this was to take too literally Newton's efforts to forestall suggestions that the mechanical philosophy could explain all phenomena without recourse to God. Being aware that atheists could appropriate to their cause a mechanical system of the world in which motions were always preserved, Newton insisted that the motions of the heavenly bodies were in gradual decay and that God's periodic intervention was required to correct this decay. Although it was not unreasonable for Leibniz to assume that Newton must have had a miraculous intervention in mind—that is to say, intervention by God's absolute power—it is clear from unpublished comments by Newton that he believed that comets were the secondary causes through which the ordained power of God operated to replenish the motions of the planets.

The Rise of Secondary Causation

Natural theology, which achieved its heyday at the end of the seventeenth and the beginning of the eighteenth centuries, was entirely based upon the traditional distinction between God and secondary causes. And, as is well known, this distinction led to the flourishing of deism, which accepted the existence of an omnipotent Creator and was willing to discuss the Creator's attributes as revealed by the intricate contrivance of his creation but denied the validity of theological and religious doctrines supposedly gleaned from revelation. When defenders of religious orthodoxy introduced reports of miracles into their attempts to defend the importance of revelation, a number of the more radical deists even went so far as to deny the possibility of miracles. Peter Annet (1693–1769), for example, used the immutability of God to argue that he could not, or would not, interrupt the normal course of nature. Deism can be seen, therefore, as an extreme version of the tradition of attributing natural efficacy to secondary causes, at the opposite end of the *potentia absoluta et ordinata* spectrum from occasionalism. It can also be seen, of course, as a major source for atheistic appropriations of explanations by secondary causes, in which the need for a primary cause is denied.

By the nineteenth century, natural philosophers were so used to developing their theories in terms of secondary causes alone—without introducing the deity—that the origin of new species of plants and animals caused some embarrassment. The fossil record seemed to suggest that new species of animals and plants had appeared on the earth at different times; creatures that were not found in earlier rock strata suddenly appear in abundance. The comparatively new science of geology was called upon to account for the changing face of both the earth and the habitat, which made it possible, perhaps for the first time, for such new creatures to thrive. But geology and paleontology could say nothing about the origins of the new creatures themselves. Secondary causation did not seem capable of extending that far. Here, then, were the limits of natural science. The origin of species became the "mystery of mysteries," to be left to the man of religion. As William Whewell (1794–1866) said, it was a problem to which "men of real science do not venture to return an answer."

Needless to say, this abdication of the rights of science did not persist for long. A group of biological scientists seeking to find the answer to this mystery developed theories of biological evolution. Once again, these theories could be presented as the workings of secondary causes established by God. As Charles Darwin (1809–82) wrote in 1842, before he became an agnostic: "It accords with what we know of the law impressed on matter by the Creator: that the creation & extinction of forms, like the birth and death of individuals, should be the effect of secondary laws. It is derogatory that the Creator of countless systems of worlds should have created each of the myriads of creeping parasites and slimy worms which have swarmed each day of life on land and water on this globe." For many believers, Darwin's theory of evolution pointed the way to a "grander view of the Creator," in which God was able to demonstrate his wisdom and omnipotence by ensuring the self-development of different life forms through the workings of secondary causes.

It is perhaps an indication of the strong links between theories of secondary causation and belief in the existence of God that anti-Darwinist Christians tried to dismiss Darwinism on the ground that it relied upon chance rather than cause and effect. This charge was vigorously rejected by "Darwin's bulldog," T. H. Huxley (1825–95), who argued that evolution involved chance no more than the scene of chaos presented at a seashore in a heavy gale: "The man of science knows that here, as everywhere, perfect order is manifested; that there is not a curve of the waves, not a note in the howling chorus, not a rainbow glint on a bubble, which is other than a necessary consequence of the ascertained laws of nature." In our postquantum age, in which Werner Heisenberg's (1901–76) uncertainty principle holds sway, it would be harder for a scientist to talk so confidently of the necessary consequences of cause and effect, but this does not mean that the uncertainty principle thwarted religious interpretations of the physical world. On the contrary, believers immediately saw Heisenberg's principle as a way of rejecting the determinism that had been all too often appropriated to the cause of atheism. It would seem that theists are ever resourceful in their use of contemporary scientific theory to support belief in God.

In later-twentieth-century physics, there has been a tendency to rely on mathematical formalism, rather than cause-and-effect accounts, to lead from one claim about the physical world to another. It has even been remarked that the word *cause* hardly appears in the discourse of modern physics. It seems unlikely, however, that Albert Einstein (1879–1955) remains unique among modern physicists in believing that there must be a real world controlled by causal mechanism underwriting the mathematical formalisms discerned in quantum physics. Moreover, causal accounts continue to be the raison d'être of most other sciences. Given the richness of the distinction between God's absolute and ordained powers and the tradition of secondary causation, it seems hardly surprising that many scientists continue to combine their science with a devout belief in God.

BIBLIOGRAPHY

Alexander, H. G., ed. *The Leibniz-Clarke Correspondence: With Extracts from Newton's* Principia *and* Opticks. Manchester, U.K.: Manchester University Press / New York: Barnes and Noble, 1956.

Beauchamp, T. L., and A. Rosenberg. *Hume and the Problem of Causation.* New York: Oxford University Press, 1981.

Brooke, John Hedley. *Science and Religion: Some Historical Perspectives.* Cambridge: Cambridge University Press, 1991.

Brown, Stuart, ed. *Nicolas Malebranche: His Philosophical Critics and Successors.* Assen: Van Gorcum, 1991.

Burns, R. M. *The Great Debate on Miracles: From Joseph Glanvill to David Hume.* London: Associated University Presses, 1981.

Courtenay, William J. "The Critique on Natural Causality in the Mutakallimun and Nominalism." *Harvard Theological Review* 66 (1973): 77–94.

Dobbs, B. J. T. *The Janus Faces of Genius: The Role of Alchemy in Newton's Thought.* Cambridge: Cambridge University Press, 1991.

Fakhry, Majid. *Islamic Occasionalism and Its Critique by Averroës and Aquinas.* London: Allen and Unwin, 1958.

Funkenstein, Amos. *Theology and the Scientific Imagination: From the Middle Ages to the Seventeenth Century.* Princeton: Princeton University Press, 1986.

Gillispie, Charles C. *Genesis and Geology: The Impact of Scientific Discoveries upon Religious Beliefs in the Decades before Darwin.* Cambridge: Harvard University Press, 1951.

Hatfield, Gary C. "Force (God) in Descartes's Physics." *Studies in the History and Philosophy of Science* 10 (1979): 113–40.

Huxley, T. H. "On the Reception of the *Origin of Species.*" In *The Life and Letters of Charles Darwin,* ed. F. Darwin. 3 vols. London: John Murray, 1887, 2:179–204.

Jolley, Nicholas. "The Reception of Descartes' Philosophy." In *The Cambridge Companion to Descartes*, ed. John Cottingham. Cambridge: Cambridge University Press, 1992, 393–423.

Kubrin, David. "Newton and the Cyclical Cosmos: Providence and the Mechanical Philosophy." *Journal of the History of Ideas* 28 (1967): 325–46.

Mackie, J. L. *The Cement of the Universe: A Study of Causation*. Oxford: Clarendon, 1974.

Oakley, Francis. *Omnipotence, Covenant, and Order: An Excursion in the History of Ideas from Abelard to Leibniz*. Ithaca: Cornell University Press, 1984.

Ospovat, Dov. *The Development of Darwin's Theory: Natural History, Natural Theology, and Natural Selection, 1838–1859*. Cambridge: Cambridge University Press, 1981.

Shanahan, Timothy. "God and Nature in the Thought of Robert Boyle." *Journal of the History of Philosophy* 26 (1988): 547–69.

Vailati, Ezio. *Leibniz and Clarke: A Study of Their Correspondence*. Oxford: Oxford University Press, 1997.

van Ruler, J. A. *The Crisis of Causality: Voetius and Descartes on God, Nature, and Change*. Leiden: Brill, 1995.

11 Mechanical Philosophy

Margaret J. Osler

The mechanical philosophy was a philosophy of nature adopted by many influential seventeenth-century natural philosophers who attempted to show that all natural phenomena could be explained in terms of matter and motion. Unwilling to accept the materialism that seemed implicit in this philosophy, mechanical philosophers developed ways of incorporating God's creation of the world, his providential care of the creation, and the immortality of an immaterial human soul into their worldview. Among the most prominent mechanical philosophers were Pierre Gassendi, René Descartes, Robert Boyle, and Isaac Newton.

Margaret J. Osler received her Ph.D. in history and philosophy of science from Indiana University. She is currently professor of history and adjunct professor of philosophy at the University of Calgary. Her publications include *Divine Will and the Mechanical Philosophy: Gassendi and Descartes on Contingency and Necessity in the Created World* (Cambridge: Cambridge University Press, 1994); "How Mechanical Was the Mechanical Philosophy? Non-Epicurean Themes in Gassendi's Atomism," in *Late Medieval and Early Modern Corpuscular Matter Theory,* edited by John Murdoch, William R. Newman, and Christoph Lüthy (Leiden: Brill, 2001); and "Whose Ends? Teleology in Early Modern Natural Philosophy," in *Osiris* Second Series 16 (2001): 151–68.

MECHANICAL PHILOSOPHY was a philosophy of nature, popular in the seventeenth century, that sought to explain all natural phenomena in terms of matter and motion without recourse to any kind of action-at-a-distance. During the sixteenth and seventeenth centuries, many natural philosophers rejected Aristotelianism, which had provided metaphysical and epistemological foundations for both science and theology at least since the thirteenth century. One candidate for a replacement was the mechanical philosophy, which had its roots in classical Epicureanism. Mechanical philosophers attempted to explain all natural phenomena in terms of the configurations, motions, and collisions of small, unobservable particles of matter. For example, to explain the fact that lead is denser than water, a mechanical philosopher would say that the lead has more particles of matter per cubic measure than water. The mechanical explanation differed from Aristotelian explanations, which endowed matter with real qualities and used them to explain the differences in density by appealing to the fact that lead has more absolute heaviness than water. A hallmark of the mechanical philosophy was the doctrine of primary and secondary qualities, according to which matter is really endowed with only a few "primary" qualities, and all others (such as color, taste, or odor) are the result of the impact of the primary qualities on our sense organs. Nature was thus mechanized, and most qualities were considered subjective. This approach enhanced the mathematization of nature at the same time that it provided an answer to the skeptical critique of sensory knowledge.

While the mechanical philosophy was attractive to thinkers working in the tradition of Galileo Galilei (1564–1642) and William Harvey (1578–1657), it posed serious problems for those holding a Christian worldview. Orthodox natural philosophers feared that the mechanical philosophy would lead to materialism or deism, resulting in the denial of Creation and divine providence. The fact that the Thomist synthesis of theology and Aristotelian philosophy had become dominant in the Catholic world, especially after the Council of Trent (1545–63), also meant that the rejection of Aristotelianism seemed to challenge the doctrine of transubstantiation (the doctrine that the bread and wine in the Eucharist were miraculously transformed into the body and blood of Christ).

Christian mechanical philosophers adopted a variety of strategies to stave off these perceived threats, including frequent appeal to the argument from design as a way of establishing God's providential relationship to the world he created, special attention to proving the existence of an immaterial, immortal human soul, and attempts to explain the real presence in the Eucharist in mechanical terms.

Background

The mechanical philosophy originated in classical times with the Greek philosopher Epicurus (341–270 B.C.), who sought to explain all natural phenomena in terms of the chance collisions of material atoms in empty space. He even claimed that the human soul is material, composed of atoms that are exceedingly small and swift. Epicurus believed that atoms have always existed and that they are infinite in number. Epicureanism, while not strictly atheistic, denied that the gods play a role in the natural or human worlds, thus ruling out any kind of providential explanation. Because of its reputation as atheistic and materialistic, Epicureanism fell into disrepute during the Middle Ages. The writings of Epicurus and his Roman disciple Lucretius (c. 99–55 B.C.) were published during the Renaissance, along with a host of other classical writings.

Following the development of heliocentric astronomy in the late sixteenth and early seventeenth centuries, many natural philosophers believed that Aristotelianism, which rests on geocentric assumptions, could no longer provide adequate foundations for natural philosophy. Among the many ancient philosophies that had been recovered by the Renaissance humanists, the atomism of Epicurus seemed particularly compatible with the spirit of the new astronomy and physics. Moreover, the mechanical philosophy often seemed easier to reconcile with Christian theology than the alternatives—Stoicism, Neoplatonism, and Paracelsianism—all of which appeared to limit the scope or freedom of God's action in the world. Early advocates of the mechanical philosophy included David van Goorle (d. 1612), Sebastian Basso (fl. 1550–1600), and various members of the Northumberland Circle, of which Walter Warner (c. 1570–c. 1642), Thomas Hariot (1560–1621), and Nicholas Hill (c. 1570–1610) were members. Although each of these men favored some version of atomism, none of them developed a systematic philosophy or addressed the theological problems associated with atomism. Isaac Beeckman (1588–1637), a Dutch schoolmaster, advocated a mechanical view of nature and wrote about it extensively in his private journal, which was not published until the twentieth century. Beeckman's personal influence was enormous, however, and he was instrumental in encouraging both Pierre Gassendi (1592–1655) and René Descartes (1596–1650) to adopt the mechanical philosophy.

Major Advocates

A number of the major figures whose names are associated with the scientific revolution adopted some version of the mechanical philosophy. Although Galileo did not write a fully articulated account of it, he implicitly adopted its major tenets. In *Il Saggiatore (The Assayer* [1623]), he employed the doctrine of primary and secondary qualities to explain perception.

Gassendi and Descartes published the first systematic and most influential accounts of the mechanical philosophy. Their treatises were not detailed accounts of particular subjects. Rather, they spelled out the fundamental terms of a mechanical philosophy and functioned as programmatic statements, describing what such a philosophy would look like in practice. While both men agreed that all physical phenomena should be explained in terms of matter and motion, they differed about the details of those explanations. Gassendi, writing in the manner of a Renaissance humanist, saw himself as the restorer of the philosophy of Epicurus. Deeply concerned with Epicurus's heterodox ideas, Gassendi, a Catholic priest, sought to modify ancient atomism so that it would be acceptable to seventeenth-century Christians. Accordingly, he insisted on God's creation of a finite number of atoms, on God's continuing providential relationship to the Creation, on free will (both human and divine), and on the existence of an immaterial, immortal human soul that God infused into each individual at the moment of conception.

Gassendi believed that God had created indivisible atoms and endowed them with motion. The atoms, colliding in empty space, are the constituents of the physical world. In his massive *Syntagma philosophicum* (published posthumously in 1658), Gassendi set out to explain all of the qualities of matter and all of the phenomena in the world in terms of atoms and the void. He argued for the existence of the void—a controversial claim at the time—on both conceptual and empirical grounds, appealing to the recent barometric experiments of Evangelista Torricelli (1608–47) and Blaise Pascal (1623–62). The primary qualities of Gassendi's atoms were size, shape, and mass. He advocated an empiricist theory of scientific method and considered the results of this method to be, at best, probable.

Gassendi insisted that God has complete freedom of action in the universe that he freely created. Indeed, God can, if he wishes, violate or overturn the laws of nature that he established. The only constraint on divine freedom is the law of noncontradiction. One consequence of divine freedom is that humans must have free will, too, for if human actions were necessarily determined, that necessity would be a restriction on God's freedom to act. Gassendi rejected as question-begging Epicurus's explanation of human freedom as a consequence of unpredictable, random swerves of the atoms composing the human. Instead, he argued for the existence of an immaterial and immortal human soul, which is the seat of the higher mental faculties. In addition to the immaterial, immortal soul, Gassendi claimed that there exists a material, sensible soul, composed of very fine and swiftly moving particles. This material soul (which animals also possess) is responsible for vitality, perception, and the less abstract aspects of understanding. The material soul is transmitted from one generation to the next in the process of biological reproduction. Gassendi's ideas were brought to England by Walter Charleton (1620–1707) and popularized in France by François Bernier (1620–88).

Although Descartes also articulated a full-fledged mechanical philosophy in his *Principia philosophiae* (*Principles of Philosophy* [1644]), his ideas were quite different from those of Epicurus. In contrast to Gassendi's atomic view of matter, Descartes claimed that matter fills all space and is infinitely divisible, thus denying the existence of both atoms and the void. He believed that matter possesses only one primary quality, geometrical extension. This belief provided foundations for his attempted mathematization of nature. Descartes drew a sharp distinction between matter and mind, considering thinking to be the essential characteristic of the mind. Like Gassendi's doctrine of the immortal soul, Descartes's concept of mind established the boundaries of mechanization in the world.

Descartes derived his mechanical philosophy directly from theological considerations. His rationalist epistemology was grounded in the conviction that, since God is not a deceiver, his existence provides a warrant for reasoning from clear and distinct ideas in our minds to knowledge about the created world. Since geometrical concepts are paradigmatic of clear and distinct ideas, we can conclude that the physical world has geometrical properties. It is on such grounds that Descartes justified the claim that matter is infinitely divisible. The divine attributes also lie at the base of Descartes's attempts to prove the laws of motion. He appealed to God's immutability to justify his law of the conservation of motion and his version of the principle of inertia, the foundations of his physics.

Like Gassendi, Descartes intended his philosophy to replace Aristotelianism. He hoped that the Jesuit colleges would adopt the *Principia philosophiae* as a physics textbook to replace the Aristotelian texts still in use. His hopes were dashed, however, when his book was condemned in 1662 and placed on the Index of Prohibited Books in 1663 in response to his attempt to give a mechanical explanation of the real presence in the Eucharist.

The differences between the mechanical philosophies of Gassendi and Descartes reflect their theological differences concerning providence, or God's relationship to the creation. Gassendi was a voluntarist, believing that the created world is utterly contingent on God's will, which is constrained only by the law of noncontradiction. The contingency of the world rules out the possibility of any kind of rationalist epistemology because it would embody some kind of necessity, such as the relationship between ideas in our minds and the world. Gassendi's empiricism and probabilism and the fact that he believed that matter possesses some properties that can be known only by empirical methods reflect his voluntarist theology. In contrast to Gassendi, Descartes believed that, although God was entirely free in his creation of the world, he freely created some things to be necessary (e.g., the eternal truths), which we are capable of knowing a priori and with certainty. Descartes's theory of matter, according to which matter possesses only geometrical properties that can be known a priori, follows from his rationalist epistemology. Both his theory of knowledge and his theory of matter are closely associated with his theological presuppositions.

Another mechanical philosopher, Thomas Hobbes (1588–1679), was the specter haunting more orthodox mechanical philosophers. Whatever the state of his religious beliefs, Hobbes's philosophy seemed—to the seventeenth-century reader—to be materialistic, deterministic, and possibly even atheistic. In *The Elements of Philosophy* (1655), Hobbes propounded a complete philosophy—of matter, of man, and of the state—according to mechanistic principles. Although the details of his mechanical philosophy were not very influential among natural philosophers, his mechanical account of the human soul and his thoroughly deterministic account of the natural world alarmed the more orthodox thinkers of his day.

Later Developments

Gassendi and Descartes were founding fathers of the mechanical philosophy in the sense that the next generation of natural philosophers, who accepted mechanical principles in general, believed that they had to choose between Gassendi's atomism and Descartes's corpuscularism. Robert Boyle (1627–91) and Isaac Newton (1642–1727), among the most prominent natural philosophers of the second half of the seventeenth century who developed their philosophies of nature in this context, were both deeply concerned with the theological implications of their views.

Boyle is best known for his attempt to incorporate chemistry within a mechanical framework. His corpuscular philosophy—which remained noncommittal on the question of whether matter is infinitely divisible or composed of indivisible atoms—was founded on a mechanical conception of matter. His reluctance to commit himself to a position on the ultimate nature of matter reflected his concern about the atheism still associated with Epicureanism, as well as his recognition that some questions lie beyond the ability of human reason to resolve. Material bodies are, according to Boyle, composed of extremely small particles, which combine to form clusters of various sizes and configurations. The configurations, motions, and collisions of these clusters produce secondary qualities, including the chemical properties of matter. Boyle conducted many observations and experiments with the aim of demonstrating that various chemical properties can be explained mechanically. He performed an extensive series of experiments with the newly fabricated air pump to prove that the properties of air—most notably its "spring"—could be explained in mechanical terms.

Boyle was a deeply religious man and discussed the theological implications of his corpuscularianism at great length. He believed that God had created matter and had endowed it with motion. God had created laws of nature but could violate those laws at will; biblical miracles provided evidence for that claim. In addition to matter, God creates human souls, which he imparts to each embryo individually. He also created angels and demons, which are spiritual, not material entities.

For Boyle—and many other natural philosophers of his day—the practice of natural philosophy was an act of worship, since it led to greater knowledge of the Creator by directly acquainting the careful observer with God's wisdom and benevolence in designing the world. God's purposes are everywhere evident to the astute observer. God is not entirely knowable, however, and neither are his purposes. Boyle was careful to acknowledge the limits of human reason in theology, and those limits extend as well to natural philosophy, in which human knowledge is limited in scope and is never certain. Boyle's ideal was that of "the Christian Virtuoso," who discovered the deep connections between natural philosophy and Christian theology.

Newton, whose reputation rests on his achievements in mathematical physics and optics, accepted the mechanical philosophy from his student days in Cambridge. A notebook written in the mid-1660s shows him thinking about natural phenomena in mechanical terms and designing thought experiments for choosing between Cartesian and Gassendist explanations of particular phenomena. A number of phenomena—gravitation, the reflection and refraction of light, surface tension, capillary action, certain chemical reactions—persistently resisted explanation in purely mechanical terms. Failing in the attempt to explain them by appeal to hypotheses about submicroscopic ethers, Newton was led to the view that there exist attractive and repulsive forces between the particles composing bodies. This idea came to him from his alchemical studies. Newton's most notable discovery, the principle of universal gravitation, which provided a unified foundation for both terrestrial and celestial mechanics and which marks the culmination of developments started by Copernicus (1473–1543) in the mid–sixteenth century, demanded a concept of attractive force. The concept of force, which seemed to some contemporaries to be a return to older theories of action-at-a-distance banished by the mechanical philosophy, enabled Newton to accomplish his stunning mathematization of physics.

In addition to physics and mathematics, Newton devoted years of intellectual labor throughout his life to the study of alchemy and theology. Recent scholarship has suggested that Newton's primary motive in all three areas was theological: to establish God's activity in the world. Theologically, Newton was an Arian, believing that Christ, while divine, was a created being and denying the doctrine of the Trinity. The transcendence of the Arian God suggested the possibility of deism, a doctrine Newton rejected. Consequently, Newton devoted himself to discovering evidence of divine activity in the world, something he found in the active matter of the alchemists, in the fulfillment of the biblical prophecies in history, and in the gravitation of matter. Because matter itself is inert, it cannot generate any motion, and it cannot deviate from uniform rectilinear motion without the action of some external mover. The orbital motions of the planets are a departure from inertial motion, which Newton explained in terms of gravitational force. Explaining gravitation had been a challenge to Newton throughout his life. Denying that it is an innate property of matter, Newton sought to explain this force in some way that was consistent with both his theology and his philosophy of nature. Early in his career, he attempted to explain gravity, along with other recalcitrant phenomena, such as surface tension, capillary action, and the reflection and refraction of light, in terms of such mechanical devices as density gradients in the ether. In the 1670s, he abandoned attempts to explain gravity in purely mechanical terms, recognizing that such explanations led to an infinite regress. Moreover, as he proved in the *Principia*, the presence of even the most subtle mechanical ether in space would resist the motions of the planets, causing the solar system to run down. Newton speculated further in several of the "Queries" to his *Opticks* (1704). At one stage, he proposed the existence of an ether that would not resist the motions of the planets and that was composed of particles endowed with both attractive and repulsive forces. At another stage, he proposed that gravitation results from God's direct action on matter. On this account, he regarded his physics and his cosmology as part of a grand argument from design, leading to knowledge of the intelligent and all-powerful Creator.

In the decades after Newton's death, the worst fears of the Christian mechanical philosophers of the seventeenth century came true. John Locke (1632–1704) argued for the reasonableness of Christianity, and his environmentalist analysis of the human mind—which grew directly from the ideas of the mechanical philosophers—implied the denial of the Christian doctrine of original sin. Deism and natural religion flourished both in England and on the Continent. Some of the French philosophes, notably Julien Offray de La Mettrie (1709–51) and Paul Henry Thiry, Baron d'Holbach (1723–89), espoused atheistic materialism and adopted vigorously anticlerical and antiecclesiastical views. David Hume (1711–76) undermined the possibility of natural religion and a providential understanding of the world by purporting to demonstrate the invalidity of the standard arguments for the existence of God, particularly the argument from design, which had played such a crucial role for the seventeenth-century mechanical philosophers. Newtonian mechanics rose to great heights, having shed the theological preoccupations of its creator. These developments culminated in the work of Pierre Laplace (1749–1827), who articulated a clear statement of classical determinism and was able to demonstrate that the solar system is a gravitationally stable Newtonian system. When asked by Napoleon what role God played in his system, Laplace is reputed to have replied: "Sire, I have no need for that hypothesis."

BIBLIOGRAPHY

Boas, Marie. "The Establishment of the Mechanical Philosophy." *Osiris* 10 (1952): 412–541.

Brandt, Frithiof. *Thomas Hobbes' Mechanical Conception of Nature.* Trans. Vaughan Maxwell and Annie Fausball. Copenhagen and London: Levin and Munksgaard / Librairie Hachette, 1928.

Cook, Margaret G. "Divine Artifice and Natural Mechanism: Robert Boyle's Mechanical Philosophy of Nature." *Osiris* 2d ser. 16 (2001): 133–50.

Des Chene, Dennis. *Spirits and Clocks: Machine and Organism in Descartes.* Ithaca: Cornell University Press, 2001.

Dijksterhuis, E. J. *The Mechanization of the World Picture.* Trans. C. Dikshoorn. Oxford: Oxford University Press, 1961.

Dobbs, Betty Jo Teeter. *The Janus Faces of Genius: The Role of Alchemy in Newton's Thought.* Cambridge: Cambridge University Press, 1991.

Funkenstein, Amos. *Theology and the Scientific Imagination from the Middle Ages to the Seventeenth Century.* Princeton: Princeton University Press, 1986.

Gabbey, Alan. "Henry More and the Limits of Mechanism." In *Henry More (1614–1687): Tercentenary Studies,* ed. Sarah Hutton. Dordrecht: Kluwer, 1990, 19–35.

Garber, Daniel. *Descartes' Metaphysical Physics.* Chicago: University of Chicago Press, 1992.

Garber, Daniel, and Michael Ayers. *The Cambridge History of Seventeenth-Century Philosophy.* 2 vols. Cambridge: Cambridge University Press, 1998.

Hutchison, Keith. "Supernaturalism and the Mechanical Philosophy." *History of Science* 21 (1983): 297–333.

Kargon, Robert. *Atomism in England from Hariot to Newton.* Oxford: Oxford University Press, 1966.

MacIntosh, J. J. "Robert Boyle on Epicurean Atheism and Atomism." In *Atoms, Pneuma, and Tranquillity: Epicurean and Stoic Themes in European Thought,* ed. Margaret J. Osler. Cambridge: Cambridge University Press, 1991, 197–219.

McGuire, J. E. "Boyle's Conception of Nature." *Journal of the History of Ideas* 33 (1972): 523–43.

Mintz, Samuel I. *The Hunting of Leviathan: Seventeenth-Century Reactions to the Materialism and Moral Philosophy of Thomas Hobbes.* Cambridge: Cambridge University Press, 1962.

Murdock, John, William R. Newman, and Christoph Lüthy, eds. *Late Medieval and Early Modern Corpuscular Matter Theory.* Leiden: Brill, 2001.

Nadler, Steven M. "Arnauld, Descartes, and Transubstantiation: Reconciling Cartesian Metaphysics and Real Presence." *Journal of the History of Ideas* 49 (1988): 229–46.

Osler, Margaret J. "The Intellectual Sources of Robert Boyle's Philosophy of Nature: Gassendi's Voluntarism and Boyle's Physico-theological Project." In *Philosophy, Science, and Religion, 1640–1700,* ed. Richard Ashcraft, Richard Kroll, and Perez Zagorin. Cambridge: Cambridge University Press, 1991, 178–98.

———. *Divine Will and the Mechanical Philosophy: Gassendi and Descartes on Contingency and Necessity in the Created World.* Cambridge: Cambridge University Press, 1994.

Redondi, Pietro. *Galileo: Heretic.* Trans. Raymond Rosenthal. Princeton: Princeton University Press, 1987.

Schaffer, Simon. "Godly Men and Mechanical Philosophers: Souls and Spirits in Restoration Natural Philosophy." *Science in Context* 1 (1987): 55–86.

Shanahan, Timothy. "God and Nature in the Thought of Robert Boyle." *Journal of the History of Philosophy* 26 (1988): 547–69.

Westfall, Richard S. *Force in Newton's Physics: The Science of Dynamics in the Seventeenth Century.* London and New York: MacDonald / American Elsevier, 1971.

Wojcik, Jan W. *Robert Boyle and the Limits of Reason.* Cambridge: Cambridge University Press, 1997.

12 Isaac Newton

Richard S. Westfall

Isaac Newton, whose invention of the calculus, theory of universal gravitation, and experimental work in optics marked the culmination of the scientific revolution that had begun with Copernicus's espousal of heliocentric astronomy, was also devoted to the study of theology. He understood his work in physics and astronomy as part of a grand argument from design. His theological views developed around Arianism or some closely related form of antitrinitarianism. He believed that Athanasius had led the Roman Church astray in the fourth century in its acceptance of the doctrine of the Trinity. He pursued biblical prophecy (particularly of the Books of Revelation and Daniel) to support his heresy. He believed that, before Christianity had been corrupted, it had consisted of belief in God and our duties to one another, the common property of all humans who are willing carefully to study nature.

Richard S. Westfall, who died in 1996, received his Ph.D. from Yale University in 1955. He twice received the Pfizer Book Prize from the History of Science Society, in 1971 for *Force in Newton's Physics* and in 1980 for *Never at Rest*, his definitive biography of Newton. He also received the History of Science Society's most prestigious award, the Sarton Medal, given for lifetime contributions to the history of science. At the time that he wrote this article, he was distinguished professor emeritus of history and philosophy of science at Indiana University, where he had taught since 1963.

Newton's Life

If the date of Isaac Newton's birth, Christmas Day 1642, suggests the important role that Christianity would play in his life, it does not even begin to hint at the heterodox opinions he would embrace as he struggled to bring his Christianity into harmony with his science.

The son of a prosperous yeoman farmer in Lincolnshire who died three months before his only child's birth, Newton (1642–1727) was educated at the grammar school in Grantham and admitted to Trinity College, Cambridge, in the summer of 1661. Trinity was his home for the following thirty-five years. Two years after commencing his B.A., he was elected to a fellowship in the college, and two years after that he received appointment as the university's Lucasian Professor of Mathematics. All of Newton's contributions to science belong to the years in Trinity. Already by the time of his B.A., his independent, untutored study of mathematics was beginning to lead him toward the calculus, and only a year later he composed what is known as the Tract of October 1666, in which he set its principles down. About the same time, he became interested in optics and phenomena of color and first entertained his insight into the heterogeneity of light. His initial series of lectures as Lucasian Professor were devoted to this topic, and in them he polished the theory that phenomena of colors arise from the analysis of heterogeneous light into its components. Although Newton published his *Opticks* only in 1704, virtually all of its content went back to the late 1660s. At this time, he also took up the science of mechanics and began to think about gravity. He was still a member of Trinity in the 1680s, when he returned to mechanics, addressed the problem of orbital dynamics, and composed *The Mathematical Principles of Natural Philosophy* (or *Principia,* from the key word in its Latin title), which was published in 1687 and established his renown in science for all time. About a decade after the *Principia,* Newton moved from Cambridge to London, where he became first warden and then master of the Royal Mint. In 1703, the Royal Society elected him president. Both of these positions he held until his death on March 20, 1727.

Newton and Theology

It was also in Cambridge that Newton began seriously to study theology. There is, of course, no reason to doubt that he had been a normally pious young man in an age when normal young men tended to be pious, but his surviving manuscripts do not indicate any sustained study of theology until about 1672. Perhaps the statutory requirement of Trinity College that fellows be ordained to the Anglican clergy within seven years of commencing the M.A. initially stimulated his active concern with theology. He had received his M.A. in 1668. The required ordination would have to take place by 1675, and Newton was never one to take an obligation lightly. Whether or not the Trinity ordinance was the cause, the manuscript remains leave no doubt that he began to study theology intensely about 1672. He devoured the Scriptures. Years later, John Locke (1632–1704) would confess that he had never known anyone with a better command of Scripture. With equal gusto, Newton devoured the work of the early fathers of the church, making himself a master of that extensive literature as well. For the following decade and more, Newton devoted very little time to the scientific and mathematical topics we associate with his name. Along with alchemy, theology was the staple of his intellectual life in those years. The *Principia* interrupted this pattern, and, during the following twenty years, the quantity of theological manuscripts that he penned greatly diminished. He returned once more to theology during the first decade of the eighteenth century, however, and, from that time until his death, it was the focus of his attention. Newton never threw a paper away, so the record of his long immersion in theology survives. The manuscript remains can only be described as immense, at least several million words. In quantity, they far exceed those from any of his other fields of study.

At the time when Newton turned to theology, a tradition of natural theology was well established among English scientists. Books by scientists bearing such titles as *The Darknes of Atheism Dispelled by the Light of Nature* and *The Wisdom of God Manifested in the Works of the Creation* appeared in constant succession. Robert Boyle (1627–91), a prolific author, included arguments to this effect in nearly every one of the many books he published, and, toward the end of his life, he summarized the message in *The Christian Virtuoso* (a title that is not distorted if translated as *The Christian Scientist*). Newton also contributed to this literature. "When I wrote my treatise about our Systeme," he stated in a letter to the theologian Richard Bentley (1662–1742), "I had an eye upon such Principles as might work with considering men for the beleife of a Deity & nothing can rejoyce me more than to find it usefull for that purpose." A passage in Query 28 at the end of the *Opticks* asserted that the "main Business of natural Philosophy is . . . to deduce Causes from Effects, till we come to the very first Cause, which certainly is not Mechanical." In perhaps his most eloquent statement of this argument, the *General Scholium* appended to the second edition of the *Principia*, he described the solar system with planets and satellites moving in the same direction in the same plane while comets course among them. "This most beautiful system of the sun, planets, and comets," he argued, "could only proceed from the counsel and dominion of an intelligent and powerful Being."

These passages have been well known ever since Newton wrote them; there is no reason whatever to doubt their sincerity. Nevertheless, they are not what one finds in his many theological manuscripts. The quoted passages repeat an inherited piety, the deposit of centuries of Christianity, that part of Newton's religion not seriously touched by the new science despite its seeming utilization of it. In the manuscripts, we find a different Newton, who brings a different attitude to bear on the established religion. The different attitude was intimately related to the new science, though not to its overt conclusions as it forged a new image of physical reality. The attitude was related, rather, to the questioning stance of science, as it rejected one received authority after another, to all that Basil Willey had in mind when he spoke (repeating a seventeenth-century phrase) of the "touch of cold philosophy" (Willey 1967, vii).

Newton's Arianism

Early in his theological reading, Newton became absorbed in the doctrine of the Trinity and in the fourth-century struggle between what became Christian orthodoxy and Arianism. He identified with Arius (250–336), who taught that Jesus was a created being and not co-eternal with God the Father, and came to hate Athanasius (c. 296–373), the principal architect of trinitarian orthodoxy. Newton did not think of Athanasius and his cohorts in the fourth century merely as mistaken. Rather, he regarded them as criminals, who had seized Christianity by fraud and perverted it as they pursued selfish ends, even tampering with the Scriptures to insert trinitarian passages that he could not find in versions earlier than the fourth century. One of his papers, "Paradoxical Questions concerning the morals & actions of Athanasius & his followers," virtually stood Athanasius in the dock and prosecuted him for an extended litany of alleged sins, both doctrinal and moral.

Newton became either an Arian or something so close to an Arian that it is difficult to distinguish them. Arianism is similar to but not identical with early unitarianism. It considered Christ to be not an eternal person in a triune God, but a created intermediary between God and man. In his manuscripts, Newton insisted on a distinction between God, the omnipotent Pantocrator, and Christ the Lord, who was always subordinate to God and whom God elevated to sit at his right hand. From this position, which he adopted in the 1670s, Newton never retreated. In his old age, he was still writing Arian statements on the nature of Christ. Trinitarianism he always treated as more an abomination than a mere error. He referred to it as the "false infernal religion" and "the whole fornication." "Idolators, Blasphemers & spiritual fornicators," he thundered at the trinitarians in the isolation of his chamber. A shrill note of iconoclasm that one does not ordinarily associate with Newton rings through these pages as he denounced "vehement superstition" and "monstrous Legends, fals miracles, veneration of reliques, charmes, the doctrine of Ghosts or Daemons, & their intercession invocation & worship & such other heathen superstitions as were then brought in."

This was hardly a safe position to adopt in England in the late seventeenth century. For a man of Newton's rigid posture, it made ordination impossible, and in 1675 he was preparing to lay down his fellowship at Trinity. Perhaps the Lucasian chair could have been held separately, but questions would inevitably have been raised in an institution that explicitly considered orthodoxy a requirement for membership. The Lucasian chair also had orthodoxy written into its statutes. It is not too much to say that Newton's career hung in the balance in 1675. At the last minute, probably through the influence of Isaac Barrow (1630–77), master of Trinity College, a royal dispensation from the requirement of ordination was granted, not to Isaac Newton, but to the holder of the Lucasian chair in perpetuity, and the crisis passed. Newton understood that absolute secrecy had now become the condition of his continued presence in Cambridge, and there is no indication that he divulged the content of his theological papers to anyone there. Though unwilling to accept ordination, he was prepared to accommodate to silence. The need to shield this important dimension of his existence from public scrutiny became a permanent part of Newton's life. Even in the greater laxity of the capital city at the turn of the century, after Newton moved there in 1696, the need for secrecy remained. The law of the land explicitly set belief in the Trinity as a requirement for public servants. It does seem clear that, in London, Newton shared his views with a restricted circle of mostly younger men, who may, indeed, have learned their heresy from him. Because some of them were less cautious, a few rumors did spread, but only in the late twentieth century, when his theological manuscripts became accessible to the scholarly public, did the extent of his heretical views become known.

Newton's Interest in Biblical Prophecy

Along with Arian theology, Newton maintained a concern with biblical prophecy, initially with the Book of Revelation and later with Daniel as well. Newton's interest in prophecy is well known; shortly after his death, his heirs published *Observations on the Prophecies*. The manuscript they published, really two different manuscripts together with three other miscellaneous chapters that they melded together, was a product of Newton's old age when, increasingly aware of his own prominence, he wanted to obscure his heterodoxy. It is quite impossible to find any point in the published book's meandering chronologies. This was not the case with the original interpretation of Revelation that Newton composed in the early 1670s. There he drew upon an established Puritan interpretation that centered on the "great Apostasy." To the Puritan interpreters, the Great Apostasy was the Roman Church. For Newton, it was trinitarianism. The purpose of the Apocalypse was to foretell the rise of the Great Apostasy and God's response to it. At the end of the sixth Seal (the Seals represented periods of time to Newton), the first six trumpets and the vials of wrath associated with them (also periods of time) prophesied the barbarian invasions of the Roman Empire, punishments poured out on a stiff-necked people who had gone whoring after false gods. In pursuit of this interpretation, Newton, who had come to doubt the accuracy of the received text of the New Testament, collated some twenty different manuscript versions of the Book of Revelation to establish the correct text. An interpretation could be correct, in his view, only if the prophecy corresponded in detail to the facts of history as they later unfolded, and he was no less vigorous in pursuing those facts in the sources for the history of the early Christian era.

In the late 1670s, Newton's interpretation acquired a new dimension as he became convinced of the role of Jewish ceremony as a "type," or figure, in prophecy. With his characteristic thoroughness, Newton plunged into Jewish literature—Josephus (c. A.D. 37–c. 100), Philo (c. 30 B.C.–A.D. 45), Moses Maimonides (1135–1204), and the Talmudic scholars. In this context, the exact shape of the temple in Jerusalem became increasingly important to Newton. He was convinced that the tabernacle of Moses and the temples of Solomon and Ezekiel had followed the same plan, although the temples were bigger. He found the best description in chapters 40–43 of Ezekiel. On its basis, he drew a detailed plan of the temple, using the exigencies of the plan when necessary to correct the text.

Early in the 1680s, Newton began to work on a new theological treatise, which never even approached final form and which he called *Theologiae gentilis origines philosophicae* (*The Philosophical Origins of Gentile Theology*) in the manuscript that most approaches a finished form. The *Origines* was the most radical of Newton's theological writings. It was also his most important. Its ideas echoed through his scientific works during the following thirty years—

in the first section of the original draft of the *Principia*'s concluding book, in the so-called *Classical Scholia* inserted in revisions of Book 3 during the 1690s but never published, in the final lines of Query 31, which he published initially in 1706, and in two footnotes to the *General Scholium* added to edition 2 of the *Principia* in 1713.

The *Origines* proposed a cyclical pattern of human history, in which the one true religion is continually perverted into idolatry. Newton argued that all of the ancient civilizations worshiped the same twelve gods, though under different names. The twelve corresponded to the seven planets, the four elements, and the quintessence, the noblest parts of nature, with which the ancient peoples had identified their divinized rulers and heroes. Kings who wanted to claim descent from divinities had every incentive to promote this religion. For Newton, gentile theology was rank idolatry. Mankind had started with the one true religion, manifest from the observation of nature, and Newton was convinced that he could see evidence of the true religion among early people before idolatry set in. Everywhere throughout the ancient world, he found, temples that he called "prytanea" had been built to a common plan. They had a central fire surrounded by seven candles, a representation of the universe—indeed, a heliocentric representation—for true philosophy had accompanied true religion. His primary evidence for the original true religion was the worship of Noah and his children. But mankind has an inherent tendency to idolatry. In fact, the common twelve gods were Noah, his children, and his grandchildren, the source of all of the ancient peoples, each of whom saw the twelve as their unique ancestors and divinized them. Egypt had led the way in idolatry; the other ancient peoples had learned it from the Egyptians.

In the twentieth century, it would be easy to overlook the radical threat of the *Origines*. We smile at the quaint themes of Noah and his children and see in the special place the treatise seems to accord to the Judeo-Christian revelation a hallmark of Christian provincialism. We might miss the message that Newton clothed in the common idiom of the day, which was the only idiom available to him. However, he treated the historical records of the other ancient peoples as sources equal in validity to the Hebrew writings. For example, he found a universal story of a flood and of a line of ten patriarchs before Noah. "Noah" was only the name most familiar to Newton's potential audience; in fact, in his view the man had been more Egyptian than Hebrew, more Hammon than Noah. Above all, the *Origines* deflated the Christian message by making Christianity another instance of a repetitive cycle. As he saw human history, God had continually sent prophets to recall mankind from idolatry. Christ was no different from the rest, and, in trinitarianism (the work of Athanasius, another Egyptian), mankind had slid again into idolatry, for what is trinitarianism but the worship of a man as God? "What was the true religion of the sons of Noah before it began to be corrupted by the worship of false Gods?" Newton set down as the title of a chapter 11 that he never, in fact, composed. "And that the Christian religion was not more true and did not become less corrupt." In turn, "the true religion of the sons of Noah" appears to have been a naturalistic religion, restricted to the acknowledgment of God and our duties to one another, the common property of all of mankind who are willing carefully to study nature.

Newton was at work on the *Origines* when the visit of Edmund Halley (1656–1742) in 1684 set him in motion toward the *Principia.* The *Principia,* in turn, marked a break in his theological endeavors, as mentioned above, and, though he later returned and devoted reams of paper to them, he was by then an old man engaged largely in reshuffling earlier ideas. He did produce a sanitized version of the *Origines,* analogous to his late manuscript on prophecy. *The Chronology of Ancient Kingdoms Amended* appeared after his death; it gives no hint of its radical provenance.

Conclusion

Different interpretations of Newton's religious odyssey are possible. It seems to offer the record of a man seeking to reconcile a spiritual heritage that was precious in his sight with a new intellectual reality. Somewhat the same can be said of all of the scientists in their exercises in natural theology. Read in isolation, each one seems to offer testimony to an unshaken faith. Read one after another, each repeating essentially the argument of its predecessors, they begin to project an uneasiness that unbelief was not so readily banished. After a lifetime of refuting atheism, Boyle left an endowment for a series of lectures to refute atheism some more. When in the previous millennium had that seemed necessary? The scientists of the late seventeenth century sensed that the ground was moving under the inherited structure of Christianity, and they sought to shore it up with a new foundation. Newton felt the same movement, but his response went further. He attempted to purge Christianity of elements, centering on the doctrine of the Trinity, that he regarded as irrational. He wrote:

> If it be said that we are not to determin what's scripture & what not by our private judgments [he wrote about the two trinitarian passages that he considered corruptions of Scripture], I confesse it in places not controverted: but in disputable places I love to take up with what I can best understand. Tis the temper of the hot and superstitious part of mankind in matters of religion ever to be fond of mysteries, & for that reason to like best what they understand least. Such men may use the Apostle John as they please: but I have that honour for him as to believe he wrote good sense, & therefore take that sense to be his which is the best.

The *Origines* arrived at a vision of true religion not far different from deism except in one, perhaps all important, respect. Where the deists could not contain their animosity toward Christianity, Newton always believed that he was restoring true Christianity. In a tract from his old age called *Irenicum,* he reduced Christianity, or true religion, to two doctrines: love of God and love of neighbor. On his deathbed, after years of compromise for the sake of appearances, he refused the sacrament of the Anglican Church.

The deist storm had broken long before Newton's death, and, in the years following his death, religious radicalism reached positions he would have abhorred. He himself contributed almost nothing to this movement. Always secretive, he had communicated his doubts only to a very narrow circle, and there is no evidence to suggest that, through them, he played a role in fomenting the storm. On the contrary, there is every reason to think that the religious rebels of the eighteenth century responded on their own to the same influences that had moved Newton. Both alike testify that, with the rise of modern science, the age of unshaken faith was forever gone in the West.

BIBLIOGRAPHY

Brooke, John H. *Science and Religion: Some Historical Perspectives.* Cambridge: Cambridge University Press, 1991.

Christianson, Gale. *In the Presence of the Creator: Isaac Newton and His Times.* New York: Free Press, 1984.

Cragg, Gerald R. *From Puritanism to the Age of Reason: A Study of Changes in Religious Thought within the Church of England, 1660 to 1700.* Cambridge: Cambridge University Press, 1950.

Dobbs, Betty Jo. *The Foundations of Newton's Alchemy: The Hunting of the Greene Lyon.* Cambridge: Cambridge University Press, 1975.

———. *The Janus Faces of Genius: The Role of Alchemy in Newton's Thought.* Cambridge: Cambridge University Press, 1991.

Force, James E., and Richard H. Popkin. *Essays on the Context, Nature, and Influence of Isaac Newton's Theology.* International Archives for the History of Ideas. Vol. 129. Dordrecht: Kluwer, 1990.

———, eds. *Newton and Religion: Context, Nature, and Influence.* Dordrecht: Kluwer, 1999.

Goldish, Matt. *Judaism in the Theology of Sir Isaac Newton.* Dordrecht: Kluwer, 1998.

Hazard, Paul. *The European Mind: The Critical Years, 1680–1715.* Trans. J. Lewis May. 1935. Reprint. New Haven: Yale University Press, 1953.

Hooykaas, R. *Religion and the Rise of Modern Science.* Edinburgh: Scottish Academic Press, 1972.

Jacob, Margaret C. *The Newtonians and the English Revolution, 1689–1720.* Hassocks, England: Harvester, 1976.

Lindberg, David C., and Ronald L. Numbers, eds. *God and Nature: Historical Essays on the Encounter between Christianity and Science.* Berkeley: University of California Press, 1986.

Manuel, Frank E. *Portrait of Isaac Newton.* Cambridge: Harvard University Press, 1968.

———. *The Religion of Isaac Newton.* Oxford: Oxford University Press, 1974.

Popkin, Richard H. "Newton's Biblical Theology and His Theological Physics." In *Newton's Scientific and Philosophical Legacy,* ed. P. B. Scheuer and G. Debrock. International Archives of the History of Ideas. Vol. 123. Dordrecht: Kluwer, 1988, 81–97.

Snobelen, Stephen D. "'God of gods and Lord of lords': The Theology of Isaac Newton's General Scholium to the Principia." In *Osiris* 2d ser. 16: 169–208.

Westfall, Richard S. *Science and Religion in Seventeenth Century England.* New Haven: Yale University Press, 1958.

———. *Never At Rest: A Biography of Isaac Newton.* Cambridge: Cambridge University Press, 1980.

Willey, Basil. *Seventeenth Century Background: Studies in the Thought of the Age in Relation to Poetry and Religion.* 1934. Reprint. New York: Columbia University Press, 1967.

13 Natural Theology

John Hedley Brooke

A recurring question in the religious history of the West has been whether anything can be known of the existence and attributes of a deity through rational inference alone. The question has sometimes acquired an urgency, as religious believers have tried to convince their critics of the reasonableness of their faith. Conversely, critics of established religion, such as the eighteenth-century deist Voltaire, have made strong claims for the superiority of a religion based on natural reason as an alternative to religions claiming their foundation in revelation. The relations between natural theology and the sciences have been particularly close because the appearance of an intelligible order and design in the natural world has provided evidence of a transcendent designer, as it did for Copernicus, Kepler, and Newton. Since the middle of the nineteenth century, however, the argument for design has lost much of its force, partly owing to the challenge from Darwinism, but also from philosophical and theological critiques.

John Hedley Brooke obtained his Ph.D. from the University of Cambridge. He is the Andreas Idreos Professor of Science and Religion at the University of Oxford. His most recent books are *Science and Religion: Some Historical Perspectives* (Cambridge: Cambridge University Press, 1991); *Thinking about Matter: Studies in the History of Chemical Philosophy* (Aldershot, Hants, England and Brookfield, Vt.: Ashgate, 1995); and (with Geoffrey Cantor) *Reconstructing Nature: The Engagement of Science and Religion* (Oxford: Oxford University Press, 2000).

NATURAL THEOLOGY is a type of theological discourse in which the existence and attributes of the deity are discussed in terms of what can be known through natural reason, in contradistinction (though not necessarily in opposition) to knowledge derived from special revelation. Routine, timeless definitions of natural theology are, however, simplistic because *natural reason* and *revelation* have been understood differently in different cultures and at different times. Since the Enlightenment, natural theology has often been characterized as the attempt to construct rational "proofs" for God's existence and attributes—a project drawing on the natural sciences but vulnerable both to philosophical critiques and to changes in scientific sensibility. By contrast, in premodern cultures, adherents of the monotheistic religions would scarcely have entertained a discourse of natural theology independent of that greater knowledge of God revealed in their sacred texts. Doctrinal disputes abounded but, within Jewish, Christian, and Islamic societies, the existence of God was rarely the issue. The psalmist had spoken of the manifold and wondrous works of God (e.g., in Psalm 19:1–6), but as an affirmation of faith in, not an attempted proof of, divine wisdom. There are many comparable examples from the history of Christendom and Islam, suggesting that to abstract what may look like "proofs" of God's existence from their contexts misses the significance that such arguments had within specific religious communities.

Order and Design in Nature

A recurring goal of natural theology has been to show the incompleteness of philosophies of nature that purport to explain the appearance of order and design without reference to supernatural agency. This was the challenge faced by Thomas Aquinas (c. 1225–74) as he engaged the philosophy of Aristotle (384–322 B.C.), for whom the world was eternal, governed by causes entirely within the cosmos. The primary controlling cause, without which there would be no change, was the "final cause," the end or purpose of the process. Such control was most evident in the development of a seed or embryo, where the inference to a goal-directed process was irresistible. Regarding the natural world as if it were a living organism, Aristotle saw these final causes as immanent within nature itself. In response, Aquinas asked why the components of nature should behave in so orderly a way. Physical bodies surely lacked knowledge and yet regularly acted in concert with others to achieve certain ends. In his *Summa Theologiae,* Aquinas insisted that such ends must, therefore, be achieved not fortuitously but designedly: Whatever lacks knowledge cannot move toward an end unless directed by some being endowed with knowledge and intelligence. This was the fifth of the "five ways" by which Aquinas affirmed the rationality of belief in a transcendent deity. But it was not so much a demonstration of the existence of God as a demonstration that the Aristotelian philosophy of nature was harmless to Christian theism, since it was not fully coherent without it.

The association of Aristotle's "final cause" with a doctrine of providence was not the only basis on which a natural theology might be constructed. It was possible to argue, as Plato (c. 427–347 B.C.) had, that the world resembles a work of craftsmanship, bearing the marks of intelligent design. Plato had ascribed the order in nature to the work of a Demiurge (Craftsman), which had molded preexisting matter according to intentions that were partly frustrated by the recalcitrance of the material at its disposal. This model of divine activity had the attraction of accounting for imperfections in nature but has often been judged defective by Christian theologians because of the restrictions it placed on the power of a God who, as Creator of all things, would not have been bound by preexisting matter.

With the increasing mechanization of nature during the seventeenth century, new images of the divine craftsman were introduced. René Descartes (1596–1650) and Robert Boyle (1627–91) compared the universe with the cathedral clock at Strasbourg, both stressing the freedom of the divine will rather than restrictions on its power. The conjunction of a corpuscular theory of matter with a voluntarist theology of Creation is often seen as propitious for the reformation of natural philosophy. It could legitimate the quest for the divinely ordained "rules" or "laws" by which the movement of passive matter was regulated; it also justified empirical methods as the only means of discovering which of the many mechanisms the deity might have chosen or that he had, in fact, employed.

Descartes himself had not developed arguments for design based on his mechanistic worldview, preferring instead variants of the ontological argument to establish beyond doubt the existence of the Perfect Being who had planted the idea of perfection in his mind. Descartes even warned against presumptuous attempts to know the designs of God. It was this prohibition that so worried Boyle who, in his *Disquisition about the Final Causes of Natural Things* (1688), accused the Frenchman of having discarded the strongest argument for a deity: that based on such wonderful contrivances as the human eye, which had so evidently been made to see with. Boyle conceded that many of God's intentions we could not presume to know; having marveled at the intricacies of nature recently disclosed by the microscope, however, he was convinced that scientific knowledge supported the Christian revelation. That the machinery necessary for life had been packed into the minutest mite was, for Boyle, more astounding evidence for a deity than the larger machinery of the macrocosm. Such a natural theology, in which evidence for divine wisdom was uncovered in scientific investigation, was to prove especially durable in the

English-speaking world. It was epitomized by Boyle's self-presentation as priest in the temple of nature.

A Higher Profile for Natural Theology

Why did natural theology gain a higher profile during the second half of the seventeenth century? Scientific discoveries could evoke a sense of wonder, as they did for Boyle. The microscope revealed finely wrought structures, vividly captured in Robert Hooke's (1635–1703) depiction of the compound eye of a fly. The microscope also provided a rhetoric for religious apologists who observed that, whereas human artifacts, when magnified, revealed all of their deformities, the works of nature, from the beauty of a snowflake to the proboscis of a flea, revealed a kind of perfection. In John Ray's (1627–1705) *Wisdom of God Manifested in the Works of Creation* (1691), this contrast was used to argue for the transcendence of nature's art over human art and, by implication, the transcendence of the Supreme Designer. Ray also expatiated on the aesthetic appeal of Copernican astronomy. A hidden beauty had been unveiled once the sun had become the focus of the planetary orbits.

Social pressures may help explain the recourse to natural theology among natural philosophers themselves. Those such as Marin Mersenne (1588–1648), Pierre Gassendi (1592–1655), Walter Charleton (1620–1707), and Robert Boyle, who favored an atomic or corpuscularian theory of matter, had to stress that they were not reviving the atheistic atomism of Lucretius (c. 99–55 B.C.). The craftsmanship discernible in nature rendered it inconceivable that a chaos of atoms, left to itself, could have produced an ordered world. The pressure to affirm an active providence was acute because mechanists such as Thomas Hobbes (1588–1679) were gaining notoriety by contending that the soul was corporeal. In England, an emphasis on the rationality of faith was also a way of dispelling the religious "enthusiasm" that had flourished during the Puritan domination (1649–60) and had led to such a proliferation of sects that Boyle had feared lest the Christian religion should destroy itself. Arguments for an intelligent Creator grounded in the realities of the natural world offered an anchor, even the prospect of consensus, amid the turbulence created by religious divisions.

It was, however, a less-than-orthodox Christian, Isaac Newton (1642–1727), who gave the most decisive impetus to the design arguments. Newton's views on the Trinity were heterodox, but his abiding interest in the fulfillment of biblical prophecy reflected his belief in a deity having dominion over both nature and history. From his correspondence with Richard Bentley (1662–1742), from the *General Scholium* that he wrote for the second (1713) edition of his *Principia*, and from the "Queries" appended to his *Opticks* (1704), it is clear that Newton saw evidence of design in the structure of the universe. Against Descartes, he argued that only an intelligent being, "very well versed in mechanics and geometry," could have calculated the correct tangential component of each planet's velocity to ensure that it went into a stable orbit.

A natural theology could even generate confidence in a providential deity whose activity had not been confined to an initial Creation. Newton was concerned about the long-term stability of the solar system, given that planets might slow down as they moved through an ether or given the loss of mass from the sun through vaporization. The necessary "reformation" of the system did not necessarily involve God's direct intervention. Secondary causes, such as comets, could serve as instruments of the divine will. But, either way, the inference to an active deity seemed to have all of the authority of Newton's science behind it. The elegance of the inverse-square law of gravity pointed to a God who had chosen to rule the world not by incessant acts of absolute power (though Newton never denied the deity that power) but through the self-limiting mediation of laws. A parallel with the constitutional monarchy that England was developing after the Revolution of 1688 has often been noted.

The Presuppositions of Natural Theology and their Exposure

In much of eighteenth-century natural theology, images of the Divine Craftsman, Geometer, or Architect were distinctly anthropomorphic and, therefore, vulnerable to the objection that the transcendence of God was being demeaned. Newton's spokesman, Samuel Clarke (1675–1729), encountered this criticism in his controversy with Gottfried Wilhelm Leibniz (1646–1716). When Newton had spoken of space as if it were the "sensorium" of God, this, for Leibniz, implied that Newton was thinking of the deity in human terms, even possessing a body. Newton had once suggested that our ability to move our limbs affords an analogy on the basis of which it may be presumed rational to believe that God, as spirit, can (even more easily) move matter. The danger, sensed by Leibniz, was that the physical universe might be identified with the body of God—a position closer to pantheism than to Christian theism. The Clarke-Leibniz controversy also shows how deep divisions could ensue in prioritizing divine attributes. Newton and Clarke emphasized the freedom and power of the divine will. If God had willed a world in which a "reformation" of the solar system was periodically required, so be it. It was not for natural philosophers to dictate to God the kind of world that God should have made. Leibniz, however, appeared to be doing just that when he argued that a perfect, rational Being would have had the foresight to make a world that did not need correction. This shows how an understanding of God's relation to nature could be deeply affected according to whether God's freedom or foresight was accentuated.

Leibniz's description of the Newtonian world-machine as second-rate clockwork highlights another feature of the design argument—one that was to engage David Hume (1711–76) in his posthumously published *Dialogues Concerning Natural Religion* (1779). Clockwork metaphors expressed an analogy between human artifacts and the natural world. In a skeptical critique, Hume exposed the fragility of analogical argument. Even if the world resembled a human artifact, one could not conclude that it had a single maker. Many hands were routinely involved in the making of machines. Consequently, polytheism was as plausible an inference as monotheism.

According to Hume, analogies other than clockwork were equally apposite for the expression of natural order. Why not regard the universe as an animal or a vegetable, in which case its cause would be an egg or a seed? The uniqueness of the universe did not mean that it was uncaused. Hume simply argued that, without experience of the creation of worlds, we can know nothing of the cause. Newton had claimed that the natural philosopher was to reason from effects to their causes, until one would finally reach an original cause that was certainly not mechanical. Hume retaliated that it was illegitimate to stop the inquiry with the introduction of an uncaused cause. To posit mental order in a divine Being as the cause of an intelligible order in nature only invited the further question: How had that mental order originated? Hume also questioned the ascription of properties such as divine, omnipotent, or omniscient to the originating cause. Claiming that it was a cardinal principle of reasoning that causes should always be proportionated to their effects, Hume saw no grounds for conferring infinite powers on the cause of a universe that, in all of its workings, was finite. Striking at the heart of natural theology, Hume also raised the problem of theodicy. If it was legitimate to infer the benevolence of the deity from beneficent features of the world, surely it was just as legitimate to infer the maleficence or, at best, the indifference of the deity from a preponderance of pain and evil.

The gist of Hume's critique was that apologists were assuming the existence of a beneficent Creator, not proving it. The dependence of the design argument on a prior, but rationally undemonstrable, belief in a Creator was also recognized by Immanuel Kant (1724–1804) in a subtle analysis that exposed other deficiencies. For example, Kant insisted that the purposive causality associated with living organisms, which allowed one to say that they were both causes and effects of themselves, could not be explained by analogy with a work of art. Among other weaknesses, Kant pointed to one that was decisive: No matter how much ingenuity and artistry might be displayed in the world, it could never demonstrate the moral wisdom that had to be predicated of God.

The Survival and Diversification of Natural Theology

In France, where a more secular culture prevailed at the time of the Revolution, and in Germany, where the full force of the Kantian critique was felt, physicotheology lost much of its appeal. In the English-speaking world, however, it continued to be visible in popular scientific and religious culture. Thus, James Hutton's *Theory of the Earth* (1795), which reported "no vestige of a beginning" or "prospect of an end," was defended from the charge of atheism by the Rev. John Playfair (1748–1819), who could point to the author's references to the wisdom of the overall design. The earth sciences, especially through Georges Cuvier's (1769–1832) proofs of extinction, raised particularly sensitive issues, but the English clerical geologists could still turn to natural theology for assistance. In Cambridge, Adam Sedgwick (1785–1873) turned the fossil record into arguments against both deists and atheists. The appearance of living forms that had once not existed confirmed a Creator who, unlike the clockmaker God of the deists, had clearly been active in the world since creation. Similarly, atheists were deprived of their solace: No longer could living things be said to have existed from eternity. A progressive pattern in the fossil record could also be used to argue for providence. Clerical scientists in Britain, such as William Whewell (1794–1866), felt the need to reassure their congregations that French science was not as dangerous as it seemed. Pierre Laplace (1749–1827) might have corrected Newton by showing that the solar system could restabilize itself without the need for a divine initiative, but, for Whewell, this only confirmed the greater skill and foresight of the Creator.

Indeed, design arguments were remarkably resilient, diversifying to meet the challenge presented by religious dissidents and new scientific perspectives. Immediately preceding the publication of Charles Darwin's (1809–82) *Origin of Species* (1859), several different styles of natural theology co-existed. William Paley's (1743–1805) argument for a divine Contriver based on the utility of anatomical structures left an almost indelible impression on Darwin himself, who later admitted that, even on his theory of natural selection, it had been difficult to relinquish the belief that every detail of organic structure must have some use. But Paley's argument based on specific contrivances was far from the only model. James Hutton (1726–97) and Joseph Priestley (1733–1804) had focused attention on the system of nature as a whole, each drawing attention to processes of replenishment that implied divine foresight—Hutton to processes of mountain building and erosion that maintained the earth's fertile soils, Priestley to the role of vegetation in maintaining the respiratory quality of the air.

A celebration of nature as a beneficent system could also be sustained by observing the harmonious manner in which many "laws" of nature combined. Whewell took this line in his *Bridgewater Treatise* (1833), arguing that a law presupposed an agent, a supreme legislator. The young Darwin was not unsympathetic to this model of God's relationship to nature. The existence of laws of nature did not exclude the existence of higher purposes, of which the production of more complex organic beings was the obvious example. Arguments based on the laws of nature could cut both ways, in that the weight of naturalistic explanation could easily exclude a personal, caring God, as it eventually did for Darwin.

Yet another style of natural theology, having resonances with Plato, was exemplified by T. H. Huxley's (1825–95) adversary Richard Owen (1804–92). The concept of a unity of skeletal structure, common to all vertebrates, had been used, especially in France, to contest the primacy given by Cuvier to teleological considerations. In response to this secular program, Owen reinterpreted the skeletal archetype as an idea in the mind of the Creator who, in the unfolding of a plan, had adapted it differently to the different needs of successive species. Still other claims for design were possible. The prolific popularizer of geology Hugh Miller (1802–56) marveled at the beauty of fossil forms, which, because they presaged all human architecture, confirmed that the divine and the human mind shared the same aesthetic sensibilities. For William Whewell, there was no way that the sciences could undermine a Christian faith, because every scientific advance pointed to the divine gift of a mind that could elicit truths about nature.

The Darwinian Challenge to Natural Theology

Darwin's theory of evolution by natural selection is seen as a watershed because it challenged so many facets of natural theology. Principally, it showed how nature could counterfeit design. In a competitive struggle for limited resources, those individual variants with advantageous characteristics would tend to survive at the expense of the norm and leave more offspring. Over innumerable generations, this process of natural selection would lead to the accumulation of favorable variations, giving rise to new and well-adapted species having all the appearance of design. No longer could finely honed organic structures constitute proof of a Designer in the manner suggested by Paley.

In Darwin's theory, the course of evolution was depicted in terms of successive branching, of divergent lines stemming from common ancestors. This jarred with the idea that *Homo sapiens* was the intended product of a divinely planned progression. In correspondence with American botanist Asa Gray (1810–88), Darwin denied that the variations on which natural selection worked were under divine supervision. They were random, in the sense that some were beneficial but some detrimental; they did not appear as if designed for a prospective use. Darwin's mechanism also highlighted the theodicy problem. The presence of so much pain and suffering in the world Darwin considered a formidable argument against a beneficent God, but it was what one would expect on the basis of natural selection. If the human mind was itself the product of evolution, there was the additional question whether it was equipped to reason profitably about the existence and attributes of the deity. Darwin's own view was that the very enterprise of natural theology was arrogant and anthropocentric. That man was descended from earlier forms of animal life taught a necessary lesson in humility.

Popular modern writers on evolution, notably Richard Dawkins, sometimes give the impression that, once Darwin had pronounced, a rational case for theism became a lost cause. Images of the divine Craftsman and the divine Magician were certainly moribund. The sufficiency of Darwin's mechanism of natural selection was, however, a contentious issue among scientists themselves. In his *Descent of Man* (1871), Darwin himself said that he had given it too great a scope in his *Origin*. Even his "bulldog," T. H. Huxley, introduced mutations to speed up the earliest stages of evolution. Consequently, despite such a considered rejection of Darwinism as that of the Princeton theologian

Charles Hodge (1797–1878), models of theistic evolution were developed that preserved a natural theology. For Asa Gray, even natural selection was compatible with natural theology because it helped rationalize suffering. If it was a necessary concomitant of a long creative process, some of its sting might be removed. In Britain, Frederick Temple (1821–1902), destined to become archbishop of Canterbury, made a not dissimilar point when he suggested that the theodicy problem was greater for those who believed in separate acts of Creation than for those who accepted the integrity of an evolutionary process. By the end of the nineteenth century, it was even possible for Aubrey Moore (1848–90) to argue that Darwin had done Christianity a service. On the basis of an Incarnational theology, in which God was immanent in the world, Moore was grateful for Darwin's destruction of semideistic schemes in which God was totally absent except for the occasional intervention. The contours of the evolutionary process as Darwin described them, coupled with distressing natural disasters, made it increasingly difficult to see any single, overriding divine purpose in the history of life on Earth. But this did not prevent sophisticated thinkers such as the Jesuit modernist George Tyrrell (1861–1909) from arguing that the universe still teemed with aims and meanings even though they could not be subsumed under a collective effort.

Natural Theology in the Twentieth Century

For much of the twentieth century, disincentives to natural theology have tended to outweigh the incentives. Traditional distinctions between "natural" and "revealed" theology proved difficult to sustain in the wake of historical criticism of the sacred texts. The meaning of *natural* as in "the natural world" has also been compromised by science-based technologies that have insinuated so many artificial products into local and global environments. The concept of a "natural reason" in all humanity has proved simplistic in the light of psychoanalytical models of the unconscious mind. Theodicies based on the evolution and prospective improvement of "human nature" were shattered by two world wars. After the first, Karl Barth (1886–1968) issued a stentorian "No" to natural theology that has resounded through the corridors of Reformed theology. Reaffirming a God of judgment and redemption, Barth insisted that such was the gulf between creatures and their Creator that no autonomous creaturely reasoning could reach a knowledge of God, who is knowable only through himself.

An interest in revised forms of natural theology has, however, never completely waned, as evidenced by the various schemes of "process theology" that have taken their inspiration from Alfred North Whitehead's (1861–1947) *Process and Reality* (1929) and the appearance of ecofeminist theologies that have also stressed the persuasive rather than the coercive agency of God. The impetus to experiment with science-based theologies has not always come from scientists themselves. Relativity theory, quantum mechanics, and, more recently, chaos theory have been used (not always circumspectly) to argue for less mechanistic, less deterministic conceptions of nature in which there might be hints of an openness to divine influence. The disclosure that the emergence of intelligent life has been possible only because of what looks like an extraordinary "fine-tuning" of the physical processes involved in the earliest moments of the big bang has encouraged those who already believe in a designing intelligence, though the apologetic force of such anthropic coincidences remains controversial, given the rejoinder that our universe may be one of a myriad possible universes—the "lucky one" that just happened to have the right parameters. In this example, as in others in which the roles of contingency and necessity in biological evolution are discussed, there is a deep paradox, in that both contingency and necessity are invoked by theists and their critics to support their respective positions. This paradox provides additional evidence for what John Hick has called the religious ambiguity of the universe.

BIBLIOGRAPHY

Barrow, John D., and Frank J. Tipler. *The Anthropic Cosmological Principle.* Oxford: Oxford University Press, 1986.

Bowler, Peter J. *The Eclipse of Darwinism: Anti-Darwinian Evolution Theories in the Decades around 1900.* Baltimore: Johns Hopkins University Press, 1983.

Brooke, John Hedley. "Indications of a Creator: Whewell as Apologist and Priest." In *William Whewell: A Composite Portrait,* ed. Menachem Fisch and Simon Schaffer. Oxford: Oxford University Press, 1991, 149–73.

———. *Science and Religion: Some Historical Perspectives.* Cambridge: Cambridge University Press, 1991.

Brooke, John Hedley, and Geoffrey Cantor. *Reconstructing Nature: The Engagement of Science and Religion.* Edinburgh: T&T Clark, 1998.

Brooke, J. H., M. J. Osler, and J. Van der Meer, eds. *Science in Theistic Contexts. Osiris* 2d ser. 16 (2001).

Brunner, Emil. *Natural Theology.* Trans. Peter Fraenkel, comprising Brunner's "Nature and Grace" and Karl Barth's reply, "No!" London: Bles, Centenary, 1946.

Buckley, Michael J. *At the Origins of Modern Atheism.* New Haven: Yale University Press, 1987.

Burbridge, David. "William Paley Confronts Erasmus Darwin: Natural Theology and Evolutionism in the Eighteenth Century." *Science and Christian Belief* 10 (1998): 49–71.

Clayton, John. "Piety and the Proofs." *Religious Studies* 26 (1990): 19–42.

Funkenstein, Amos. *Theology and the Scientific Imagination from the Middle Ages to the Seventeenth Century.* Princeton: Princeton University Press, 1986.

Gascoigne, John. "From Bentley to the Victorians: The Rise and Fall of British Newtonian Natural Theology." *Science in Context* 2 (1988): 219–56.

Gaskin, J. C. A. *Hume's Philosophy of Religion.* London: Macmillan, 1978.

Gillespie, Neal C. "Divine Design and the Industrial Revolution: William Paley's Abortive Reform of Natural Theology." *Isis* 81 (1990): 214–29.

Gillispie, Charles C. *Genesis and Geology.* New York: Harper, 1959.

Glacken, Clarence J. *Traces on the Rhodian Shore: Nature and Culture in Western Thought from Ancient Times to the End of the Eighteenth Century.* Berkeley: University of California Press, 1967.

Gregory, Frederick. *Nature Lost? Natural Science and the German Theological Traditions of the Nineteenth Century.* Cambridge: Harvard University Press, 1992.

Harsthorne, Charles. *A Natural Theology for Our Time.* La Salle, Ill.: Open Court, 1967.

Hick, John. *An Interpretation of Religion.* London: Macmillan, 1989.

Hurlbutt, Robert H. *Hume, Newton, and the Design Argument.* Lincoln: University of Nebraska Press, 1965.

Kenny, Anthony. *The Five Ways: Saint Thomas Aquinas' Proofs of God's Existence.* London: Routledge, 1969.

Mackie, John L. *The Miracle of Theism: Arguments for and Against the Existence of God.* Oxford: Oxford University Press, 1982.

Manson, N. A. "There Is No Adequate Definition of 'Fine-tuned for Life.'" *Inquiry* 43 (2000): 341–52.

Moore, James R. *The Post-Darwinian Controversies: A Study of the Protestant Struggle to Come to Terms with Darwin in Great Britain and America.* Cambridge: Cambridge University Press, 1979.

Ospovat, Dov. *The Development of Darwin's Theory: Natural History, Natural Theology, and Natural Selection.* Cambridge: Cambridge University Press, 1981.

Roberts, Jon H. *Darwinism and the Divine in America: Protestant Intellectuals and Organic Evolution, 1859–1900.* Madison: University of Wisconsin Press, 1988.

Ruse, Michael. *Can a Darwinian Be a Christian?* Cambridge: Cambridge University Press, 2001.

Russell, Robert J., Nancey Murphy, and Arthur R. Peacocke, eds. *Chaos and Complexity: Scientific Perspectives on Divine Action.* Vatican City: Vatican Observatory Press, 1995.

Stenmark, Mikael. *Rationality in Science, Religion, and Everyday Life.* Notre Dame: University of Notre Dame Press, 1995.

Wildman, W. J. "Evaluating the Teleological Argument for Divine Action." In *Evolu-*

tionary and Molecular Biology: Scientific Perspectives on Divine Action, ed. R. J. Russell, W. R. Stoeger, S.J., and F. J. Ayala. Vatican City: Vatican Observatory Press; Berkeley, Calif.: Center for Theology and the Natural Sciences, 1998, 117–50.

Young, Robert M. *Darwin's Metaphor: Nature's Place in Victorian Culture*. Cambridge: Cambridge University Press, 1985, 126–63.

Part IV :: Transformations in Geology, Biology, and Cosmology, 1650–1900

14 Geology and Paleontology

Nicolaas A. Rupke

The alliterative coupling of "Genesis and geology" stands in the historiography of science for two seemingly opposing views about the history of the earth and for the debates that occurred during the late eighteenth and early nineteenth centuries. During this period, the study of rocks and fossils became an independent scientific discipline, with its own methods of fieldwork and its specialized textbooks that defined the subject's contents and boundaries, wresting away from humanistic scholarship, including theology, the intellectual authority to establish historical facts about the earth. Geological discoveries and theories opened up a vista of prehuman earth history that vastly exceeded any perception of the past based on a literal reading of the Genesis account of Creation and the Deluge. In the Anglo-American world, where geology was extensively patronized by Anglican and other church-controlled institutions, a great deal of effort was made by the geologists to harmonize their subject with Genesis. This was done by interpreting the paleontological-stratigraphic record in terms of the functionalist design argument of natural theology, as well as by means of various reinterpretations of the biblical account. On the European continent, by contrast, there existed less concern with the issues of Genesis and geology, as in many instances geologists had found secular professional niches and were under no pressure to deal with any real or apparent discrepancies.

Nicolaas A. Rupke received his Ph.D. from Princeton University. He is professor of the history of science and director of the Institute for the History of Science in Goettingen University, Germany. Among his publications are *The Great Chain of History: William Buckland and the English School of Geology, 1814–1849* (Oxford: Clarendon Press, 1983) and *Richard Owen: Victorian Naturalist* (London and New Haven: Yale University Press, 1994).

From Buffon to Darwin

During the late eighteenth and early nineteenth centuries, geology opened up a vast and unfamiliar vista of earth history. The study of rocks and fossils showed that the history of the earth had not covered the same stretch of time as the history of mankind but extended back immeasurably before the advent of *Homo sapiens.* It also appeared that prehuman earth history had been not a single period of continuity but a great chain of successive worlds (i.e., of periods of geological history), each with its own distinctive flora and fauna. Moreover, it emerged that the nature of the historical succession had been progressive: Successive worlds increasingly resembled our present world, with respect to both its inhabitants and the environmental conditions under which they had lived. This new vista of earth history equaled the Copernican revolution in its intellectual implications, reducing the relative significance of the human world in time just as early modern astronomy had diminished it in space.

The discoveries and theories of this new historical geology dominated the discourse of science with religion during the years 1780–1860, a period that began with the *Époques de la nature* (1778) by the Parisian naturalist Georges Louis Leclerc, Comte de Buffon (1707–88), whose book made a large and international readership familiar with the notion of periods of prehuman earth history, and that ended with Charles Darwin's (1809–82) *Origin of Species* (1859). The Darwinian theory of organic evolution, which incorporated "geological time" and "progressive development," started a new chapter in the science-religion debates by denying the special creation of organic species, including humans. Until then, the question of organic origins was, to a large extent, kept off the research agenda of the geological community.

The main sticking points of the geological theories with religion were those in which the new geology no longer fitted a literal interpretation of Genesis. These points were the age of the earth, the geological effect of the Deluge, the impact of original sin on the natural world, and the question of whether or not earth history is an eschatological process. Geology and paleontology triggered controversies over these issues; yet it should be emphasized that, in some ways, geology and religion interacted fruitfully, with reciprocal stimulation, mediated through the concepts and institutions of natural theology.

Dating the Past

Traditional sacred chronology calculated the age of the earth at some six thousand years, but there existed no consensus among chronologists. No fewer than 140 different estimates for the earth's birth year were put forward, ranging from 3616 to 6484 years B.C. A widely used figure was 4004 B.C., worked out by the Rev. James Ussher (1581–1656), archbishop of Armagh, and published in his scholarly *Annales Veteris et Novi Testamenti* (*Annals of the Old and New Testament* [1650–54]). The principal source for these figures was the Pentateuch—in particular, the genealogies of the patriarchs recorded in Genesis. In the course of the seventeenth and eighteenth centuries, sacred chronology gained considerable scholarly weight, being made a subject of study by such giants of early modern humanistic scholarship as the Leyden professor Joseph Justus Scaliger (1540–1609) and his Jesuit competitor Dionysius Petavius (1583–1652). One of the assumptions of sacred chronology was that the earth, having been created for humans, had no fundamentally different or longer history than its human inhabitants and that, therefore, the earth's history and human history were effectively of identical duration.

Historical geology subverted biblical chronology by uncoupling earth history from world (or human) history and assigning a much greater antiquity to the earth than could be accommodated by the Old Testament genealogies. Initially, evidence for a long stretch of earth history prior to man was qualitative. The absolute age of the earth and the duration of the periods of its history could not be reliably estimated until the full sequence of rock formations and fossil communities, the very entities to be dated, had been determined. However, the known cumulative thickness of the sedimentary strata of the stratigraphic column, which acquired the essential outline of its modern form by the early 1840s, amounted to several kilometers, indicating the immensity of geological time. The Oxford geologist William Buckland (1784–1856) spoke of "millions of millions of years."

In one respect, sacred chronology and the new geological time scale were in agreement—namely, that the earth, irrespective of the length of its history, had a distinct beginning in time. Historical geology thus helped counter the eternalism of Enlightenment deism and, in particular, the theory of an "eternal present" expressed in the famous maxim of the Scottish naturalist James Hutton (1726–97): "We find no vestige of a beginning,—no prospect of an end."

An early quantitative method for estimating the age of the earth was based on the belief that the earth had originated as an incandescent blob and had undergone a process of secular cooling. The stretch of time from the moment that the earth had acquired a solid crust until the present day could be estimated, given a figure for the rate of heat loss. Buffon carried out ingenious refrigeration experiments and dated the earth at some seventy-six thousand years of age, the earth thus having been in existence for seventy thousand years prior to the appearance of its human inhabitants. During the 1850s and subsequent decades, this method was perfected, and the figure much increased, by the British physicist William Thomson (Lord Kelvin [1824–1907]), who calculated that the earth was maximally four hundred million years old and minimally twenty million, the minimum figure being his preferred estimate. Such numbers were regarded by the Darwinians as too conservative, however, and by this time, Kelvin's figures were used less to invalidate sacred chronology than to undercut Darwin's theory of organic evolution.

The Deluge

Some naturalists and scholars were not swayed by the stratigraphic evidence for an immense stretch of geological time. These men, the Mosaical geologists, who were not part of the leadership of the new geology that was being institutionalized in geological societies and university chairs, continued the sacred-cosmogony tradition of authors such as the Gresham College (London) professor of physic John Woodward (1667–1728), who, in his *Essay towards a Natural History of the Earth* (1695), had attributed the entire sedimentary column to the Deluge. Others who had referred many of the sedimentary and tectonic features of the earth's crust to the Deluge included the London cleric Thomas Burnet (c. 1635–1715), the Cambridge mathematician William Whiston (1667–1752), the Swiss naturalist Johann Jakob Scheuchzer (1672–1733), and such internationally less renowned figures as the Bristol vicar Alexander Catcott (1725–79), a follower of the anti-Newtonian John Hutchinson (1674–1737).

Mosaical geology survived in leading English-language geological treatises until the early nineteenth century and can be found in the first volume of the *Organic Remains of a Former World* (1804), written by the London physician and amateur paleontologist James Parkinson (1755–1824). However, by the time Parkinson published the third volume of his trilogy (1811), he had changed his mind and adopted the Cuvierian view of earth history. Mosaical geology was undermined by, among others, the French geologists Georges Cuvier (1769–1832) and Alexandre Brongniart (1770–1847), who demonstrated that successive geological formations are characterized by distinctive assemblages of fossils, concluding that these formations are the record of separate periods of geological history. Cuvier presented his synthesis of earth history in a *Discours préliminaire* (*Preliminary Discourse* [1811]) to his *Ossemens fossiles.* The English translation of the earlier work was edited by the Scottish geologist Robert Jameson (1774–1854). Not only the two periods of biblical cosmogony, ante- and postdiluvial, were recognized, but several more prior to the creation of man. These had been terminated by geological upheavals, causing the extinction of many forms of life. The last of these upheavals was generally believed to be identical with the Genesis Flood.

This perception of the history of the earth as a concatenation of prehistoric periods was sensationally confirmed by Buckland's cave researches. In his classic *Reliquiae Diluvianae* (*Diluvial Remains* [1823]), Buckland developed his so-called hyena-den theory of caves: Assemblages of bones and teeth, found fossilized in the floor of caves, are the product not of a cataclysmal event but of a gradual accumulation over a long period of time, during which the caves were the den and larder of cave hyenas. The extinction of the cave carnivores, however, had been caused by the Deluge, which simultaneously had scooped out valleys and deposited surface detritus, the so-called Diluvium. Buckland's hyena-den theory undermined traditional diluvialism in that it limited the effect of the biblical Flood to the emplacement of loose surface sediment; the massive rock strata below it had been deposited during earlier periods of geological history.

The appearance of the *Reliquiae Diluvianae* was followed by a stream of articles and books in which Buckland's hyena-den theory and its diluted diluvialism were fiercely attacked. Most famous were *Scriptural Geology* (1826–27), written by the Oxford-educated Granville Penn (1761–1844) expressly in refutation of Cuvier and Buckland, and *New and Conclusive Physical Demonstrations, Both of the Fact and the Period of the Mosaic Deluge* (1837), by George Fairholme, who, in sticking to traditional Mosaical catastrophism, put forward ingenious arguments in support of the view that the entire sedimentary column was emplaced in a single global cataclysm. The writings of these men form the intellectual roots of twentieth-century flood geology, although there is little evidence of real continuity.

Buckland's diluted diluvialism did not survive for long. By around 1830, the Geological Society of London formed the stage for a classic debate, pitching the catastrophist-diluvialists, headed by the Oxford clergymen-geologists Buckland and William Daniel Conybeare (1787–1857), against the Scottish geologist Charles Lyell (1797–1875) and his supporters. A central issue was whether valleys are diluvial or fluvial in origin. The Cambridge professor of geology Adam Sedgwick (1785–1873), initially a confirmed Bucklandian diluvialist, in his 1831 anniversary address as president of the society dramatically recanted, admitting that Noah's Deluge had left no appreciable geological traces. The Bible, Sedgwick more comprehensively maintained, contains information for our moral conduct, not for scientific instruction.

Sedgwick's public retraction was followed by Buckland's. In his Bridgewater Treatise, Buckland disentangled earth history from human history by arguing that the last geological catastrophe had been not Noah's Deluge but an earlier event that had taken place just before the creation of man; the Genesis Flood, though real and historical, had been a geologically quiet event. Many of the phenomena that Buckland had attributed to the Deluge he now began seeing as the result of glacial action, and Buckland led the way in Britain when he introduced the glacial theory of the Swiss (and later Harvard) geologist Jean Louis Rodolphe Agassiz (1807–73). In other words, the Deluge was now interpreted not as a nonhistorical myth but as a nongeological event, a view that had been championed before by Scottish Presbyterians, such as the zoologist John Fleming (1785–1857) and the Edinburgh professor of physics and mathematics John Playfair (1748–1819). This vindication of the Lyellian position on the origin of "diluvial" phenomena and the removal of the Mosaical Deluge from the research agenda of London's Geological Society did not mean, however, that Lyell's anticatastrophism was generally adopted: Catastrophic occurrences continued to be postulated by the Cuvierians.

By this time, two basic schemes to reconcile Genesis with geology had become current. The first, the "day-age" exegesis, was advocated by Parkinson, the nonconformist minister Joseph Townsend (1739–1816), the Oxford chemistry professor John Kidd (1775–1851), the surgeon-paleontologist Gideon Algernon Mantell (1790–1852), and various other geologists, who suggested that the days of the Creation Week of Genesis should be understood as periods of geological time. They engaged in the complex task of demonstrating that the sequence of events of the Creation Week are exactly paralleled by the essential features of the stratigraphic column.

The second scheme, advocated by Buckland and followed by many others, such as the nonconformist divine and naturalist John Pye Smith (1774–1851) in his *On the Relation between the Holy Scriptures and Some Parts of Geological Science* (1839), made room for geology with the following exegesis. The first verse of Genesis, "In the beginning God created the heaven and the earth," is not a prospective summary of the Creation Week that follows but a retrospective reference to the primordial creation of the universe, including the earth; the first part of the second verse, "And the earth was without form and void," takes up the history of the earth following an indefinite and possibly very long interval, after the destruction of the last geological world and just before the appearance of humans.

Buckland's scriptural interpretation had the sanction of leading Anglican theologians, including the low-church evangelical John Bird Sumner (1780–1862), bishop of Chester and later archbishop of Canterbury, and E. B. Pusey (1800–1882), Regius Professor of Hebrew at Oxford and leader of the high-church Tractarian movement. The same exegesis had previously been put forward by the Scottish Presbyterian Thomas Chalmers (1780–1847), author of the Bridgewater Treatise on *The Adaptation of External Nature to the Moral and Intellectual Constitution of Man* (1833).

The advantage of this reconciliation scheme was that the earth's history was completely extricated from biblical history; the Bible covered only the history of mankind, while sacred chronology applied exclusively to the period of human existence. This perception seemed corroborated by Cuvier's observation that there were no fossil humans and by Buckland's failed attempt to find human remnants in diluvial deposits. However, this particular view required a further reconciliatory adjustment when, by the end of the 1850s, it became undeniable that humans had been contemporaneous with extinct mammals and, as Lyell showed in his *Geological Evidences of the Antiquity of Man* (1863), were of much greater age than allowed for by biblical chronology.

There is little or no evidence to suggest that the advocates of these reconciliation schemes were influenced by higher criticism of the Bible. Higher criticism did not become a topic of major public debate in Britain before *Essays and Reviews* (1860) appeared and the bishop of Natal, John William Colenso (1814–83), published *The Pentateuch and the Book of Joshua Critically Examined* (1862), questioning the Mosaic authorship and therewith the historicity of the Pentateuch.

Extinction, Death, and Sin

The fact of extinction became established by around 1800, largely from a comparative anatomical study of fossil mammals, a study that culminated in Cuvier's monumental *Recherches sur les ossemens fossiles* (*Researches on Fossil Bones* [1812]). Initially, some naturalists regarded extinction as incompatible with their belief in a plenitude of forms. Each species was believed to represent a necessary link in the Chain of Being, an integral part of creation as a whole, which contributed to its perfection. Destruction of a single link would lead to the dissolution of the entire chain. Divine providence would not let this happen, a belief theologically supported by the story of Noah's ark, which had preserved representatives of all species. Ironically, the notion of extinction was frowned upon also by Enlightenment "eternalists" such as Hutton because of its historicist connotation.

The eighteenth-century language of providence and the Chain of Being was gradually adjusted to the discoveries of historical geology. Some (such as Parkinson) argued that extinction was part of divine superintendence of earth history; others (such as Buckland) argued that fossils are missing links that, when added to the array of living forms, fill gaps and produce a complete Chain of Being. Hence, plenitude became a historical notion; the Chain of Being had no deficiencies when considered as a chain of history. In this way, paleontology significantly added to, and refreshed, the argument from design.

The very existence of fossils also represented a new problem: It indicated that death had occurred long before the appearance of humans on Earth. Moreover, death had taken place not only by natural means but also violently, inflicted by individuals of one animal species on those of another. This was apparent from the carnivorous anatomy of certain vertebrate fossils. Animal aggression in the geological past was depicted with savage realism in the various reconstructions of ancient landscapes and was expressed by the English poet Alfred Tennyson (1809–92) in the famous lines from *In Memoriam*, "Dragons of the prime, That tare each other in their slime." The problem posed by this discovery of carnivorousness and death in the geological past derived from the traditional belief that such phenomena had not existed in the Garden of Eden and had entered the world because of, and subsequent to, the Fall of man. Old and New Testament texts formed the basis for this belief, in particular St. Paul's letter to the Romans: "Wherefore, as by one man sin entered into the world, and death by sin" (5:2). Mosaical geologists, opposed to the new geology, such as the English clergyman-naturalist George Young, were quick to point out the apparent discrepancy.

The geologists, Buckland in his Bridgewater Treatise prominently among them and also several Scottish Presbyterians, responded by using the argument of Paleyan utilitarianism: Carnivorous animals function as a "police of nature," eliminating the sick and the old who would otherwise suffer as a result of pain and a lingering death; the aggregate of animal enjoyment is, thus, increased and that of pain diminished; moreover, carnivorous animals are a check on excess numbers that would have produced a shortage of food and starvation among herbivores; therefore, carnivorousness is a "dispensation of benevolence."

The utilitarian argument did not provide a solution, however, to the problem of the prehuman existence of death. How could death be a punishment for man's sin if it already existed in prehuman geological history? In an attempt to solve this conundrum, St. Paul's letter to the Corinthians was cited: "For since by man came death, by man came also the resurrection of the dead" (15:21). The passage was interpreted to mean that, just as resurrection exclusively applies to humans, so death has been a punishment only to man and not to the creation as a whole.

Designer Fossils

An area in which the interaction of religion (in the form of a belief in providence) with the new geology proved particularly fruitful was the functionalist study of fossils. The design argument (natural theology, physicotheology), prevalent throughout early modern times, attracted new popularity with the *Natural Theology* (1802) of the Cambridge theologian William Paley (1743–1805), cresting during the 1830s when the Bridgewater Treatises appeared. One of the most successful of these treatises was Buckland's *Geology and Mineralogy Considered with Reference to Natural Theology* (1836). The nonconfessional, general nature of the argument from design made it suitable as an instrument of interdenominational cooperation in the furtherance of science, and the use of nature as a source of design arguments made it possible that ecclesiastical sinecures were awarded to men who devoted their time to science. Both Buckland and Sedgwick, for example, lived on church incomes. In this way, natural theology became a vehicle for the introduction of geology at Oxford and Cambridge, for the promotion of this new subject at the British Association for the Advancement of Science, and for "geologizing" by many a clergyman across the British Isles.

Paley had made the human body the main source of evidence for design, though he also used plants and animals. Geology now provided a new range of facts that exemplified adaptation and design. In particular, paleontology, with its extinct and unfamiliar forms of life, enriched the canon of design examples by adding new, in some instances bizarre, contrivances from the geological past. The megatherium, for example, one of several genera of extinct sloths, became a cause célèbre of natural theology when its seemingly monstrous frame was successfully interpreted in terms of functional anatomy. Its grotesque-looking "claws" were a perfect adaptation to the environmental conditions of the South American pampas, where the giant sloths had dug up roots or, in a modified interpretation, had wrenched tree trunks out of the ground. Natural theology did not allow for imperfections in nature and sharpened the interpretative faculties of its practitioners in perceiving functional adaptation. Many geologists, their scientific colleagues and patrons across the diluvialist-fluvialist divide, applauded this work and added to it.

The historical dimension of paleontology provided natural theology with an altogether novel argument. Design in the world indicated a supreme designer, but this argument could be used only as a refutation of atheists. Deists, who reduced the operations of nature to those of an autonomous machine that had been designed and set in motion only at the moment of its origin, were not threatened by the argument from design. Their mechanistic worldview seemed invalidated, however, by historical geology, with its evidence of not just one single beginning but of a series of successive worlds, each with a fresh beginning and a new creation. Hence, geology served to refute not only atheists, but also deists—a function of the new science that was highlighted by its clerical practitioners.

Eschatology

In the course of the first few decades of the nineteenth century, a consensus emerged that the relationship of successive fossil worlds was one of progress or progressive development. One criterion for this was taxonomic: The lower and earlier forms of life are simpler, and the higher, or late, ones are more complex. For example, an age of fishes had preceded an age of reptiles, and this, in turn, had been followed by the rule of mammals; human beings, at the top of the taxonomic ladder, had come last. A second criterion was ecological: Progress was a matter of the improvement in the habitability of the earth to humans, and taxonomic progress was reduced to a subsidiary effect of environmental change.

The ecological criterion connected paleontology to the study of the earth as a planet. At this time, there was a revival of the old notion of a central heat, which stated that the earth had originated as an incandescent mass, that this mass had cooled down gradually and acquired a solid crust, and that it still retained a core of primeval heat. The central heat was believed to have influenced the climate of the earth, especially during its early stages of thermal evolution. The dominant form of life during a particular period of earth history had been the one most perfectly adapted to contemporary environmental conditions.

The progressivist synthesis strengthened the biblical, Christian notion of time as a directional phenomenon against the cyclical notion of time found in Enlightenment eternalism. It undercut the uniformitarian, steady-state model of earth history, prominently defended by Lyell in his *Principles of Geology* (3 vols., 1830–33). Subsequently, in two famous anniversary addresses to the Geological Society of London (1850, 1851), Lyell persisted in his denial that the stratigraphic record showed a progressive trend. The vertebrate paleontologist Richard Owen (1804–92) accepted the challenge and, point by point, refuted Lyell's arguments. Yet, in adopting progressivism, the natural theologians took in a Trojan horse because the progressivist synthesis became the hard core of the argument for organic evolution in the anonymously published *Vestiges of a Natural History of Creation* (1844), written by the Edinburgh publisher Robert Chambers (1802–71). Sedgwick, one of the fiercest critics of *Vestiges*, in the famous fifth edition of his *Discourse on the Studies of the University of Cambridge* (1850), backtracked on his earlier commitment to progressivism. Ironically, some of the antiprogressionist arguments used by Lyell have become part of the armamentarium of modern-day creationists in their opposition to the Darwinian theory.

Both the taxonomic and the environmental criteria of geological progress were anthropocentric, defining progress in relation to humans. Buckland ended his Bridgewater Treatise with illustrations of how the composition and structure of the earth's crust had been purposefully designed for the benefit of mankind, particularly for nineteenth-century society and most generously for Great Britain in support of industry and empire. Geological progressionism merged seamlessly with the Victorian belief in sociopolitical and economic progress.

The perception of an anthropocentric design in earth history went hand in hand with the biblical notion of history as a directional and a teleological process. Yet, to many geologists, progressive development seemed open-ended, and, as such, historical geology weakened part of the Christian view of history, namely the belief that history follows an eschatological course toward an apocalyptic conflagration and the Second Advent of Christ and his millennial reign. A fine example of an earlier historical study of the earth, couched in eschatological language, is the *Three Physico-theological Discourses, Concerning (1) the Primitive Chaos and Creation of the World, (2) the General Deluge, Its Causes and Effects, (3) the Dissolution of the World and Future Conflagration* (1693), written by the Cambridge-educated naturalist John Ray (1627–1705).

Apocalyptic theology, especially millenarianism, had been strong as late as the seventeenth and eighteenth centuries and, in the course of the first half of the nineteenth century, experienced a new wave of popularity, especially in England, where Warburtonian lecturers (the Warburtonian Lecture at Lincoln's Inn, London, had been established to prove the truth of the Christian religion from the completion of prophecy in the Old and the New Testaments) and such popular authors as the prolific millenarian writer William Cuninghame (d. 1849) added their voices to the apocalyptic chorus, lambasting the progressivist synthesis of earth history. Among the scientists themselves, those who adhered to the "day-age" exegesis were inclined to believe that, with the reign of man, Earth's history had come to an end. The geological present, by corresponding to the seventh day of the Creation Week, marks the end of God's creative work. The Presbyterian stonemason-geologist Hugh Miller (1802–56) felt uneasy about the emphasis on progress, pointing instead to instances of degeneration in the fossil record. In his *Footprints of the Creator* (1847), he saw the final culmination of the earth's vast history in an eschatological future kingdom of Christ.

National Context

Concern with the "Genesis and geology" issue was not uniformly spread across national boundaries. For the most part, it was a British preoccupation or, more precisely, a preoccupation in the English-speaking world, as leading figures of the North American scientific community also took part in attempts to reconcile the Bible with science. Prominent among them was the Congregationalist Edward Hitchcock (1793–1864), president of Amherst College and professor of natural theology and geology. In *The Religion of Geology and Its Connected Sciences* (1851), he advocated the "day-age" reconciliation scheme, although in the second edition of his major book (1859) he let go of the need for exact correspondences, emphasizing instead that a study of nature's laws would lead us to the divine lawmaker. Many of Hitchcock's fellow American naturalists shared his interest in harmonizing the findings of geology with the biblical account of Creation, among them Hitchcock's teacher Benjamin Silliman (1779–1864), professor of chemistry and natural history at Yale University and founder-editor of the *American Journal of Science and Arts.*

On the European mainland, original literature on the "Genesis and geology" issue was relatively scarce, yet major contributions were not altogether lacking. Already the Genevan and Calvinist naturalist Jean André Deluc (1727–1817) had presented a famous reconciliation scheme in, for example, his *Lettres sur l'histoire physique de la terre* (*Letters on the Physical History of the Earth* [1798]), arguing that the Creation days were to be understood as geological epochs and that the present epoch of earth history, conformable in length to biblical chronology, began with Noah's Deluge, which had occurred when large cavities below the antediluvial continents had collapsed, draining the oceans and exposing their floors to become our dry land. In the Netherlands, the Calvinist polymath and Romantic poet Willem Bilderdijk (1756–1831) followed Deluc's scheme in his treatise on *Geologie* (1813). A rare German contribution to the genre was *Geschichte der Urwelt, mit besonderer Berücksichtigung der Menschenrassen und des mosaischen Schöpfungsberichtes* (*History of the Antediluvian World, with Special Reference to the Races of Men and the Mosaic Creation Accounts* [1845]), written by the Lutheran professor of zoology at the University of Munich Andreas Wagner (1797–1861).

One could make a case for the thesis that the issue was of interest primarily in Protestant communities. There were exceptions to such a rule, however. The bishop of Hermopolis and minister for ecclesiastical affairs and public instruction Denis Antoine Luc de Frayssinous (1765–1841) wrote an enormously popular *Défense du Christianisme; ou, conférences sur la réligion* (*Defense of Christianity; or, Lectures on Religion* [1825]; 17 eds.), in which he argued that the days of Creation were indeterminate periods of time and that Cuvier's work had demonstrated that the Mosaical order of the creation of living beings matched their geological occurrence. Moreover, one of the most substantial reconciliation books of the period was the *Cosmogonie de Moise comparée aux faits géologiques* (*Cosmogony of Moses Compared with Geological Events* [1838–59]), written by the Catholic naturalist and magistrate Marcel Pierre Toussaint de Serres de Mesplès (1780–1862).

It is true that, on the Continent, "Genesis and geology" never became as prominently controversial as it was in Britain and North America. In Germany, for example, none of the leading names of the geological community, such as

Leopold von Buch (1774–1853), bothered to address the issue. The writings of Alexander von Humboldt (1769–1859) lacked all biblical concern. In Cuvier's *Preliminary Discourse,* too, the Pentateuch was treated not as Holy Writ but as one of several histories of nations. Such absence of reverence for the text of Genesis, however, was less the result of major differences in belief with British colleagues than of the circumstances in which Continental Europeans wrote. French and German science had nonecclesiastical, professional niches—for example, in the Parisian Musée d'Histoire Naturelle and in the many secular universities of Germany. By contrast, English cultural and religious life was dominated by the two ancient universities of Oxford and Cambridge, in essence Anglican seminaries. This had the dual effect that science in England was supported and cultivated by the church but also had much greater difficulty in acquiring autonomy and had to be discussed in terms that were directly relevant to the education of the clergy. Both diluvialism and the geological design argument were examples of this phenomenon. In the United States, too, institutions of higher learning were, not uncommonly, denominational foundations.

Hence, the science-religion discourse developed a professional dimension: At Oxbridge, the clergymen-geologists, speaking to a largely church-destined audience, stressed the relevance of geology to Genesis, whereas in London, at meetings of the Geological Society, the same Oxbridge dons mingled with nonecclesiastical and increasingly professional geologists, keeping Genesis as much as possible off the agenda and not infrequently taking scientific cues from their Continental colleagues.

By and large, mainstream Christian geologists and paleontologists succeeded in coming to terms with the new geology. Their reconciliation schemes provided space for scientific inquiry as well as for religious belief. Traditional Flood geology, with its tenets of a young earth and a geologically effective, cataclysmal deluge, became regarded as incorrect and antiquated. Interestingly, it has reemerged in the twentieth century, primarily among American fundamentalist scientists.

BIBLIOGRAPHY

Appel, Toby A. *The Cuvier-Geoffroy Debate: French Biology in the Decades before Darwin.* Oxford: Oxford University Press, 1987.

Brooke, John H. "The Natural Theology of the Geologists: Some Theological Strata." In *Images of the Earth*, ed. Ludmilla Jordanova and Roy Porter. Chalfont St. Giles: British Society for the History of Science, 1979, 39–64.

———. *Science and Religion: Some Historical Perspectives.* Cambridge: Cambridge University Press, 1991.

Burchfield, Joe D. *Lord Kelvin and the Age of the Earth.* London: Macmillan, 1975.

Cannon, Susan Faye. *Science in Culture: The Early Victorian Period.* New York: Dawson/Science History Publications, 1978.

Cohn, Norman. *Noah's Flood: The Genesis Story in Western Thought.* New Haven: Yale University Press, 1996.

Conser, Walter H. *God and the Natural World: Religion and Science in Antebellum America.* Columbia: University of South Carolina Press, 1993.

Corsi, Pietro. *Science and Religion: Baden Powell and the Anglican Debate, 1800–1860.* Cambridge: Cambridge University Press, 1988.

Gillispie, Charles C. *Genesis and Geology: The Impact of Scientific Discoveries upon Religious Beliefs in the Decades before Darwin.* 1951. Reprint. New York: Harper and Row, 1959.

Haber, Francis C. *The Age of the World: Moses to Darwin.* Baltimore: Johns Hopkins Press, 1959.

Herbert, Sandra. "Between Genesis and Geology: Darwin and Some Contemporaries in the 1820s and 1830s." In *Religion and Irreligion in Victorian Society: Essays in Honor of R. K. Webb*, ed. R. W. Davis and R. J. Helmstadter. London: Routledge, 1992, 68–84.

Hooykaas, Reijer. *Natural Law and Divine Miracle: A Historical-Critical Study of the Principle of Uniformity in Geology, Biology, and Theology.* Leiden: Brill, 1959

Livingstone, David N. *Darwin's Forgotten Defenders: The Encounter between Evangelical Theology and Evolutionary Thought.* Grand Rapids, Mich.: Eerdmans, 1987.

Millhauser, Milton. "The Scriptural Geologists: An Episode in the History of Opinion." *Osiris* 11 (1954): 65–86.

Moore, James R. "Geologists and Interpreters of Genesis in the Nineteenth Century." In *God and Nature: Historical Essays on the Encounter between Christianity and Science*, ed. David C. Lindberg and Ronald L. Numbers. Berkeley: University of California Press, 1986, 322–50.

North, John D. "Chronology and the Age of the World." In *Cosmology, History, and Theology*, ed. Wolfgang Yourgrau and Allen D. Breck. New York: Plenum, 1977, 307–33.

Numbers, Ronald L. "Science and Religion." *Osiris* 2d ser. 1 (1985): 59–80.

———. *The Creationists: The Evolution of Scientific Creationism.* New York: Knopf, 1992.

Page, Leroy E. "Diluvialism and Its Critics in Great Britain in the Early Nineteenth Century." In *Toward a History of Geology*, ed. Cecil J. Schneer. Cambridge: MIT Press, 1969, 257–71.

Porter, Roy M. "Creation and Credence: The Career of Theories of the Earth in Britain, 1660–1820." In *Natural Order: Historical Studies of Scientific Culture*, ed. Barry Barnes and Steven Shapin. Beverley Hills, Calif.: Sage, 97–123.

Rappaport, Rhoda. "Geology and Orthodoxy: The Case of Noah's Flood in Eighteenth-Century Thought." *British Journal for the History of Science* 11 (1978): 1–18.

Rudwick, Martin J. S. "The Shape and Meaning of Earth History." In *God and Nature:*

Historical Essays on the Encounter between Christianity and Science, ed. David C. Lindberg and Ronald L. Numbers. Berkeley: University of California Press, 1986, 296–321.

———. *Georges Cuvier, Fossil Bones, and Geological Catastrophies: New Translations and Interpretations of the Primary Texts*. Chicago: University of Chicago Press, 1997.

Rupke, Nicolaas A. *The Great Chain of History: William Buckland and the English School of Geology, 1814–1849*. Oxford: Clarendon, 1982.

———. "Caves, Fossils, and the History of the Earth." In *Romanticism and the Sciences*, ed. Andrew Cunningham and Nicholas Jardine. Cambridge: Cambridge University Press, 1990, 241–62.

———. "A Second Look: C. C. Gillispie's Genesis and Geology." *Isis* 85 (1994): 261–70.

———. "'The End of History' in the Early Picturing of Geological Time." *History of Science* 36 (1998): 61–90.

Stiling, Rodney. "The Diminishing Deluge: Noah's Flood in Nineteenth-Century American Thought." Ph.D. diss., University of Wisconsin, 1991.

Turner, Frank M. *Contesting Cultural Authority: Essays in Victorian Intellectual Life*. Cambridge: Cambridge University Press, 1993.

Van Riper, A. Bowdoin. *Men among the Mammoths: Victorian Science and the Discovery of Human Prehistory*. Chicago: University of Chicago Press, 1993.

Young, Davis A. *The Biblical Flood: A Case Study of the Church's Responses to Extrabiblical Evidence*. Grand Rapids, Mich.: Eerdmans, 1995, chaps. 7–9.

15 Natural History

Peter M. Hess

Natural history has since antiquity been entwined in intricate ways with the theological and philosophical assumptions of the diverse cultures and periods in which it developed. Its interpretation has been profoundly influenced by the gradual transformation of the classical and early Christian view of an immutable nature into the modern conception of a dynamic and evolving biological world. The Christian tradition endorsed the study of creation as appropriate to the glorification of God, and natural history in turn supported physicotheology. But the relationship did not always remain benign. Developing sophistication in the understanding of flora, fauna, and their related ecologies would gradually reinforce a natural theology independent of Christianity. As the world and its biological processes came to be seen as subject to historical development, a methodological naturalism began to challenge traditional theistic assumptions about design.

Peter M. Hess earned his M.A. from Oxford University and his Ph.D. in historical theology from the Graduate Theological Union in Berkeley, California. He has taught theology and history at the University of San Francisco since 1985 and serves as associate program director with the Center for Theology and the Natural Sciences in Berkeley. His research focuses on the interaction between religion and science in early modern Europe, particularly in the field of natural theology.

Introduction

Throughout much of Western intellectual history, the relationship between religion and the study of natural history was relatively serene, untroubled by the spectacular displays that were provoked by advances in astronomy. Indeed, during the two millennia after Aristotle (384–322 B.C.), there was little to spark significant controversy until the secular implications of biological evolution became apparent following the publication of Charles Darwin's (1809–82) *Origin of Species* (1859). Nevertheless, the many theories of natural history and the rationales underlying its practice were perennially intertwined in complex ways with the theological assumptions of the cultures in which it developed.

The term *natural history* is not susceptible to easy and consistent definition in the multiple cultural and temporal contexts through which it developed in the centuries covered by this essay. While natural history always dealt with the study of living animals and plants, during a substantial portion of its history it also included fossils, minerals, and geological formations. In the Middle Ages, it included the study of rocks and fossils, but they became detached from natural history during the eighteenth century, as recognition of the deep history of time propelled paleontology and geology toward the status of autonomous disciplines. Natural history in 1850 hardly resembled the eclectically comprehensive body of knowledge it had been in Pliny the Elder's (c. A.D. 23–79) *Natural History,* which included everything from the study of anatomy and physiology to the anthropological treatment of comparative cultures. These subjects had become separated from the topic by the mid–nineteenth century. In this essay, *natural history* will be understood as referring primarily to the study of the organic world as manifested in species of plants and animals.

While making collections of curiosities was a significant dimension of natural history after 1700, the discipline never consisted merely of cataloging the elements of nature. It always involved, however unsystematically, the construction of (or at least the possibility of) philosophical interpretations of the things found. Nevertheless, natural history needs to be distinguished from *natural philosophy* not only by virtue of its subject matter but because the latter—especially from the seventeenth century onward—designated the elaboration of causal interpretations of measurable phenomena, particularly in physics and astronomy.

Underlying the relationship between religious thought and natural history in the Western tradition, one crucial paradigm shift stands out: the great sea change in thought from the classical view of nature as the immutable foundation of human affairs, which served in medieval Christendom as the backdrop for the great drama of salvation, to the modern view in which the earth, as well as biological life and the human species, have become thoroughly historicized. The natural history pursued by Aristotle and Pliny, as well as by patristic writers and medieval bestiarists, was qualitatively different from the nineteenth-century biological science practiced by Charles Darwin and integrated by such contemporary theologians as Frederick Temple (1821–1902) into a Christian, or more generally, a theistic, worldview.

The Dual Heritage

The intersection between religion and natural history in the Western tradition naturally has plural roots, three of which, at the risk of undue simplification, can be identified as being of primary significance: Greek rationality, Roman pragmatism, and the Judeo-Christian appraisal of the world as intrinsically good.

Natural history was not foremost among the interests of the pre-Socratic Greek philosophers of the sixth and fifth centuries B.C., whose speculations were primarily cosmological, nor was it particularly important for Plato (c. 427–347 B.C.). The latter's major contribution to natural history was his epistemology: His theory of Forms served to undergird the concept of species as an entity that truly existed within the divine mind, a theory that was not seriously challenged until the work of Georges Leclerc, Comte de Buffon (1707–88), in the eighteenth century. Of even greater significance was Aristotle, who made seminal contributions in at least three areas. First, his doctrine of the tripartite soul, which took vegetative, animate, and rational forms, established a crucial principle of continuity within the biological realm. However, his assertion that the source of the rational soul comes from outside the physical process of reproductive transmission set up later tensions by encouraging Christian theologians to assert that the soul was separately created by God. Second, his careful attention to firsthand observation—whether in his own work or in the reports of others—was conducive to what would become a tradition of empirical research that was pursued only intermittently until its firm establishment as a methodological approach by Francis Bacon (1561–1626). Third, Aristotle's pervasive teleology ensured that final causes would serve, at least until 1800, as a fundamental organizing principle in natural history, governing our understanding of how bodies are organized, of the purposes served by individual organs, and of the relationships between plant and animal members of the medieval "Great Chain of Being." Religion and natural history were, at best, only tangentially (and perhaps negatively) related in Aristotle's thought, however: His remote Prime Mover was unconcerned with individual plants or animals or even human beings.

Among other classical sources for natural history, one of the most influential was the work of Pliny the Elder, whose voluminous *Historia naturalis* (*Natural History*) exercised a profound influence on the West well into the Renaissance. The encounter of natural history and religion in his thought takes place against the background of his central concern to catalogue and exhibit nature in all of its strange and wonderful variety. Skeptical of the existence of the gods, Pliny proceeded from the assumption that "the world is the work of nature and the embodiment of nature itself" (2.1). Hence, his collection of facts about everything imaginable can almost literally be said to articulate his "theology." Early medieval encyclopedists drew heavily from this eclectic, unsystematic, and uncritical work and even patterned their own works after it.

The clear relationship between religion and the world in Hebrew thought is expressed in the Genesis Creation story. The sixfold divine affirmation of creation as "good" in Genesis 1 reflects the Jewish view that God's revelation does not merely designate the communication of divine truth to humankind but involves the whole of the natural world. The sharp distinction between God and nature that was characteristic of Judaism ensured that the Creator and his creation could not be conflated, but it also opened the door to potential scientific investigation of the world. Of course, the Hebrew authors were in no sense scientists and offered no systematic account of the organic world around them. The theological structure of the first eleven chapters of Genesis clearly shows that fact. However, the positive appraisal of nature found throughout Hebrew literature would hold great significance for the development of natural history. "The heavens declare the glory of God, and the firmament shows forth his handiwork" (Psalm 19:1 Revised Standard Version). If all creatures are called upon to praise God for his creation (as in Psalm 148), the investigation of natural history becomes a worthy enterprise provoking awe and, ultimately, worship.

The Christian New Testament follows the Hebrew Scriptures in lacking any systematic interest in natural history. But the Gospels and the Pauline literature are positive in their view of nature in contrast, for example, with contemporary Gnosticism. The view of the Old Testament that the world is good is echoed in the Pauline declaration that "ever since the creation of the world his invisible nature, namely his eternal power and deity, has been clearly perceived in the things that have been made" (Romans 1:20). This passage would exert a powerful influence on the development of natural theology in Christian Europe well into the eighteenth century and, thus, serve as one rationale for the study of natural history.

From the Fathers to the Middle Ages

The portion of patristic literature that deals with natural history tends to reflect the dual heritage of Greco-Roman science and the Hebrew-Christian affirmation of the world as God's handiwork. Especially in the writings of the Greek fathers, nature was widely regarded as sacramental and was, therefore, considered a proper subject of investigation. As bishops and theologians, the Fathers were only amateur naturalists, but their appreciation of God's creation went far beyond merely using it as a convenient source for theological metaphors. Basil of Caesarea (c. 330–79), Gregory of Nyssa (c. 330–c. 395), and Nemesius (late fourth century) made objective observations about the plants and animals around them. These men had studied natural history as part of their general education. For example, they studied botany not merely through written sources, but by a close observation of nature that was frequently directed to practical ends, such as perfecting the cultivation of fruit trees and staple crops. On the other hand, the Fathers demonstrated an eclectic and unscientific knowledge of zoology and showed (by modern standards) considerable credulity in mixing fact with fancy. Natural history in their hands was not experimental. They diligently mined the florilegia and manuals at their disposal for scientific knowledge, which they pursued for its theological and moral use. In this respect, they demonstrated a closer affinity to Plato than to Aristotle. Hence, the revival of natural history as a constructive attempt to understand the animate world in itself would have to await the high-medieval recovery of the latter's biological works.

Natural history shared in the precipitous decline of Western intellectual culture in the period (c. 500–c. 1000) following the collapse of Roman civilization. For much of the early Middle Ages up to the twelfth century, scholarly knowledge about nature and natural phenomena was largely lost in the Christian West, and, although works from this period record considerable familiarity with local plants and their medicinal uses, no significant developments were made in the direction of a systematic understanding of natural history. Such knowledge as existed was preserved and transmitted by the medieval encyclopedias, such as Isidore of Seville's (c. 560–636) *Etymologies* and *De proprietatibus rerum* (*On the Properties of Things*) by the thirteenth-century Englishman Bartholomeus Anglicus (fl. 1230–50). Ultimately, all such works descended from the anonymous Greek *Physiologus* (c. A.D. 200), and their construction reflected a continuous literary tradition more than it relied on empirical observation, which was used only infrequently to supplement excerpts made from classical texts. The knowledge displayed in these works—especially the illuminated books of animals, known as *bestiaries*—was clearly oriented to didactic and nonscientific purposes. For example, a central theological doctrine governing the interpretation of natural history was that the Fall of Adam and Eve had effected a dramatic physical transformation in nature, including the initiation of carnivorous behavior among animals. Treatment of animals in bestiaries could not simply recount their ecological circumstances or life histories but had also to pay attention to the natural symbolism of spiritual truths, such as the role of the serpent in the Garden of Eden or of the whale in the story of Jonah, not to mention moral qualities, such as the fox's cunning or the dog's fidelity.

With the reintroduction of the full Aristotelian corpus into the West in the thirteenth century, studies of botany and zoology began gradually to move beyond the bestiary tradition. The rediscovery of the biological works of Aristotle (and of those spuriously attributed to him) was an important catalyst to new thought, especially in the cases of Albertus Magnus (1193–1280) and Roger Bacon (1213–91). However, Pliny continued to be widely read well into the sixteenth century, and natural history remained oriented toward the dominant theological paradigm, fleshing out particular details in the Chain of Being between God and prime matter, in which humanity constituted the vitally important link.

From the Renaissance to Physicotheology

The early modern treatment of the natural-historical tradition bequeathed to it by the Middle Ages was marked by both important continuities and significant critiques. On the one hand, the ponderous weight of received tradition is illustrated by the printed floral works of the early sixteenth century, which, in some respects, merely recapitulate the themes of medieval herbals. Likewise, Conrad Gesner's (1516–65) handsomely printed *Historia animalium* (*History of Animals* [1551–8]) engaged the reader in the systematic study of animals but also evidenced a credulity that would be unacceptable a century later. Among the nicely drawn exempla of known species to appear in Gesner and in Edward Topsell's *History of Four-Footed Beasts* (1608) were mythical creatures such as the manticora, which was part man and part lion.

On the other hand, the invention of printing and the new humanistic scholarship together exercised a profound effect on the study of natural history, with the result that, by the fifteenth century, some biologists were beginning to examine Pliny with a decidedly critical eye. Printing offered the advantage of conveying information in the form of images, which moved well beyond a crude iconographic tradition and served to educate naturalists uniformly in far flung parts of Europe. Albrecht Durer's (1471–1528) detailed drawings of plants, for example, paid careful attention to their ecologies. The emblematic worldview—in which to know a creature was to know all of its literary associations—would gradually give way to the inductive methodology championed by Francis Bacon (1561–1626), in which physical observation was paramount.

Important humanist critics of the medieval tradition included Thomas Browne (1605–82), a scholar who was both deeply respectful of the authorities of the past and a scientifically minded naturalist. In his *Pseudodoxia epidemica,* or *Enquiries into . . . Vulgar and Common Errors* (1646), Browne submitted the reliability of many past writers on natural history to careful examination according to the three Anglican determinants of truth: sense, authority, and reason. His presupposition of the truth of the divinely ordained laws of nature became a touchstone of the Royal Society (founded in 1660), for whose members doing good science was a deeply religious activity. It is no accident that Nehemia Grew's (1641–1712) important empirical researches in botany found their highest expression in his *Cosmologia sacra* (*Sacred Cosmology* [1701]). Genuine advances in scientific method and a critical understanding of phenomena fostered the empirical research of naturalists such as Robert Hooke (1635–1703) and Anton van Leeuwenhoek (1632–1723) and established biology on a firm footing by the end of the seventeenth century. Still, natural historians continued to view the evidence they accumulated in the light of received theological assumptions. As a context for interpreting his careful observations about fossils, Nicolaus Steno (1638–86), for example, posited the Noachian Flood as the source of the fossils that he meticulously described.

In the latter half of the seventeenth century, the English physicotheology movement developed the closest intertwining that natural history would perhaps ever enjoy with religion. The Cambridge divine and scientist John Ray (1627–1705) saw clear evidence of divine planning in the complex adaptations of plants and animals to their environments. His treatise *The Wisdom of God Manifested in the Works of the Creation* (1691) not only restated the design argument but also offered a vast compendium of natural history that was characterized by the observation of important nuances within species and the incipient recognition of ecological relationships. The Boyle Lectures (founded in 1692) institutionalized this approach by providing a public forum for the articulation of the new science in support of traditional religious belief. The most influential course of lectures dealing with natural history that provided support for the design argument was delivered by William Derham (1657–1735) and published as *Physico-Theology; or, A Demonstration of the Being and Attributes of God from the Works of Creation* (1716). The establishment of the physicotheology tradition stands as both the embodiment of an ancient tradition wrapped in the mantle of the scientific revolution and the point from which natural history would split into two streams: one professional and increasingly secular, the other popular and persistently religious.

Eighteenth- and Nineteenth-Century Developments

If religion and natural history enjoyed a close relationship in the seventeenth century, as exemplified by the edifying structure of the physicotheology tradition, cultural and intellectual factors would erode its foundations through a variety of channels and bring about its collapse in the next century. In one degree or another, the Enlightenment exaltation of reason at the expense of revelation was implicated in most of them, but factors in the social construction of science also played an important role. In particular, four significant shifts in pre-Darwinian natural history contributed to this erosion.

First, it is important to recognize that natural history existed in two parallel traditions. The physicotheology tradition persevered with elegance until Victorian times, nourished by the religious impulses of reverence and awe at the divine wisdom revealed in the works of Creation. A classic example is the collection of observations on local flora and fauna made by the English clergyman Gilbert White (1720–93) and published in 1788 as *The Natural History of Selborne.* A far more philosophically rigorous contribution in this vein was William Paley's (1743–1805) *Natural Theology* (1802), which continued to influence a generation of students of impeccable Christian orthodoxy, including (as a young man) Charles Darwin. This popular tradition extended into the 1860s, embodying an approach to natural history that was solidly grounded in the assumptions of natural theology.

However, the tradition of natural theology had already begun to face strenuous competition in the mid–eighteenth century from professional natural historians. Although Carolus Linnaeus (1707–78), in creating his system of taxonomic nomenclature, regarded himself as a recorder of God's creation, in his work he avoided as consistently as possible any appeal to God for causal explanation of natural-historical phenomena. And from 1700 onward, the increasing sophistication of research tools and instrumentation, together with the establishment of endowed chairs in European universities, led inevitably to the professionalization of natural history. A field in which amateur collectors of specimens could still make respectable contributions was on its way to becoming the largely secular professional discipline of biology of the nineteenth century. Impelled by the research of such towering figures as Linnaeus in Sweden, Buffon in France, and Albrecht von Haller (1708–77) in Germany, natural history on the professional level discarded in theory (if not in fact) the religious assumptions of physicotheology. As scientific sophistication spread to the wider culture, discoveries that had provoked awe and reverence in the early physicotheologians were now regarded as merely commonplace.

A second shift may be seen in the eclipse of teleology and the growth of scientific naturalism. One of the most important Aristotelian legacies had been the adoption of teleology as a fundamental organizing principle in science, and it played an especially crucial role in natural history. Organisms were thought to develop according to a preconceived plan, just as organs were assumed to have been designed to serve specific purposes and animals to thrive within particular habitats, all for the service of humans. However, in the eighteenth century this basic principle began to falter, and, while it would be anachronistic to suggest that by 1859 teleological thinking already lay shipwrecked on the shoals of naturalism, the importance of its piecemeal dismantling cannot be underestimated.

Whatever personal and methodological differences there may have been among Linnaeus, Buffon, and Haller, as scientists they shared basic assumptions about the existence of final causes and immutable plans regarding the objects of their study. In contrast, their successors in the next generation of natural historians uniformly relied upon the assumptions of Enlightenment science, discarding as useless tools the teleologies and immutable plans that had served such a vital role from Aristotle to Ray and Derham. Their intentionally nonteleological approach found philosophical legitimation in Immanuel Kant's (1724–1804) *Critique of Teleological Judgment* (1790).

Another impetus to the dissociation of natural history from a religious interpretation of nature was the extension of the seventeenth-century mechanistic cosmological model into the biological sphere. The reintroduction of Lucretius's (c. 99–55 B.C.) atomic theory of matter, purged of its atheistic elements, had reduced physical reality to matter in motion under the influence of Galileo (1564–1642) and René Descartes (1596–1650) and had already excluded the vast continuum of "vital" powers intrinsic to the Aristotelian universe. The logical sequel was to extend this reductionism to life itself and to provide a purely naturalistic explanation of life from a mechanicochemical perspective. Even if the mechanistic interpretation of life was not ultimately successful, its presence caused some wear on the supports for a religious interpretation of nature.

A third eighteenth-century challenge to the received tradition of natural history came in the form of a secularizing "historicization" of natural history. Two growing mountains of evidence—one temporal and the other geographical—suggested that the biblical cosmogony could not be accepted literally. First, almost from the moment of the European discovery of the New World, an endless stream of information about previously unknown plants and animals began inundating the minds of natural historians. Whereas John Ray listed fifteen hundred species of animals, Linnaeus knew of fifty-six hundred species of quadrupeds alone, and further geographical exploration offered no end to this explosion of knowledge. It began to seem difficult to reconcile such an abundance of species with Creation from a central point or with the story of Noah having saved two pairs of each species in his ark, thus forcing natural historians into the increasingly uncomfortable position of having to choose between their empirical evidence and the dictates of theological tradition. The temporal factor influencing the historicization of natural history was the gradual discovery of the "deep history" of time. The evidence being gathered by the young science of geology suggested that the sedimentary strata of the earth and the fossils they contained were far older than the few thousand years that a literal reading of Genesis would allow. Consequently, by the 1830s the natural historian had every good reason to believe that the history of creation was not coterminous with human history and that species, over time, might, indeed, have come into existence and become extinct.

Nevertheless, even in these developments the importance of religious factors to natural history must not be gainsaid. It had been recognized in classical times that fauna and flora varied considerably with location, and the biblical description of a universal flood and subsequent diffusion of species from

Mount Ararat served as a powerful organizing idea. Even after eighteenth-century natural historians had come to think of the Flood as a local event, the entrenched idea of the radiation of life-forms from a central point was only gradually abandoned. Likewise, the rationale for establishing botanic gardens in Europe was initially, at least in part, theological. In the sixteenth century, it had been hoped that the Garden of Eden—which, according to one tradition, had survived the destructive ravages of Noah's Flood—might be rediscovered in the course of European exploration. Disappointment in this sphere was tempered by excitement at the idea of re-creating Paradise, by reassembling from the farthest corners of the globe the many species dispersed by the Flood.

Finally, the story of pre-Darwinian natural history would not be complete without a comment on its gradual assimilation of elements of what would eventually become the new paradigm of biological evolution. From the time of Ray and Derham onward, the physicotheological tradition had paid increasing attention to the adaptations of organisms to particular environments, interpreting them through the lens of divine design. William Paley's diligent natural historical studies in *Natural Theology* masterfully rearticulated the argument for a divinely designed and providentially arranged world. Parallel to the secular work of professional natural historians, the theological interpretation of ecological adaptation persevered, both in the Bridgewater Treatises of the 1830s and in Charles Babbage's (1792–1871) argument in 1838 that the very laws governing the extinction and creation of new forms of life suitable for particular environments could be interpreted as further evidence of the benevolent, if inscrutable, purposes of God.

But, with the Darwinian synthesis, the hallowed and familiar relationship between religion and natural history became decidedly more ambiguous. Darwin's carefully substantiated case for natural selection in the *Origin of Species* found a mixed reception, with some theologians accepting it and some scientists opposing it. While Darwin himself only gradually abandoned Christianity, after 1859 professional biologists would refer to the classic metaphor of God's "Book of Nature" with increasing rarity.

Conclusion

In the two millennia separating Aristotle from Darwin, both the assumptions underlying natural history and the express rationale for its practice were profoundly influenced by theologies in the Judeo-Christian tradition. This influence was reciprocal, as sophistication in natural-historical studies in the later medieval and early modern periods reinforced a natural theology that had become quasi-independent of Christianity. It is perhaps significant that Aristotle and Darwin—the two thinkers who most clearly frame the period in which natural history developed into the science of biology—each operated out of a theological framework that was, in important respects, incompatible with Christianity. In the interim, the historical and metaphysical assumptions of the Judeo-Christian West provided fertile ground in which the seeds of the modern understanding of flora, fauna, and their related ecologies could take root.

The period from 1750 onward witnessed an increasingly secular approach to the practice of natural history, less and less determined by the agenda of a literal adherence to biblical dogma. The growing belief that Scripture was a historically conditioned document was paralleled by a radical historicization of the natural world and biological processes, and, since the late eighteenth century—and in a greatly accelerated fashion since Darwin—traditional theistic assumptions have been largely replaced by an underlying methodological and metaphysical naturalism. Relating theology and natural history meaningfully to each other has not become impossible, but, since the mid–nineteenth century, practitioners of each discipline have been compelled to become acutely aware of the limitations of the competencies of both as they relate to each other.

BIBLIOGRAPHY

Ashworth, William B. "Natural History and the Emblematic World View." In *Reappraisals of the Scientific Revolution,* ed. David C. Lindberg and Robert S. Westman. Cambridge: Cambridge University Press, 1990, 302–32.

Barber, Lynn. *The Heyday of Natural History, 1820–1870.* New York: Doubleday, 1980.

Bates, Marston. *The Nature of Natural History.* Princeton: Princeton University Press, 1950.

Brooke, John. *Science and Religion: Some Historical Perspectives.* Cambridge: Cambridge University Press, 1991.

Browne, Janet. *The Secular Ark: Studies in the History of Biogeography.* New Haven: Yale University Press, 1983.

Clark, Willene B., and Meradith T. McMunn. *Beasts and Birds of the Middle Ages: The Bestiary and Its Legacy.* Philadelphia: University of Pennsylvania Press, 1989.

Crombie, A. C. *Augustine to Galileo.* Cambridge: Harvard University Press, 1961.

Findlen, Paula. *Possessing Nature: Museums, Collecting, and Scientific Culture in Early Modern Italy.* Berkeley: University of California Press, 1994.

Gillespie, Neal C. "Natural History, Natural Theology, and Social Order: John Ray and the 'Newtonian Ideology.'" *Journal of the History of Biology* 20 (1987): 1–49.

Glacken, Clarence. *Traces on the Rhodian Shore: Nature and Culture in Western Thought from Ancient Times to the End of the Eighteenth Century.* Berkeley: University of California Press, 1967.

Goerke, Heinze. *Linnaeus.* Trans. Denver Lindley. New York: Scribner's, 1973.

Hoeniger, F. D., and J. F. M. Hoeniger. *The Growth of Natural History in Stuart England from Gerard to the Royal Society.* Charlottesville: University Press of Virginia for the Folger Shakespeare Library, 1969.

Jardine, N., J. A. Secord, and E. C. Spary, eds. *Cultures of Natural History.* Cambridge: Cambridge University Press, 1996.

Knight, David M. *Natural Science Books in English 1600–1900.* London: Batsford, 1989.

Larson, James L. *Interpreting Nature: The Science of Living from Linnaeus to Kant.* Baltimore: Johns Hopkins University Press, 1994.

Lenoir, Timothy. *The Strategy of Life: Teleology and Mechanics in Nineteenth-Century German Biology.* Dordrecht: Reidel, 1982.

Lindberg, David C. *The Beginnings of Western Science: The European Scientific Tradition in Philosophical, Religious, and Institutional Context, 600 B.C. to A.D. 1450.* Chicago: University of Chicago Press, 1992.

Ospovat, Dov. *The Development of Darwin's Theory: Natural History, Natural Theology, and Natural Selection, 1838–1859.* Cambridge: Cambridge University Press, 1981.

Pelikan, Jaroslav. *Christianity and Classic Culture: The Metamorphosis of Natural Theology in the Christian Encounter with Hellenism.* New Haven: Yale University Press, 1993.

Prest, John. *The Garden of Eden: The Botanic Garden and the Recreation of Paradise.* New Haven: Yale University Press, 1981.

Rehbock, Philip F. *The Philosophical Naturalist.* Madison: University of Wisconsin Press, 1983.

Roger, Jacques. "The Mechanistic Conception of Life." In *God and Nature: Historical Essays on the Encounter between Christianity and Science,* ed. David C. Lindberg and Ronald L. Numbers. Berkeley: University of California Press, 1984, 277–95.

Rudwick, Martin J. S. *Scenes from Deep Time: Early Pictorial Representations of the Prehistoric World.* Chicago: University of Chicago Press, 1992.

Shteir, Ann B. *Cultivating Women, Cultivating Science: Flora's Daughters and Botany in England, 1760 to 1860.* Baltimore: Johns Hopkins University Press, 1996.

Sloan, Phillip R. "John Locke, John Ray, and the Problem of the Natural System." *Journal of the History of Biology* 5 (1972): 1–53.

———. "Natural History, 1670–1802." In *Companion to the History of Modern Science,* ed. R. C. Colby et al. London: Routledge, 1990, 295–313.

Stannard, Jerry. "Natural History." In *Science in the Middle Ages,* ed. David C. Lindberg. Chicago: University of Chicago Press, 1978, 429–60.

Wallace-Hadrill, D. S. *The Greek Patristic View of Nature.* Manchester, England: Manchester University Press, 1968.

Welch, Margaret. *The Book of Nature: Natural History in the United States, 1825–1875.* Boston: Northeastern University Press, 1998.

16 Charles Darwin

James Moore

For more than a century, the agelong "warfare" between science and religion has functioned as a creation myth to explain the world as seen in the light of modern research. This "New Nature" (as Thomas Huxley called it) emerged—according to the myth—from the relentless conflict of reason versus faith, inquiry versus authority, truth versus error. And the greatest Victorian cause of this war was a heroic evolutionist, Charles Darwin, whose theory of evolution by natural selection finally "disproved" the Book of Genesis. But modernity's creation myth, too, tends to be undermined by familiarity with historical evidence. As a spate of recent scholarship has shown, Darwin's own Christian background, his lingering religious beliefs, and his attitude toward the faith of others does not harmonize well with the old myth of scientific origins.

James Moore is reader in history of science and technology at the Open University in England and also teaches at Cambridge University. His books include *The Post-Darwinian Controversies: A Study of the Protestant Struggle to Come to Terms with Darwin in Great Britain and America, 1870–1900* (Cambridge: Cambridge University Press, 1979) and (with Adrian Desmond) *Darwin* (London: Michael Joseph, 1991). He is working toward a biography of Alfred Russel Wallace.

EACH AGE fashions nature in its own image. In the nineteenth century, the English naturalist Charles Darwin (1809–82) recast the living world in the image of competitive, industrial Britain. He abandoned the Bible as a scientific authority and explained the origin of living things by divinely ordained natural laws. Once destined for the church, he became the high priest of a new secular order, proclaiming a struggling, progressive, and law-bound nature to a struggling, improving, and law-abiding society. For his devotion to science and his exemplary life, he received England's highest religious honor when scientists joined churchmen and politicians of all parties to inter his mortal remains in Westminster Abbey.

Darwin was born at Shrewsbury in 1809, the second son of a wealthy Whig household. His father was a freethinking physician; his mother, a conservative Unitarian. Upright and respectable, they had Charles christened in the local Anglican Church. As a boy, he attended chapel with his mother and was first educated by the minister. After her death in 1817, he sat under a future bishop at Shrewsbury School and learned to despise the classics. Chemistry was more to his taste, and his first experiments were conducted in a garden shed with his brother, Erasmus. Five years older, Erasmus followed their father into medicine at Edinburgh University, leaving Charles in the care of his sisters. He fretted and his lessons suffered, so, in 1825, his father sent him to study medicine with Erasmus.

Edinburgh was liberal and cosmopolitan, full of brash freethinkers. Charles struck up a friendship with one of them, Dr. Robert E. Grant (1793–1874), Britain's leading invertebrate zoologist. Together they scoured the coast for exotic sea life and attended the university's Plinian Society, in which students and staff debated hot topics in natural history. Grant was an atheist and evolutionist, following the French naturalist Jean Baptiste de Monet de Lamarck (1744–1829); the Plinian served as a platform for his and other members' radical, materialist ideas. Here Charles first saw scientific heresy punished when a fellow student's remarks on the identity of mind and brain were struck from the minutes.

Charles dropped out of medicine, unable to stomach surgery. To cure his indirection, his father prescribed a stint at Cambridge University to train for the Church of England. A country parish would make few demands on his son's faith; he would have a respectable social role, a guaranteed income, and, above all, the leisure to indulge his Edinburgh interest in natural history. Charles read a few divinity books and decided there was nothing in them he could not say he believed. In 1828, he went up to Christ's College to study for the B.A. and ordination.

Cambridge was strict and feudal, a market town dominated by a medieval university. Here the professors were untainted by French radicalism. They included clergymen like John Stevens Henslow (1796–1861), who taught Charles botany, and Adam Sedgwick (1785–1873), who introduced him to geology. These men believed that living species and society alike were kept stable by God's will. This was the reigning orthodoxy, enshrined in required textbooks by the Rev. William Paley (1743–1805). Everyone conformed to it, more or less. Unbelievers were unwelcome. In 1829, when a renegade Cambridge graduate, the Rev. Robert Taylor, attempted an "infidel mission" to the university, he was hounded out of town. Charles never forgot the example of this apostate priest, dubbed "the Devil's Chaplain."

In August 1831, after a geological fieldtrip with Sedgwick, Darwin found a letter at home from Henslow offering him a place as captain's companion aboard H.M.S. *Beagle.* This was the turning point of his life. The church could wait. His path to a country parish was now diverted via a voyage around the world. For five years, Darwin collected specimens, kept a diary, and made countless notes. He dreamed of becoming a parson-naturalist, and his religious beliefs and practices remained conventional. Like his professors, he did not take Genesis to be a literal account of creation, but he quoted Scripture as a supreme moral authority. He carried a copy of Milton's *Paradise Lost* with him and, on Henslow's recommendation, the first volume of Charles Lyell's (1797–1875) *Principles of Geology* (1830–33). Lyell, a Unitarian, argued that the earth's crust had been laid down over countless ages according to natural law. Darwin was convinced. More and more, he saw himself as a geologist, and he began to theorize about the formation of islands and continents and the causes of extinction. The *Beagle*'s aristocratic captain, Robert FitzRoy (1805–65), disagreed. He held to the literal interpretation of Genesis, and his faith became a foil for Darwin's developing science. Equally, it was a reminder of Tory-Anglican prejudice. FitzRoy's defense of slave-owning colonial Catholics outraged Darwin's Whig abolitionist morals, although in 1836 they jointly published an article vindicating the moral influence of missionaries in Tahiti.

Nothing on the voyage prepared Charles for the political sea change at home. The ferment was palpable in March 1837, when he took lodgings in London near Erasmus to seek expert help with his *Beagle* collections. Successive Whig governments had tackled corruption, extended the franchise, and opened public offices to non-Anglicans. Angry radicals and nonconformists, unappeased, demanded further concessions, including the disestablishment of the Church of England. A national movement was already under way, leading to a general strike in August 1842. For Britain, these were the century's most turbulent years; for Darwin, they were the most formative.

He entered scientific society, his fame as the *Beagle*'s naturalist preceding him. Here materialism and evolution were debated as in Edinburgh, though, again, he had little to prepare him—only his copy of Lyell's *Principles,* with its refutation of Lamarck. Evolution had been taken up by radical naturalists and medical men, not just as a true theory of life but as a political weapon for attacking miracle-mongering creationists—Oxbridge professors and Tory placement—who kept scientific institutions in a stranglehold. To the radicals, evolution meant material atoms moving themselves to ever higher states of organization, just as social atoms—humans—could. It was nature's legitimation of democracy in science and society alike.

Darwin himself was rising fast. Within months he had a huge government grant to publish his *Beagle* research. Lyell became his patron at the elite Geological Society and saw him on to the governing council. Here Darwin read papers before the Oxbridge dons, and one of them, the Rev. William Whewell (1794–1866), the president, asked him to become a secretary. All in all, the young man was a paragon of public respectability. But, in private, the voyage, the political ferment, and specialist reports on his collections had shattered his orthodoxy; he became a closet evolutionist. In a series of pocket notebooks started in 1837, Darwin began working out a theory that would transform the study of life. His aim was to explain the origin of all plants and animal species, including the human mind and body, by divinely ordained natural laws. Such a theory was dangerous—"oh you Materialist!" he jotted half in jest (C. Darwin 1987, 291)—and it was sure to be damned as atheistic by those he least wished to offend. So secrecy was vital.

About this time, Darwin became unwell, with headaches and stomach troubles. Insomnia and nightmares plagued him, and once he even dreamed of public execution. He felt like a prisoner in London, tied down by his *Beagle* work, theorizing about evolution, and dreading the consequences. In his notebooks, he devised protective strategies lest he should ever publish. He would pitch his theory to Anglican creationists by emphasizing its superior theology. A world populated by natural law was "far grander" than one in which the Creator interferes with himself, "warring against those very laws he established in all organic nature." Just think—Almighty God personally lavishing on Earth the "long succession of vile Molluscous animals!" "How beneath the dignity of him, who is supposed to have said let there be light & there was light" (C. Darwin, 1987, 343).

In mid-1842, Darwin took up the theme again in a pencil sketch of his theory, which he now called "natural selection." It seemed so obvious: Nature alone "selects" the best-adapted organisms, those celebrated in Paley's *Natural Theology* (1802) as proofs of a designing Providence. They survive the constant struggle for food described in the Rev. Thomas Malthus's (1766–1834) *Essay on the Principle of Population* (1798), passing on their adaptive advantage to offspring. In this way, Darwin believed, the laws governing "death, famine, rapine, and the concealed war of nature" bring about "the highest good, which we can conceive, the creation of the higher animals." Good from evil, progress from pain—this was a boost for God. "The existence of such laws should exalt our notion of the power of the omniscient Creator" (Darwin 1909, 51–52).

Darwin might have sounded like a parson, but the church was now the last thing on his mind. He knew that his theory undermined the "whole fabric" of Anglican orthodoxy. Let one species alter, he noted tartly, and the whole creationist edifice "totters & falls" (C. Darwin 1987, 263). With such ideas, he was plainly unfit to seek ordination, quite apart from his devotion to geology and his bad health. In 1838, Charles's father had opened the family purse to endow him as a gentleman naturalist. Months later, Charles married his first cousin Emma Wedgwood and began making plans to escape from London. In September 1842, they moved out fifteen miles to the Kentish village of Downe, where Charles fulfilled his old ambition to be a parish naturalist. His new home was the former parsonage, Down House. Here his clerical camouflage was complete.

Emma became his full-time nurse and the mother of ten. She was a sincere Christian, like all Wedgwoods of her generation: Unitarian by conviction, Anglican in practice. Charles differed from her painfully. Ever since their engagement, when he revealed his evolution heresy to her, she had feared that in death they would be separated, and he would suffer eternal torments. Emma's anxiety remained a sad undercurrent throughout the marriage, her heartache and prayers increasing with his illness.

Darwin's own feelings sometimes showed, as on the rare occasions he mooted his theory to friends. It was criminal, "like confessing a murder," he confided to a colleague, Joseph Hooker. In 1845, when Sedgwick damned the anonymous evolutionary potboiler *Vestiges of the Natural History Creation* (1844) for being subversive and unscientific, Darwin read his old professor's review with "fear & trembling." He had just finished a draft of his own theory and given Emma instructions for publishing it "in case of my sudden death" (Burkhardt et al. 1985–97, 3:2, 43, 258).

Events came to a head when he had a serious breakdown after his father's death in 1848. For the first time, he felt sure that he himself was about to die. Four months at a spa worked wonders, but he returned home only to see his eldest daughter taken ill. When Annie died tragically in April 1851, at age ten, he found no comfort in Emma's faith. After years of backsliding, he finally broke with Christianity. His father's death had spiked the faith; Annie's clinched the point. Eternal punishment was immoral. He would speak out and be damned.

Down House was now his pulpit; evolution, the new "gospel." He pressed on through sickness and sorrow, polishing his theory, extending it, finding illustrations everywhere. Finally, in 1856, he was ready to write it up. His confidants—Lyell now, as well as Hooker and Thomas Henry Huxley (1825–95)—egged him on. Huxley, angry and anticlerical, baited him with juicy tidbits, like the indecency of jellyfish cross-fertilizing through the mouth. Darwin, about to start the *Origin of Species* (1859), shared the lewd jest with Hooker: Good grief, he spouted, "What a book a Devil's Chaplain might write on the clumsy, wasteful, blundering low & horridly cruel works of nature!" (Burkhardt et al. 1985–97, 4:140, 6:178).

But he was the apostate now, touting not treachery but a "grander" theology than Anglican creationism. His book would be a hymn to the Creator's immutable laws by which the "higher animals" had evolved. The *Origin of Species* did not once use the word *evolution*, but *creation* and its cognate terms appeared more than one hundred times. At the front, opposite the title, stood a quotation from Francis Bacon (1561–1626) on studying God's works as well as his Word and another from Whewell on "general laws" as God's way of working. On the last page, Darwin rhapsodized about the "grandeur" of viewing nature's "most beautiful and most wonderful" diversity as the product of the "several powers . . . originally breathed into a few forms or into one." From start to finish, the *Origin* was a pious work, "one long argument" against miraculous creationism but equally a reformer's case for creation by law (Darwin 1959, 719, 759).

There was doublethink in it and a certain subterfuge. The book was the man, after all—ambiguous, even contradictory. In the end, the *Origin* held multiple meanings; it could become all things to everyone. Radicals like brother Erasmus loved it, the theology notwithstanding. Anglican diehards loathed it, and some, like Sedgwick, muttered about Darwin's eternal destiny. Emma now worried more about her husband's present suffering, his anxiety and illness, as the *Origin* went into the world. But she still prayed that these pains would make him "look forward . . . to a future state" in which their love would go on forever (Burkhardt et al. 1985–97, 9:155).

Not all Anglicans damned Darwin. The "celebrated author and divine" quoted in later editions of the *Origin* was the Rev. Charles Kingsley (1819–75), novelist, amateur naturalist, and professor of history at Cambridge. His plug for Darwin's theology—it seemed "just as noble" as miraculous creationism (Darwin 1959, 748)—was timely but timid, a mere "yea" to the hearty "amen" from the Oxford geometry professor, the Rev. Baden Powell (1796–1860). Writing in the Broad Church manifesto *Essays and Reviews* (1860), he declared that the *Origin* "must soon bring about an entire revolution of opinion in favour of the grand principle of the self-evolving powers of nature" (Powell 1860, 139). For such remarks, Powell and his fellow authors were hounded for heresy and two of them eventually prosecuted. In 1861, when a private petition was got up in their defense, Darwin rallied to the cause, adding his signature. He welcomed the essayists' efforts to "establish religious teaching on a firmer and broader foundation" (Burkhardt et al. 1985–97, 9:419).

Worse heretics embarrassed the church from without, and, during the 1860s and 1870s, Darwin was repeatedly asked to back them. But although the *Origin* became all things to everyone, he found this impossible. He steered clear of public support for religious heretics—in Great Britain. Only in the United States did he allow freethinkers to use his name. They called themselves the Free Religious Association, and their creed, printed as *Truths for the Times,* augured "the extinction of faith in the Christian Confession" and the development of a humanistic "Free Religion" (Abbot 1872, 7). Darwin wrote that he agreed with "almost every word" and allowed his remark to be published (Desmond and Moore 1991, 591).

Meanwhile, at Downe, his dual life went on. For years he had worked closely with the incumbent, the Rev. John Brodie Innes. Together they started a benefit society for the local laborers, with Darwin as guardian, and Innes made his friend treasurer of the parish charities and the village school. But in 1871, a boorish new vicar took over and soon fell out with the Darwins. Charles cut his ties with the charities; Emma left the church for one a few miles away. The "great folks" in Down House continued to be parish paternalists, tending the social fabric. With Emma's help, Charles started a temperance reading room in an old hall, where, for a penny a week, working men could smoke, play games, and read "respectable" literature without resorting to the pub.

In 1871, his long-awaited *Descent of Man* came out bearing the imprimatur of his daughter Henrietta. Parts, he had feared, would read like an infidel sermon—"Who w[oul]d ever have thought I sh[oul]d turn parson"! (Burkhardt et al. 1994, 7:124)—and he asked her to tone them down. Emma, too, had jogged the family censor, reminding her that, however "interesting" the book's treatment of morals and religion might be, she would still "dislike it very much as again putting God further off" (Litchfield 1915, 2:196). Henrietta dutifully preened the proofs, and the *Descent* caused few commotions. For her good work, she was given a free hand in Charles's biographical sketch of his grandfather. These proofs she pruned. *Erasmus Darwin* appeared in 1879 shorn of everything religiously risqué.

No one curbed Darwin's candor in his own biography, written between 1876 and 1881. But, then, it was intended for the family, not publication. Here he gave his fullest statement ever on religion (Darwin 1958, 85–96). At first he had been unwilling to abandon Christianity and had even tried to "invent evidence" to confirm the Gospels, which had prolonged his indecision. But just as his clerical career had died a slow "natural death," so his faith had withered gradually. There had been no turning back once the deathblow fell. His dithering had crystallized into a moral conviction so strict that he could not see how anyone—even Emma—"ought to wish Christianity to be true." If it were, "the plain language" of the New Testament "seems to show that the men who do not believe, and this would include my Father, Brother and almost all my best friends, will be everlastingly punished. And this is a damnable doctrine."

These hard, heartfelt words recalled the bitter months and years after his father's death. Since then, Darwin's residual theism had weakened, worn down by controversy. Now, as one with "no assured and ever present belief in the existence of a personal God or of a future existence with retribution and reward," he confessed, "I . . . must be content to remain an Agnostic." An unbeliever, yes, but still an upright man, living without the threat of divine wrath. "I feel no remorse from having committed any great sin," he assured Emma and the children. "I believe that I have acted rightly in steadily following and devoting my life to science."

Charles entrusted the autobiography to Francis, the son who shared his biological interests. William, the eldest, was asked to tackle a more sensitive matter. He had married a relative of one of the Free Religious Association's founders, so was well placed to ask in 1880 for his father's endorsement of *Truths for the Times* to be stopped. He did not explain why this was necessary, but the Americans complied.

William's intervention, like Henrietta's editing, served to conceal Charles's identity and restore it to the family. As his anxious life drew to a close, he was his own man again, safe at Downe, guarded by loved ones. They knew him in different ways, for he had shown them his separate sides. To the daughters, he was the respectable evolutionist, careful not to offend; to his sons, he was the radical unbeliever whose worst heresies were tucked away in the autobiography (as they once had been in pocket notebooks). Only Emma knew him as he knew his own divided self, and he was desperate that she should survive him. With her guidance, the world would know only the "Darwin" the family chose to reveal.

Not that no one pried. Within weeks of his brother Erasmus's death in August 1881, Darwin was, it seems, visited by the dowager Lady Hope, an evangelical temperance worker who read the Bible from door to door among the poor, the sick, and the elderly. She later claimed to have found Darwin himself reading the Bible, and this story, first published in 1915, became the basis of a deathbed-conversion legend.

About the same time, Edward Aveling (1851–98), a young medical doctor and militant secularist, came to lunch at Down House. It was he (not Karl Marx, as was long believed) to whom Darwin had written, refusing permission for an atheist primer, *The Student's Darwin* (1881), to be dedicated to him. Books like Aveling's and current secularist agitation had, in fact, probably made Darwin cautious about his exposure in the United States, and after lunch he remained coy. Aveling tried to extract an atheistic confession. Darwin insisted on calling himself an agnostic. Only one subject could they agree on, Christianity. Darwin admitted that it was not "supported by evidence" but pointed out that he had reached this conclusion very slowly. "I never gave up Christianity until I was forty years of age" (Aveling 1883, 4–5). It had taken his father's and Annie's death to make him shake off the last shreds. And even then he had refused to speak out or to assail people's faith. He never was a comrade at arms.

In this period, Darwin thought much on the eternal questions—chance and design, providence and pain—and struggled with despondency, feeling worn out. He saw his last book, on earthworms, published and resigned himself to joining them. On April 19, 1882, he succumbed to a massive heart attack. Emma and the daughters were present to hear him whisper, "I am not in the least afraid to die" (F. Darwin 1887, 3:358). The family had planned for a funeral at Downe, but it was not to be. In London, Darwin's scientific friends lobbied for a public funeral in Westminster Abbey. Churchmen joined in, heralding the event as a visible sign of "the reconciliation between Faith and Science" (Moore 1982, 103). On April 26, at high noon, Darwin's body was borne up the nave at Westminster as white-robed choristers sang, "I am the resurrection." Behind them in the procession came the Darwin children, followed by the elders of science, state, and church. After the service, the coffin was carried to the north end of the choir screen, where the floor was draped with black cloth that dropped into the grave. Anglican priests rubbed shoulders with agnostic scientists; the Tory leaders closed ranks with Liberal lords. The coffin was lowered, and the choristers sang: "His body is buried in peace, but his name liveth evermore." Emma stayed at Downe.

BIBLIOGRAPHY

Abbot, Francis Ellingwood. *Truths for the Times.* Ramsgate, Kent: Scott, [1872].

Aveling, Edward B. *The Religious Views of Charles Darwin.* London: Freethought, 1883.

Brooke, John Hedley. "The Relations between Darwin's Science and His Religion." In *Darwinism and Divinity: Essays on Evolution and Religious Belief,* ed. John R. Durant. Oxford: Blackwell, 1985, 40–75.

Brown, Frank Burch. *The Evolution of Darwin's Religious Views.* Macon, Ga.: Mercer University Press, 1986.

Browne, Janet. "Missionaries and the Human Mind: Charles Darwin and Robert FitzRoy." In *Darwin's Laboratory: Evolutionary Theory and Natural History in the Pacific,* ed. Roy MacLeod and Philip F. Rehbock. Honolulu: University of Hawaii Press, 1994, 263–82.

———. *Charles Darwin.* Vol. 1: *Voyaging.* London: Cape, 1995.

Burkhardt, Frederick et al., eds. *The Correspondence of Charles Darwin.* 12 vols. to date. Cambridge: Cambridge University Press, 1985.

———. *A Calendar of the Correspondence of Charles Darwin, 1821–1882, with Supplement.* Cambridge: Cambridge University Press, 1994.

Colp, Ralph, Jr. "The Myth of the Darwin-Marx Letter." *History of Political Economy* 14 (1982): 461–82.

Darwin, Charles. *The Foundations of the Origin of Species: Two Essays Written in 1842 and 1844.* Ed. Francis Darwin. Cambridge: Cambridge University Press, 1909.

———. *The Autobiography of Charles Darwin, 1809–1882, with Original Omissions Restored.* Ed. Nora Barlow. London: Collins, 1958.

———. *The Origin of Species by Charles Darwin: A Variorum Text.* Ed. Morse Peckham. Philadelphia: University of Pennsylvania Press, 1959.

———. *Charles Darwin's Notebooks, 1836–1844: Geology, Transmutation of Species, Metaphysical Enquiries.* Ed. Paul H. Barrett et al. Cambridge: Cambridge University Press and British Museum (Natural History), 1987.

Darwin, Francis, ed. *The Life and Letters of Charles Darwin, Including an Autobiographical Chapter.* 3 vols. London: John Murray, 1887.

Desmond, Adrian. *Huxley: From Devil's Disciple to Evolution's High Priest.* London: Penguin, 1998.

Desmond, Adrian, and James Moore. *Darwin.* London: Michael Joseph, 1991.

Gillespie, Neal C. *Charles Darwin and the Problem of Creation.* Chicago: University of Chicago Press, 1979.

Keynes, Randal. *Annie's Box: Charles Darwin, His Daughter, and Human Evolution.* London: Fourth Estate, 2001.

Kohn, David. "Darwin's Ambiguity: The Secularization of Biological Meaning." *British Journal for the History of Science* 22 (1989): 215–39.

———. "The Aesthetic Construction of Darwin's Theory." In *The Elusive Synthesis: Aesthetics and Science,* ed. A. I. Tauber. Dordrecht: Kluwer, 1996, 13–48.

Litchfield, Henrietta. *Emma Darwin: A Century of Family Letters, 1792–1896.* 2 vols. London: John Murray, 1915.

Moore, James. "Charles Darwin Lies in Westminster Abbey." *Biological Journal of the Linnean Society* 17 (1982): 97–113.

———. "Darwin of Down: The Evolutionist as Squarson-Naturalist." In *The Darwinian Heritage,* ed. David Kohn. Princeton: Princeton University Press, 1985, 435–81.

———. "Freethought, Secularism, Agnosticism: The Case of Charles Darwin." In *Religion in Victorian Britain.* Vol. 1: *Traditions,* ed. Gerald Parsons. Manchester, England: Manchester University Press, 1988, 274–319.

———. "Of Love and Death: Why Darwin 'gave Up Christianity.'" In *History, Humanity, and Evolution: Essays for John C. Greene,* ed. James Moore. New York: Cambridge University Press, 1989, 195–229.

———. *The Darwin Legend.* Grand Rapids, Mich.: Baker, 1994.

Ospovat, Dov. *The Development of Darwin's Theory: Natural History, Natural Theology, and Natural Selection, 1838–1859.* Cambridge: Cambridge University Press, 1981.

Powell, Baden. "On the Study of the Evidences of Christianity." In *Essays and Reviews.* Oxford: Parker, 1860, 94–144.

Secord, James. *Victorian Sensation: The Extraordinary Publication, Reception, and Secret Authorship of "Vestiges of the Natural History of Creation."* Chicago: University of Chicago Press, 2000.

17 Evolution

Peter J. Bowler

Evolutionary theory has always been seen as a major area in which science and religion interact because of its influence on our ideas about the origin of the world and of the human mind and because it affects our beliefs about God's interaction with his creation. The opposition of American creationists encourages a polarized image of the interaction reminiscent of that proposed in the Victorian era: Evolution is seen as a battleground in the war between science and religion. But just as many modern religious thinkers look for a common ground in which evolution can be seen as God's method of creation, so modern historians have shown the extent to which similar efforts at reconciliation have been prominent since the time of Darwin.

Peter J. Bowler obtained his Ph.D. from the University of Toronto and is professor of the history of science in the Queen's University of Belfast, Northern Ireland. He is the author of numerous books on the history of evolutionism, including *The Eclipse of Darwinism: Anti-Darwinian Evolution Theories in the Decades around 1900* (Baltimore: Johns Hopkins University Press, 1983) and *The Non-Darwinian Revolution: Reinterpreting a Historical Myth* (Baltimore: Johns Hopkins University Press, 1988). His latest book is a survey of the interaction of science and religion in early-twentieth-century Britain.

OF ALL OF THE TOPICS that have fueled the antagonism between science and religion, evolutionism remains perhaps the only one with power to stimulate debate even today. Following on from the impact of geology and paleontology in the early nineteenth century, evolutionary theories challenged the story of human origins recounted in sacred texts. By rendering humankind a product of nature, evolutionism broke down the barrier between human spirituality and the mentality of the "brutes that perish." Equally seriously, some of the more materialistic theories of evolution undermined the traditional belief that nature itself is a divine construct. In the Darwinian theory of natural selection, struggle and suffering are the driving forces of natural development and, hence, the root cause of our own origin. Not surprisingly, there are many who still think that the human species must be the product of a more purposeful mode of development and some who wish to retain the traditional view that we are divinely created.

Despite the ongoing sources of conflict, historians have shown that the conventional image of nineteenth-century Darwinism sweeping aside religious belief is an oversimplification. There were, indeed, great controversies, and Darwinism was supported by liberal intellectuals, who had good reason to be suspicious of the ways in which the image of a divinely created universe had been used to sustain a conservative model of the social order. Conservative thinkers quite correctly pointed out the materialistic implications of Charles Darwin's (1809–82) theory. It was Darwin's supporter Thomas Henry Huxley (1825–95) who coined the term *agnosticism* to denote the critical state of unbelief generated by a scientific approach to nature. Historians have shown, however, that the materialistic aspects of Darwin's theory were suppressed by many of the first-generation evolutionists; even Huxley did not accept the theory of natural selection. In the so-called Darwinian revolution, evolutionism was popularized only by linking it to the claim that nature is progressing steadily (if a little irregularly) toward higher mental and spiritual states and by making the human species both the goal and the cutting edge of that progressive drive. A sense of purpose—and, for many, it was still a divine purpose—was built into the operations of nature itself. The more materialistic implications of Darwin's thinking became widely accepted only in the twentieth century, when biologists at last became convinced that natural selection was the driving force of evolution. As scientists began to insist that we must learn to live with the idea that we are the products of a purposeless and, hence, morally neutral natural world, so the modern creationist backlash began.

Early Evolutionism

In the seventeenth century, naturalists believed that the world was created by God only a few thousand years ago. Books such as *The Wisdom of God Manifested in the Works of Creation* by John Ray (1627–1705) argued that each species was perfectly adapted to its environment because it had been created by a wise and benevolent God. This view was repeated in the *Natural Theology* (1802) of William Paley (1743–1805) and was still popular in conservative circles in the early nineteenth century. In the eighteenth century, however, the worldview of what would now be called simple creationism was challenged. In part, this was a product of the discoveries made by geologists and paleontologists. The world was clearly much older than a literal interpretation of the Genesis story would suggest. There was increasing evidence from the fossil record that some species had become extinct in the course of geological time and had been replaced by others. Following the work of Georges Cuvier (1769–1832), these conclusions became inescapable.

Even before this, however, materialist thinkers such as Georges Leclerc, Comte de Buffon (1707–88), and Denis Diderot (1713–84) had begun to suggest that life could be created on the earth by natural processes (spontaneous generation) and that the species thus produced might change in response to natural forces. By the end of the eighteenth century, Erasmus Darwin (1731–1802) and Jean Baptiste Lamarck (1744–1829) were beginning to suggest comprehensive theories of transmutation in which life had advanced slowly from primitive origins to its present level of development. The adaptation of species to their environments was explained by supposing that individual animals modified their behavior in response to environmental change, and any resulting changes in their bodily structure were inherited (the inheritance of acquired characteristics, now often called Lamarckism). Paley's defense of the claim that adaptation indicated design by a benevolent God was a reaction to these new ideas.

Historians used to assume that Lamarck's views were dismissed by most of his contemporaries. Cuvier may have demonstrated that new species must appear from time to time, but he and his followers did not believe that natural evolution was the source of new species. In Britain especially, conservative geologists invoked the image of a series of divine creations spread through geological time, thereby accounting for the evident discontinuity of the fossil record. Recent work by Adrian Desmond (1989) has shown that radical anatomists, especially in the field of medicine, were using materialistic theories such as Lamarckian transformism to attack the image of a static, designed universe that sustained the traditional social structure. Evolutionism became firmly linked to materialism, atheism, and radical politics. In Britain, the anatomist Richard Owen (1804–92) gained his reputation by holding back the demands of the radicals. Owen modernized the view that all species are divinely created by stressing the underlying unity of structure among all of the members of each animal group: The Creator had instituted a rational plan for his universe that could be deciphered by the comparative anatomist. Unity of design, rather than a list of particular adaptations, offered the best illustration of the Creator's handiwork.

In 1844, an effort to make evolutionism acceptable to a middle-class audience was made in an anonymously published book, *Vestiges of the Natural History of Creation,* actually written by Robert Chambers (1802–71). The book proclaimed a message of progress through nature and human history but attempted to circumvent the charge that transmutationism was atheistic by arguing that progress represented the unfolding of a divine plan programmed into nature from the beginning. *Vestiges* clearly explored the implications of the view that the human mind was a product of nature by linking transmutationism to phrenology (the belief that the brain is the organ of the mind). For Chambers, the human mental and moral faculties were generated by the enlarged brain produced by progressive evolution in the animal kingdom. On this count, his book was roundly condemned as materialistic by conservative scientists such as Hugh Miller (1802–56).

By the 1850s, however, the possibility that the divine plan might unfold through the operations of natural law, rather than by a sequence of miracles, was being taken increasingly seriously even by conservative naturalists such as Owen. As science grew more powerful, it became necessary to bring the operation of creation under the control of law, provided the law was seen as having been instituted for a purpose by the Creator of the universe. The mathematician and philosopher Baden Powell (1796–1860) argued that design was seen more obviously in the operation of law, rather than in capricious miracles, and noted that one implication of this view was that the introduction of new species would have to be seen as a lawlike process. The threat to the status of the human mind remained, however, a potent check to full exploration of this idea.

Darwinism

In 1859, the situation was changed dramatically by the publication of *Origin of Species* by Charles Darwin. Darwin proposed new lines of evidence to show how evolutionism could explain natural relationships, but he also suggested a new and potentially more materialistic mechanism of evolution. He had developed his theory of natural selection in the late 1830s, following his voyage of discovery around the world aboard the survey vessel H.M.S. *Beagle.* In the Galapagos Islands, Darwin found evidence that new species were produced when populations became separated in isolated locations and subject to new conditions: In these circumstances, several different "daughter" species could be produced from the parent form. Darwin then went on to develop the theory of natural selection to explain how the separated populations might change to adapt to their new environment. In this theory, it was assumed that the individuals making up a population differ among themselves in various ways; the differences have no apparent purpose and are, in that sense, "random." Following the principle of population expansion suggested by Thomas Malthus (1766–1834), Darwin deduced that there must be a "struggle for existence," in which any slight advantage would be crucial. Those individuals with variant characters that conferred such an advantage would survive and reproduce, passing the character on to their offspring. Those with harmful characters would be eliminated. This process of natural selection would, thus, gradually adapt the species to any changes in its environment. The philosopher Herbert Spencer (1820–1903) called it the "survival of the fittest." Darwin almost certainly began as a progressionist but gradually lost his faith in the idea that evolution moved toward a morally significant goal. As understood by modern biologists, his theory implied a branching model of relationships, in which there could be no single goal toward which life has tended to evolve and no inevitable trend toward higher levels of organization.

Darwin had delayed publication of his theory, partly to wait for a change in the climate of opinion. An intense controversy followed the publication of *Origin of Species* (Ellegard 1958). Radical scientists such as Huxley proclaimed that Darwin's new insights at last opened up the subject to rational investigation. Conservative opponents labeled the theory as the most extreme manifestation of the atheistical tendency inherent in the basic idea of evolution. Not only were humans reduced to the status of animals, but also the natural world that produced us was reduced to a purposeless sequence of accidental changes. It is clear that the evolutionists carried the day. By the 1870s, the vast majority of scientists and educated people had accepted the basic idea of evolution. The question that historians now ask is: In what form did they accept the theory? Was it the radical materialism of the theory of natural selection, or was it a less threatening version of evolutionism, a compromise in which some form of purpose was retained by assuming that natural developments tended to progress toward higher states?

The most comprehensive accounts of the religious debate (Livingstone 1987, Moore 1979) suggest that, in the long run, there was as much compromise as confrontation. This does not mean that passions were not aroused: The issues were important, and the conflict between conservatives and radicals was intense. Huxley and Spencer hated the way in which the idea of design was used to block aspirations for social change and wanted to see humankind firmly embedded in a universe subject to change under natural law. But, in the end, both sides came to accept evolution, and neither wanted a worldview based on nothing but chance and suffering. Faced with the even greater threat of natural selection, conservatives took up the (once radical) argument that evolution represented the unfolding of a divine plan. They concentrated their efforts not on blocking the case for evolution but on showing that the process could not be driven by a purely haphazard mechanism such as natural selection. Radicals wanted a changing universe based on natural law but assumed that the changes would, in the end, be beneficial and moral. They were more willing to let individual effort determine success in this world but were comforted by the fact that success depended not on brute force, but on the old Protestant virtues of thrift and industry. As James Moore (1985b) has noted, this allowed liberal Protestants to accept Spencer's philosophy of cosmic evolutionism, in which the old human values were now built into nature itself as the driving force of progress. Since the agnostics also argued for a purposeful universe, those religious thinkers who wanted to keep up with the times could accept that the new cosmology was not antithetical to their beliefs. The old image of Spencer and Darwin destroying all moral values and sweeping Western culture immediately into an age in which the only measure of worth was brute force is, thus, a myth. It took many decades for the full implications of Darwin's thinking to become apparent, and much "social Darwinism" has its sources in other models of nature.

Human Origins

The most controversial issue at first was the evolutionary origin of the human race. Huxley was engaged in an intense debate with Owen over the degree of relationship between humans and apes. He is popularly supposed to have demolished Samuel Wilberforce (1805–73), bishop of Oxford, at the 1860 meeting of the British Association for the Advancement of Science by declaring that he would rather be descended from an ape than from a man who misused his position to attack a theory he did not understand. It has now been suggested that the popular story of this confrontation is also a myth: Huxley certainly did not convert his audience to Darwinism overnight. But the animal ancestry of man was increasingly taken for granted over the next decade, and everyone had to grapple with the implications of it.

Darwin had been aware from the start of his theorizing that evolutionism would affect our ideas about human nature in a way that would undermine the traditional concept of the soul. His mature views on this issue were eventually presented in his *Descent of Man* (1871). He believed that many aspects of human behavior are controlled by instincts that have been shaped by natural selection. Our moral values are merely rationalizations of social instincts built into us because our ape ancestors lived in groups. Spencer had already proposed an evolutionary psychology, and evolutionists such as George John Romanes (1848–94) built upon Darwin's work to propose evolutionary sequences by which the various mental faculties had been added in the progress toward mankind.

A few evolutionists, including the co-discoverer of natural selection, Alfred Russel Wallace (1823–1913), were so concerned that they refused to endorse such views, holding that some supernatural intervention was still required to explain the appearance of the human mind. The Roman Catholic anatomist St. George Jackson Mivart (1827–1900) argued that, while the evolution of the human body might be explained naturally, the soul must be a divine creation. Mivart defined what would eventually become the Roman Catholic position on this issue, although it would take some time for this compromise to become widely accepted by the church hierarchy.

Most Darwinists believed that an ad hoc discontinuity marking the advent of the human spirit violated the logic of the evolutionary program, and the image of a distinct human spiritual character was abandoned. The situation was made bearable by assuming that traditional moral values were not at variance with nature but were built into nature in a way that ensured their emergence in the human mind. For religious thinkers such as Henry Drummond (1851–97), the highest moral value, altruism, was the foundation of the evolutionary process. Drummond's *Ascent of Man* (1894) presented cooperation, not competition, as the driving force of progressive evolution and implied that the human race, with its expanded sense of altruism, was the inevitable culmination of the development of life. Another way of minimizing the emotional shock of the idea of human evolution was adopted by paleontologists such as Henry Fairfield Osborn (1857–1935). He attempted to block early creationist attacks (discussed below) by suggesting that we had evolved not from the disgusting apes but from some more remote (and, hence, less immediately threatening) animal ancestor.

The potentially disruptive implications of the integration of humankind into nature became apparent only as early-twentieth-century thinkers began to explore the possibility that the world might not, after all, be evolving toward ever-higher states. Sigmund Freud (1856–1939) built on the idea of evolution to argue that our subconscious thoughts are shaped by instincts from our animal past. The loss of faith in progress precipitated by World War I also helped create a framework in which the more pessimistic implications of Darwinism might be explored.

Design in Nature

If the traditional gulf between humankind and the animals was bridged, it was made possible for most thinkers by rejecting the Darwinian theory of natural selection in favor of a more purposeful or morally acceptable process. From the start, there were many objections to the selection theory by conservatives who wanted to believe that nature must still exhibit evidence of design by God, even if individual species were produced by natural law. But radicals also found natural selection hard to accept: Huxley was never happy with the theory, and Spencer was an avowed Lamarckian. Non-Darwinian evolutionary mechanisms allowed everyone to believe that there was something more to natural development than mere trial and error. The Lamarckian theory seemed to imply a more purposeful evolutionary process because it allowed individual self-improvement to be inherited (the main point in Spencer's social philosophy) and implied that purposeful changes in animals' behavior was the directing agent of evolution.

It was the opponents of natural selection who correctly identified its materialistic implications. They saw that, in a universe governed solely by random variation and the survival of the fittest, the existing state of nature must be the outcome of trial and error, not of purposeful intention. In a letter to Charles Lyell (1797–1875), Darwin reported that Sir John Herschel (1792–1871) had called natural selection "the law of higgledy-piggledy" (quoted in Burkhardt and Smyth 1991, 423). Herschel certainly expressed his preference for the view that the history of life must be under the control of divinely planned laws of development. The biologist William Benjamin Carpenter (1813–85), while accepting evolutionism, argued that the exquisite structures of the single-celled

Foraminifera could only be the product of design. The duke of Argyll (George Douglas Campbell [1823–1900]) claimed that the beauty of many birds was intended by their Creator for us to appreciate, and he saw rudimentary organs as structures being prepared for future use.

The most effective collection of antiselectionist arguments was assembled by St. George Jackson Mivart in his *Genesis of Species* (1870). His strategy was to demonstrate that evolution was under divine control. He argued that a wide range of characters cannot be explained by mere utility to the individual; they are the products of trends built into evolution by the God who established the laws of development. As evidence for the existence of such trends, he pointed to many cases of parallel evolution, in which several branches of the animal kingdom seem to have moved independently in the same direction. Mivart's arguments were linked to his claim that the origin of the human spirit could not be explained by natural evolution: both formed part of his strategy for reconciling the Roman Catholic Church to aspects of the new biology. He argued that the writings of the church fathers did not rule out the natural evolution of the body. Although his efforts were at first welcomed by the church, by the end of the century he encountered increasing hostility and was excommunicated.

Darwin rejected the claim that there were aspects of the evolutionary process that were not susceptible to a natural explanation. For him, bright colors were developed by sexual selection because they conferred an advantage in the struggle to obtain a mate, and rudimentary organs were merely relics of what had once been useful in the past. The disparity between his theory and what has become known as theistic evolutionism (evolutionism under the control of a divine plan) became evident in a controversy with the American botanist Asa Gray (1810–88). Gray defended Darwin vigorously against those who rejected evolution, but, in a series of papers collected in his *Darwiniana* (1876), his views on design forced him to express doubts about natural selection. Having tried to defend the position that any form of lawlike production of species was compatible with belief in a Creator who established the laws, Gray was forced to admit that selection based on random variation seemed to eliminate any real sense of design in nature. He suggested that the variation within each population was somehow led along beneficial lines, thus removing the need for the elimination of unfit variants. Darwin protested that all of the evidence from plant and animal breeders proved that variation was purposeless. If someone builds a wall by picking out useful pieces of stone fallen from a cliff, the design is in the selection of the stones: No one would suggest that nature was set up in such a way that stones split from the rock with useful shapes (Darwin 1882, 2:427–28).

For many evolutionists wishing to retain the belief that nature is somehow the expression of the divine will, Lamarckism seemed to solve the problem highlighted by Gray. Animals acquire useful characters by learning new habits that encourage them to use their bodies in different ways. The fact that the animals can make a deliberate choice of a new behavior pattern in a new environment seems to imply a kind of creative input by the organisms themselves. American neo-Lamarckians such as Edward Drinker Cope (1840–97) used this aspect of the inheritance of acquired characters to argue that the Creator had delegated his creative power to life itself. This position also accepts a continuity between animal and human mental faculties: Even the most primitive animals have rudimentary mental powers, which enable them to make conscious choices when faced with an environmental challenge.

In Britain, similar points were made by the novelist Samuel Butler (1835–1902) in a series of antiselectionist books beginning with his *Evolution Old and New* (1879). Butler recognized the force of Mivart's arguments but thought that the answer was to invoke not design built into the laws of nature but a nature that was itself creative. Natural selection was a "nightmare of waste and death," but Lamarckism made life self-creative in a way that fit a more general belief in the purposeful character of nature. Butler's arguments were taken up by many literary figures who had moral objections to natural selection, including the playwright George Bernard Shaw (1856–1950) and the author Arthur Koestler (1905–83). Butler himself alienated the scientific community by his personal attacks on Darwin, but, by the end of the nineteenth century, his views were becoming more acceptable. Many early-twentieth-century scientists participated in a widespread moral reaction against Darwinism based on the belief that evolution must be a purposeful process designed to enhance mental and moral progress.

The Roman Catholic Church began to adopt a different position, accepting that evolution might be an indirect mechanism of creation for the body but refusing to extend the argument to include the creation of the soul. By the 1920s, a significant body of opinion began to build up among Roman Catholic scholars in favor of the position that Mivart had defined in response to Darwinism. Works such as Ernest Messenger's *Evolution and Theology* (1931) argued that there was nothing in the writings of the church fathers that prevented acceptance of evolution, provided the process was seen as a manifestation of divine creativity. This movement paved the way for the modern Roman Catholic view of evolution, in which natural processes (assumed to be of divine origin) have formed the human body, while the soul has been introduced by direct divine intervention.

Modern Darwinism

Although many nonscientists felt strongly that natural selection was unacceptable for moral and philosophical reasons, the biologists themselves gradually began to believe that Darwinism might, after all, be the most promising theory. To a large extent, this was a product of the emergence of modern genetics, which undermined the credibility of Lamarckism by showing that acquired characters cannot be reflected in the organism's genes. Genetic mutation also supplied a plausible source of the random variation that Darwin had noticed in populations. By the 1930s, the "modern synthesis" of genetics and Darwinism was being constructed, a theory that has remained the dominant force in scientific evolutionism. Some modern Darwinians, including Julian Huxley (1887–1975), have tried to defend the view that evolution is progressive in a way that reflects human values. Julian Huxley even endorsed the theistic evolutionism of Pierre Teilhard de Chardin (1881–1955), according to which the development of life is tending toward an "omega point" of spiritual unification (Teilhard de Chardin 1959). But other founders of the modern synthesis, especially George Gaylord Simpson (1902–84), argued that Darwinism is essentially materialistic: There is no purpose in nature and no goal toward which evolution is striving (Simpson 1949). The human race simply has to grow up and realize that the values it cherishes are not respected by nature. Such a position had, in fact, been anticipated by Thomas Henry Huxley at the end of his career. His 1893 lecture on "Evolution and Ethics" had denied progress in nature and insisted that moral values are not products of the evolutionary process. Indeed, he had proclaimed, nature may be actively hostile to our moral feelings. In such a view, we are, indeed, the products of a cosmic accident.

Not surprisingly, these developments in science have been resisted both by religious thinkers and by those who want to see human moral values as having some natural foundation. Two very different stands of protest can be identified. The best known is that leading to what is now known as creationism, in effect the return to a preevolutionary worldview in which species (especially the human species) have been directly created by God. In its most extreme manifestation, creationism has led to a complete repudiation of the geological time scale and a renewed acceptance of a literal interpretation of the Genesis account of Creation. Less well known is a current of anti-Darwinian thought emanating from both religious and philosophical critics of Darwinism who unite around the claim that the development of life cannot have been brought about by a process as purposeless as natural selection. This movement generates continued support for non-Darwinian evolutionary mechanisms in defiance of the geneticists.

The creationist reaction has received more publicity in recent years. It began in America during the 1920s, when many ordinary people, especially in the rural South, began to see Darwinism as a symbol of the moral corruption that was undermining traditional values. To treat humans as animals, they claimed, was to invite the evils of hedonism and social Darwinism. In 1925, the state of Tennessee passed legislation forbidding the teaching of evolution in the public schools. This was challenged in the famous "monkey trial" of John Thomas Scopes (1900–70), who was convicted after a much-publicized court case in which he was prosecuted by William Jennings Bryan (1860–1925) and defended by Clarence Darrow (1857–1938). Although much of the resulting publicity ridiculed the creationists, evolutionists prudently kept their subject out of biology textbooks for several decades.

In the 1960s, the now-confident Darwinians again tried to reintroduce evolutionism into the American school curriculum, precipitating the most recent outburst of creationism. Fundamentalist Protestantism now had a much wider power base in American society, and efforts had been made to establish a "creation science," in which the earth's geological structure was explained as the result of Noah's Flood. In several states, creationists urged legislation requiring equal time in high school classes for what they presented as a scientific alternative to evolutionism. In a series of much-publicized trials, this legislation was banned by the courts as unconstitutional on the ground that creation science was little more than an attempt to claim the literal truth of the Genesis story.

A far less visible campaign continued throughout the twentieth century by religious thinkers and moralists who accept evolution but argue that a more purposeful process than Darwinism must be involved. In Britain, this campaign was sparked by the inclusion of a Darwinian view of evolution in *The Outline of History* (1920) by H. G. Wells (1866–1946). The Catholic writer Hilaire Belloc (1870–1953) challenged Wells, proclaiming that Darwinism was by now dead even in science. Spirit could play a role in human history because spiritual factors were involved in evolution itself (although Belloc did not accept the evolutionary account of human origins). Unfortunately for Belloc, and for moralists such as George Bernard Shaw, who shared his distaste for selectionism, the biologists were by this time beginning to argue that theories of purposeful evolution were untenable and that selectionism would have to be accepted as the principal mechanism of evolution. Biologists such as J. B. S. Haldane (1892–1964) responded to Belloc and the anti-Darwinians and may even have been prompted to think more carefully about Darwinism by the challenge.

The rise of scientific Darwinism in the last half of the twentieth century has, however, been matched by a continued reluctance on the part of outsiders to admit that the theory can offer a complete explanation of the development of life. The popularity of Teilhard de Chardin's evolutionary mysticism is but one example of an ongoing rejection of the selection theory by those who think that nature must be based on principles that guarantee progress toward a spiritually significant goal and that the human race itself must be the highest product of such a process. For some, however, it is the origin of the human spirit that remains the chief stumbling block. As a Jesuit, Teilhard had been refused permission to publish during his lifetime because his vision included human origins, illustrating a tension within the Roman Catholic Church's position. Modern religious opposition to Darwinism thus runs the whole gamut from a creationism that rejects the orthodox scientific explanation of the geological record through more sophisticated versions in which occasional creations are required to establish the main groups of animals and, of course, the human species. Even more liberal are those who accept a completely evolutionary worldview, so long as the Darwinian mechanism is marginalized in favor of something that allows for progress and purpose in nature. This latter position is maintained by many who would claim that their concerns are motivated by philosophical or moral, rather than purely religious, principles.

BIBLIOGRAPHY

Bowler, Peter J. *The Eclipse of Darwinism: Anti-Darwinian Evolution Theories in the Decades around 1900*. Baltimore: Johns Hopkins University Press, 1983.

———. *Theories of Human Evolution: A Century of Debate, 1844–1944*. Baltimore: Johns Hopkins University Press, 1986.

———. *The Non-Darwinian Revolution: Reinterpreting a Historical Myth.* Baltimore: Johns Hopkins University Press, 1988.

———. *Evolution: The History of an Idea.* 2d ed. Berkeley: University of California Press, 1989.

Burkhardt, Frederick, and Sydney Smyth, eds. *The Correspondence of Charles Darwin.* Vol. 7. Cambridge: Cambridge University Press, 1991.

Chambers, Robert. *Vestiges of the Natural History of Creation.* 1844. Reprint. Chicago: University of Chicago Press, 1994.

Corsi, Pietro. *Science and Religion: Baden Powell and the Anglican Debates, 1800–1860.* Cambridge: Cambridge University Press, 1988.

Darwin, Charles. *The Variation of Animals and Plants under Domestication.* 2 vols. 2d ed. London: Murray, 1882.

Desmond, Adrian. *The Politics of Evolution: Morphology, Medicine, and Reform in Radical London.* Chicago: University of Chicago Press, 1989.

Ellegard, Alvar. *Darwin and the General Reader: The Reception of Darwin's Theory of Evolution in the British Periodical Press, 1859–1872.* 1958. Reprint. Chicago: University of Chicago Press, 1990.

Gillispie, Charles Coulston. *Genesis and Geology.* New York: Harper, 1959.

Greene, John C. *The Death of Adam: Evolution and Its Impact on Western Thought.* Ames: Iowa State University Press, 1959.

———. "The Interaction of Science and World View in Sir Julian Huxley's Evolutionary Biology." *Journal of the History of Biology* 23 (1990): 39–55.

Jensen, J. Vernon. "Return to the Wilberforce-Huxley Debate." *British Journal for the History of Science* 21 (1988): 161–79.

Koestler, Arthur. *The Ghost in the Machine.* New York: Macmillan, 1967.

Larson, Edward J. *Summer for the Gods: The Scopes Trial and America's Continuing Debate over Science and Religion.* New York: Basic Books, 1997.

Livingstone, David. *Darwin's Forgotten Defenders: The Encounter between Evangelical Theology and Evolutionary Thought.* Edinburgh: Scottish Universities Press and Grand Rapids, Mich.: Eerdmans, 1987.

McQuat, Gordon, and Mary P. Winsor. "J. B. S. Haldane's Darwinism in Its Religious Context." *British Journal for the History of Science* 28 (1995): 227–31.

Messenger, Ernest. *Evolution and Theology: The Problem of Man's Origin.* London: Burnes, Oates, and Washbourne, 1931.

Moore, James R. *The Post-Darwinian Controversies: A Study of the Protestant Struggle to Come to Terms with Darwin in Great Britain and America, 1870–1900.* New York: Cambridge University Press, 1979.

———. "Evangelicals and Evolution: Henry Drummond, Herbert Spencer, and the Naturalization of the Spiritual World." *Scottish Journal of Theology* 38 (1985a): 383–417.

———. "Herbert Spencer's Henchmen: The Evolution of Protestant Liberals in Late Nineteenth-Century America." In *Darwinism and Divinity,* ed. John Durant. Oxford: Blackwell, 1985b, 76–100.

———. "Of Love and Death: Why Darwin 'gave Up Christianity.'" In *History, Humanity, and Evolution,* ed. J. R. Moore. Cambridge: Cambridge University Press, 1989, 195–230.

Numbers, Ronald L. *The Creationists: The Evolution of Scientific Creationism.* New York: Knopf, 1992.

———. *Darwinism Comes to America.* Cambridge: Harvard University Press, 1998.

Ospovat, Dov. *The Development of Darwin's Theory: Natural History, Natural Theology, and Natural Selection, 1838–1859.* Cambridge: Cambridge University Press, 1981.

Richards, Robert J. *Darwin and the Emergence of Evolutionary Theories of Mind and Behavior.* Chicago: University of Chicago Press, 1987.

Simpson, George Gaylord. *The Meaning of Evolution.* New Haven: Yale University Press, 1949.

Sulloway, Frank. *Freud: Biologist of the Mind.* London: Burnett, 1979.

Teilhard de Chardin, Pierre. *The Phenomenon of Man.* London: Collins, 1959.

Turner, Frank Miller. *Between Science and Religion: The Reaction to Scientific Naturalism in Late Victorian England.* New Haven: Yale University Press, 1974.

18 Cosmogonies

Ronald L. Numbers

In the mid–seventeenth century René Descartes proposed one of the first naturalistic theories of the origin of the solar system. While the creation of the world in six days was still widely accepted, even those who rejected a completely naturalistic cosmogony (or explanation of the origin of the solar system), such as Isaac Newton, developed quasi-naturalistic theories. English cosmogonists, such as Thomas Burnet and William Whiston, were careful to develop theories that corresponded with the Genesis account of Creation. By contrast, continental naturalists, such as Georges Louis Leclerc, Comte de Buffon, were unconcerned to harmonize their explanations with the biblical narrative and repudiated any appeal to the supernatural in their theories. Not until Pierre Laplace formulated his nebular hypothesis, however, was there a scientifically respectable account of the origin of the solar system as a product of natural law rather than of special creation.

Ronald L. Numbers is Hilldale and William Coleman Professor of the History of Science and Medicine at the University of Wisconsin, Madison. He holds a Ph.D. from the University of California at Berkeley. His numerous publications include *The Creationists: The Evolution of Scientific Creationism* (New York: Knopf, 1992) and *Darwinism Comes to America* (Cambridge: Harvard University Press, 1998). This essay is extracted from his *Creation by Natural Law: Laplace's Nebular Hypothesis in American Thought* (Seattle: University of Washington Press, 1977). Complete documentation, including citations to sources quoted, can be found in the original volume.

LATE IN THE SEVENTEENTH CENTURY, a young English divine named William Whiston (1667–1752) criticized his contemporaries for habitually "stretching [the Six Days Work] beyond the Earth, either to the whole System of things, as the most do, or indeed to the Solar System, with which others are more modestly contented in the case." At the time, it was customary in Western science to accept the Mosaic story of Creation found in the first chapter of the Book of Genesis as a literal cosmogony, or account of how the universe began. For most, this meant that the entire cosmos, or at least the solar system, had been created by God's fiat in six successive twenty-four-hour periods, approximately four thousand years before the birth of Christ. As long as Western natural philosophers were willing to tolerate supernatural explanations within the domain of science, there was little motivation to discard this traditional cosmogony. The desire for a natural history of Creation became acute only after those pursuing science in Europe began to view their task as explaining the workings of nature without recourse to direct supernatural activity.

Descartes, Newton, and Buffon

Modern attempts to explain the origin of the solar system naturalistically date from the mid–seventeenth century. As the new science resolved one after another of nature's mysteries, the temptation to formulate a purely naturalistic cosmogony became increasingly great. One of the first modern Europeans to yield to this temptation was the French philosopher René Descartes (1596–1650). His theory of the origin of the solar system, sketched in the *Principia philosophiae* (*Principles of Philosophy* [1644]) as well as in his suppressed treatise *Le Monde* (*The World* [completed in 1633 but published posthumously]), followed logically from his twin beliefs in the constancy of the laws of nature and the sufficiency of these laws to explain the phenomena of nature. Using vortices as a creative mechanism, he showed how the solar system could have been formed by the God-ordained laws of nature operating on a primitive chaos. Then, undoubtedly prompted by the recent experiences of Galileo Galilei (1564–1642), he cautiously added that he considered this hypothesis to be "absolutely false" and asserted his belief in the orthodox doctrine of the creation of the world in the beginning in a fully developed state.

This thinly veiled attempt to eliminate God as a necessary element in the creation of the world brought Descartes considerable notoriety. In relegating God to the remote and seemingly minor task of establishing the laws of nature, he had overstepped the bounds of seventeenth-century tolerance. His fellow countryman Blaise Pascal (1623–62) could never forgive such blatant impiety. "In all his philosophy," wrote Pascal, Descartes "would have been quite willing to dispense with God. But he had to make Him give a fillip to set the world in motion; beyond this, he has no further need of God."

Descartes fared no better with Isaac Newton (1642–1727), who consistently rejected suggestions that the solar system had been created by the "mere Laws of Nature." In the *General Scholium* at the end of the second edition of his *Principia* (1713), Newton summarized his reasons for believing in the necessity of divine action:

> The six primary planets are revolved about the sun in circles concentric with the sun, and with motions directed towards the same parts, and almost in the same plane. Ten moons are revolved about the earth, Jupiter, and Saturn, in circles concentric with them, with the same direction of motion, and nearly in the planes of the orbits of those planets; but it is not to be conceived that mere mechanical causes could give birth to so many regular motions, since the comets range over all parts of the heavens in very eccentric orbits. . . . This most beautiful system of the sun, planets, and comets, could only proceed from the counsel and dominion of an intelligent and powerful Being.

Newton did not deny that natural causes had been employed in the production of the solar system, but he insisted that it could not have been made by the laws of nature alone. He stated his position clearly in a letter to Thomas Burnet (c. 1635–1715): "Where natural causes are at hand God uses them as instruments in his works, but I do not think them alone sufficient for ye creation." In particular, he could discover no cause for the earth's diurnal motion other than divine action.

Newton's own theory of the creation of the inanimate world, which he confided to Burnet, illustrates his preference for natural explanations. He regarded the Mosaic account as a description of "realities in a language artificially adapted to ye sense of ye vulgar," not as a scientifically accurate record of events. The Genesis story of Creation related primarily to developments on this globe; thus, the creation of the sun, moon, and stars had been assigned to the fourth day because they first shone on the earth at that time. Newton imagined that the entire solar system had been formed from a "common Chaos" and that the separation of the planets into individual "parcels" and their subsequent condensation into solid globes had been effected by gravitational attraction, though this possibly was the work of "ye spirit of God." Since the earth's diurnal motion probably had not begun until the end of the second day, at which time it had first become a terraqueous globe, it seemed as if the first two days of Creation week could be made "as long as you please" without doing violence to the language of Genesis.

Neither the private speculations of Newton nor the quasi-naturalistic cosmogonies offered by Burnet and Whiston succeeded in breaking the hold of a static Creation on the collective mind of the seventeenth century. Newton's widely circulated condemnation of hypothesizing about Creation by natural law and his insistence on the necessity of divine intervention left the distinct impression, in the words of David Kubrin, that "Newtonianism and cosmogony were absolutely incompatible." Meanwhile, the general public continued to follow "those Divines" who, in the words of Burnet, "insist upon ye hypothesis of 6 dayes as a physical reality." Into the eighteenth century, even such well-informed cosmogonists as Immanuel Kant (1724–1804) continued to view Newton as a biblical literalist who "asserted that the immediate hand of God had instituted this arrangement [i.e., the solar system] without the intervention of the forces of nature."

Although Newton and other English cosmogonists readily utilized natural laws to assist in interpreting the events of Creation, they always did so within the context of the biblical record. They might speak of the days of Creation as long periods of time and discuss the possible role of gravitational attraction in the formation of the solar system, but they did not discard the basic features of the Mosaic story. It was in France, among the scientific disciples of Newton and the spiritual heirs of Descartes, that totally secular cosmogonies, free from all scriptural influence, first gained a foothold.

Georges Louis Leclerc, Comte de Buffon (1707–88), was one of the first French admirers of Newton; he was also one of the sternest critics of Newton's cosmogony. Whereas Newton actively encouraged the union of science and theology, Buffon demanded a complete separation. Those studying physical subjects, he argued, "ought, as much as possible, to avoid having recourse to supernatural causes." Philosophers "ought not to be affected by causes which seldom act, and whose action is always sudden and violent. These have no place in the ordinary course of nature. But operations uniformly repeated, motions which succeed one another without interruption, are the causes which alone ought to be the foundation of our reasoning." Whether or not such explanations were true was of no consequence. What really mattered was that they appear probable. Buffon acknowledged, for example, that the planets had been set in motion originally by the Creator—but he considered the fact of no value to the natural philosopher.

Buffon's repudiation of the supernatural in science led him to search for a natural history of the solar system. He was far too good a Newtonian to consider the discredited Cartesian theory of vortices, and Newton's cosmogony was out of the question, so he had no choice but to formulate his own naturalistic hypothesis. The first description of it appeared in 1749 in his *Theorie de la terre* (*Theory of the Earth*). Thirty years later he gave a somewhat modified version of his original ideas in *Les Epoques de la nature* (*The Epochs of Nature* [1778]).

The numerous uniformities in the solar system persuaded Buffon that a common cause was responsible for all planetary motions. All of the planets revolved around the sun in the same direction, and, according to his calculations, the probability is 64 to 1 that this was the product of a single cause. In addition, the planes of the planetary orbits are inclined no more than 7½ degrees from the ecliptic, and the probability is 7,692,624 to 1 that this could not have been produced by accident. Such a high degree of probability, "which almost amounts to a certainty," seemed to be conclusive evidence "that all the planets have probably received their centrifugal motion by one single stroke."

Though his calculations did not indicate whether the stroke had come from the hand of God, as Newton had assumed, or from some natural heavenly body, Buffon arbitrarily limited his search for an explanation to the latter type of cause. Since "nothing but comets [seemed] capable of communicating motion to such vast masses" as the planets, he confidently turned to them for the solution to his cosmogonical problem. The hypothetical production of the solar system he described in the following way:

> The comet, by falling obliquely on the sun . . . must have forced off from his surface a quantity of matter equal to a 650th part of his body. The matter being in a liquid state, would at first form a torrent, of which the largest and rarest parts would fly to the greatest distances, the smaller and more dense, having received only an equal impulse, would remain nearer the sun; his power of attraction would operate upon all the parts detached from his body, and make them circulate round him; and, at the same time, the mutual attraction of the particles of matter would cause all the detached parts to take on the form of globes, at different distances from the sun, the nearer moving with greater rapidity in their orbits than the more remote.

Buffon believed the diurnal motion of the planets to have resulted from the oblique blow of the comet, which would have caused the matter detached from the sun to rotate. And he imagined that a very oblique blow would have given the planets a rotation so great that small quantities of matter would have been thrown off to form the satellites. The fact that the satellites "all move in the same direction, and in concentric circles round their principal planets, and nearly in the place of their orbits" appeared to be a striking confirmation of this theory.

Laplace's Nebular Hypothesis

Although Buffon's cosmogony seemed "extremely probable" to him, it failed to win widespread acceptance. Nevertheless, it remained the most serious challenger to the Mosaic account of Creation through the latter half of the eighteenth century. Indeed, Pierre Simon Laplace (1749–1827), the leading Newtonian scientist in postrevolutionary France, could think of no one besides Buffon "who, since the discovery of the true system of the world, has endeavored to investigate the origin of the planets, and of their satellites."

Laplace applauded Buffon's efforts to fashion a naturalistic cosmogony but faulted him for a number of scientific errors, such as erroneously assuming that an oblique blow by a comet would necessarily impart a rotation to the planets in the same direction as their revolutions around the sun. In 1796, in a lengthy note appended to his *Exposition du systeme du monde* (*Explanation of a World System*), Laplace sketched out an alternative cosmogony that came to be known as the nebular hypothesis. According to his view, the planets had been created from the atmosphere of the sun, which, because of its heat, had originally extended beyond the orbit of the most distant planet. As this atmosphere condensed, it abandoned a succession of rings—similar to those of Saturn—in the plane of the sun's equator, and these rings then coalesced to form the various planets. In a similar way, the satellites developed from the planetary atmospheres. In a later edition of *Systeme du monde,* Laplace speculated that the primitive condition of the solar system closely resembled a slowly rotating hot nebula, like the cloudy bodies recently discovered by the British astronomer William Herschel (1738–1822).

Laplace's *Systeme du monde* had appeared in two English editions by 1830, but popular knowledge of the nebular hypothesis in the English-speaking world came largely from other sources, such as *Views of the Architecture of the Heavens* (1837) by the Scottish astronomer John Pringle Nichol (1804–59). Nichol congratulated Laplace for putting humans virtually "in possession of that primeval Creative Thought which originated our system and planned and circumscribed its destiny." No fewer than three of the Bridgewater Treatises on the Power, Wisdom, and Goodness of God as Manifested in the Creation, published in the 1830s, discussed Laplace's cosmogony. In *Astronomy and General Physics Considered with Reference to Natural Theology* (1833), one of the most popular volumes in the series, William Whewell (1794–1866) disarmed potential critics by pointing out that the nebular hypothesis in no way affected the much cherished argument from design: "If we grant . . . the hypothesis, it by no means proves that the solar system was formed without the intervention of intelligence and design. It only transfers our view of the skill exercised, and the means employed, to another part of the work. . . . What but design and intelligence prepared and tempered this previously existing element, so that it should by its natural changes produce such an orderly system?" If the motions and the arrangement of the planetary bodies were the inevitable result of the operation of natural laws on nebulous matter, the design of the solar system no longer gave evidence of God and his wisdom; it revealed only what happens to nebulous matter under the influence of natural laws. But that raised the question of who established those laws. To Whewell and other like-minded Christians, the obvious answer was that they had been instituted by God and were evidence of his existence and wisdom.

Although the Scottish divine Thomas Chalmers (1780–1847) in his Bridgewater Treatise condemned the nebular hypothesis for giving aid to atheism, the theory encountered little religious opposition before 1844, when it had the misfortune of being included in the scandalous little volume *Vestiges of the Natural History of Creation*, written anonymously by the Scottish publisher Robert Chambers (1802–71). In this widely read work, Chambers brought together the nebular hypothesis, developmental geology, and Lamarckian evolution in an attempt to show how all of nature had originated as a product of natural law. Many Christian critics regarded the *Vestiges* as blatantly atheistic and roundly condemned it in both the scientific and the religious press. However, some reviewers softened their attack when dealing with the nebular hypothesis. As one writer in the *American Review* put it, he would not have rejected the nebular hypothesis "so rudely, founded as it is upon excellent proofs, if it had not come attended by a load of false conclusions, as of . . . men originating by slow degrees from monkeys."

Nevertheless, its association with the *Vestiges* left the nebular hypothesis tainted with atheism. Thus, widespread enthusiasm greeted the announcement by William Parsons (1800–1867) in the mid-1840s of resolutions of nebulae. With his "leviathan" telescope, this Irish astronomer, the earl of Rosse, showed that many of the heavenly objects Herschel had formerly classed as nebulae were nothing but dense clusters of individual stars. This discovery deprived Laplace's hypothesis of what many considered to be its most convincing evidence. "The Nebular Hypothesis," wrote one American author, "vanishes as a pleasant dream, profitable though we believe it has been; and with it various systems of cosmogony, the fear of timid Christians, and the hopes of Atheistical philosophers."

The Nebular Hypothesis in America

The demise of the nebular hypothesis, however, proved to be only temporary, at least in the United States, for new evidence soon came to light suggesting that it might be true after all. In the late 1840s, an academy principal in the backwoods of Pennsylvania, Daniel Kirkwood (1814–95), discovered a simple equation, based on the nebular hypothesis, relating the rotations of the planets to their spheres of attraction. When his discovery was announced at the 1849 meetings of the American Association for the Advancement of Science, it almost single-handedly restored the faith of American scientists in the nebular hypothesis, and they went home generally agreed "that Laplace's nebular hypothesis, from its furnishing one of the elements of Kirkwood's law, may now be regarded as an established fact in the past history of the solar system." Across the Atlantic, the analogy created a different response among men who derived no nationalistic pride from Kirkwood's accomplishment. Though some European scientists expressed admiration for what the American had done, they often remained skeptical of the analogy's scientific value and its bearing on the nebular hypothesis.

Some clever experiments by the Belgian scientist Joseph Plateau (1801–83) on globules of oil rotating in a mixture of water and alcohol also seemed to confirm the nebular theory. The spheres of oil, when rotated, abandoned rings, which ruptured and formed rotating gloves circling around the central mass. This experiment was allegedly performed in a fashionable New York City church to show how modern science supported the biblical story of Creation. At its conclusion, the congregation reportedly voted unanimously to thank the demonstrator "for this perfect demonstration of the exact and literal conformity of the statements given in Holy Scripture with the latest results of science."

This anecdote illustrates the ease with which many Christians accommodated the nebular hypothesis to their understanding of the Bible. By midcentury, many Bible believers had abandoned the notion that the Mosaic account of Creation limited the history of the world to six thousand years. One American observer estimated that, by the early 1850s, only about half of the Christian public still believed in a young earth. Orthodox Christians were not repudiating the authenticity of Genesis, but many were reinterpreting it in the light of modern science. Some chose to read the "days" of Genesis 1 as representing vast geological ages; others argued for inserting a chronological gap between the original Creation of the earth mentioned in the first verse of Genesis 1 and the allegedly much later six-day Creation described in the following verses.

The first explicit attempt to harmonize the nebular cosmogony with Genesis 1 came from the Swiss-American geographer Arnold Guyot (1807–84), who viewed the days of Genesis as six great epochs of creative activity. As he saw it, the nebular development of the solar system occurred during the first three of these epochs. The formless "waters" mentioned by Moses symbolized gaseous matter. The light of the first "day" was generated by chemical action as this gas concentrated into nebulae. The dividing of the waters on the second "day" corresponded to the breaking up of the nebulae into various planetary systems, of which ours was only one. On the third "day," the earth condensed to form a solid globe; on the fourth, the nebulous vapors surrounding our globe dispersed to allow the light of the sun to shine on the earth. This striking correlation between the cosmogonies of Moses and Laplace moved one writer to ask rhetorically: "If Moses had actually, in prophetic vision, seen the changes contemplated in this theory taking place, could he have described them more accurately, in popular language, free from the technicalities of science?" During the second half of the nineteenth century, no one did more to popularize Guyot's exegesis than the Yale geologist James Dwight Dana (1813–95).

The widespread acceptance of cosmogonical evolution before 1859 contributed significantly to the willingness of some Christians to accept organic evolution after that date. Like historical geology, the nebular hypothesis argued for an ancient world. It also promoted an interpretation of Genesis 1 congenial to theories of organic development. If the creation of the solar system resulted from an extended process rather than an instantaneous act, it seemed to increase the likelihood that the organic world also arose from a process. Advocates of biological evolution were quick to use the acceptance of the nebular hypothesis as an argument for embracing their views. As the Harvard botanist Asa Gray (1810–88) pointed out, "the scientific mind of an age which contemplates the solar system as evolved from a common revolving fluid mass . . . cannot be expected to let the old belief about species pass unquestioned." The experience of reconciling the nebular hypothesis with a theistic view of nature and with the Mosaic account of Creation gave comfort to those who might otherwise have viewed biological evolution as a threat to natural and revealed religion. After all, Laplace's theory had once been condemned by some as heretical, but time had proved them wrong.

The nebular hypothesis convinced many nineteenth-century Christians that the solar system was a product of natural law rather than divine miracle. For such persons, its acceptance permanently erased all notions of supernaturally created planetary bodies. When the scientific inadequacies of the Laplacian theory became evident around the turn of the twentieth century and cosmogonists turned increasingly to competing views (such as the planetesimal hypothesis that the solar system owed its origin to matter drawn off from our ancestral sun when a passing star approached close enough to produce a large-scale tidal effect), no one with any scientific pretensions gave consideration to miraculous explanations. Among those who expressed themselves on the subject, few seemed to care much anymore whether or not the proposed substitutes harmonized with the Mosaic story of Creation or the once cherished doctrines of natural theology. For all but the most conservative Christians, the nebular hypothesis had established natural law in the heavens.

BIBLIOGRAPHY

Brooke, J. H. "Nebular Contraction and the Expansion of Naturalism." *British Journal for the History of Science* 12 (1979): 200–211.

Brush, Stephen G. *A History of Modern Planetary Physics.* 3 vols. Cambridge: Cambridge University Press, 1996.

Collier, Katharine Brownell. *Cosmogonies of Our Fathers: Some Theories of the Seventeenth and Eighteenth Centuries.* New York: Columbia University Press, 1934.

Jaki, Stanley L. *Planets and Planetarians: A History of Theories of the Origin of Planetary Systems.* New York: Wiley, 1978.

Kubrin, David. "Newton and the Cyclical Cosmos: Providence and the Mechanical Philosophy." *Journal of the History of Ideas* 28 (1967): 325–46.

———. "Providence and the Mechanical Philosophy: The Creation and Dissolution of the World in Newtonian Thought: A Study of the Relations of Science and Religion in Seventeenth Century England." Ph.D. diss., Cornell University, 1969.

Lawrence, Philip. "Heaven and Earth: The Relation of the Nebular Hypothesis to Geology." In *Cosmology, History, and Theology,* ed. Wolfgang Yourgrau and Allen D. Breck. New York: Plenum, 1977, 253–81.

Numbers, Ronald L. *Creation by Natural Law: Laplace's Nebular Hypothesis in American Thought.* Seattle: University of Washington Press, 1977.

Ogilvie, Marilyn Bailey. "Robert Chambers and the Nebular Hypothesis." *British Journal for the History of Science* 8 (1975): 214–32.

Schaffer, Simon. "The Nebular Hypothesis and the Science of Progress." In *History, Humanity, and Evolution: Essays for John C. Greene,* ed. James R. Moore. Cambridge: Cambridge University Press, 1989, 131–64.

Schweber, Silvan S. "Auguste Comte and the Nebular Hypothesis." In *In the Presence of the Past: Essays in Honor of Frank Manuel,* ed. Richard T. Bienvenu and Mordechai Feingold. Dordrecht: Kluwer, 1991, 131–91.

Part V :: The Response of Religious Traditions

19 Roman Catholicism since Trent

Steven J. Harris

The condemnation of Galileo did much to give the Roman Catholic Church the modern reputation of opposing scientific research that challenged theological dogma. In fact, the church had long encouraged freedom of speculation in natural philosophy. A more nuanced historical reading of the condemnation of Galileo would include a complex variety of factors that reflected the papal reaction to the Protestant Reformation and the clash of powerful personalities, as well as the church's endorsement of a geocentric cosmos as an embodiment of the Latin Christian worldview. The church continued to patronize scientific endeavors, and some religious orders, particularly the Jesuits, were known for their work in astronomy and mathematics. In the nineteenth century a renewal of this patronage led to the founding of the Vatican Observatory and the Pontifical Academy of Sciences. The tradition of clerical science carried over into the nineteenth and twentieth centuries with such notable researchers as Gregor Mendel, whose work led to the foundation of modern genetics; Pierre Teilhard de Chardin, whose philosophical speculation on human evolution was highly controversial; and Georges Lemaître, who formulated the big bang cosmology.

Steven J. Harris holds a Ph.D. in the history of science from the University of Wisconsin, Madison. He is currently a visiting scholar in the Department of History of Science at Wellesley College. He has published several articles and essays on Jesuit science and on science in early modern Europe, including "Confession-Building, Long-Distance Networks, and the Organization of Jesuit Science," *Early Science and Medicine* 1, no. 3 (1996): 287–318.

THE ATTEMPT to characterize the relationship between the Roman Catholic Church and science has often suffered from two broad assumptions: first, that the Roman Catholic Church has been monolithic in regard to its institutions and opinions; and second, that there has existed a fundamental—perhaps inevitable—conflict between the aims and methods of the Catholic faith and those of modern science. These assumptions are nowhere more strongly in evidence than in the literature on the trial and condemnation of Galileo Galilei (1564–1642), in which the church is portrayed as a univocal, authoritarian, and dogma-bound institution that invoked the inviolability of Scripture to suppress an essentially correct theory of the world (heliocentrism) and the mathematical and empirical methods upon which it rested. Although this reading of the Galileo affair has gained wide acceptance, there are several difficulties with the conflict thesis as a general characterization of the last four hundred years of Roman Catholic interaction with science. First, the Catholic hierarchy has rarely been of one mind regarding controversial scientific theories. Second, the church's strong tradition of conservatism has not precluded accommodation to novel astronomical, evolutionary, and cosmological theories. Third, despite the implication of a "fundamental conflict" found in the Galileo affair, post-Galilean episodes fail to reveal evidence of a uniform, deliberate, and sustained attack on the methods of modern science. And, finally, an unqualified conflict thesis is difficult to reconcile with the long tradition of support of scientific activity within the church itself. Perhaps most surprising in this regard is the fact that the greatest levels of clerical activity in science are to be found in the two hundred years following the Council of Trent.

The Council of Trent, an ecumenical gathering of bishops, cardinals, and prominent theologians who met in three sessions between 1545 and 1563, marked the beginning of a concerted effort on the part of the Roman Catholic Church to counter the advances made in the previous fifty years by the breakaway reform churches of the Lutherans, Calvinists, and Anglicans. Protestant challenges to papal authority, profound theological disagreement regarding matters of doctrine and faith, and an acknowledgment by the church hierarchy of indiscipline within its own ranks led the council to issue a series of decrees and institutional reforms that initiated what came to be called the Catholic Counter Reformation. It was within this "era of restrictive orthodoxies" (both Protestant and Catholic) that the so-called Copernican revolution unfolded, a revolution that would result in the abandonment of the earth-centered worldview and the beginnings of modern astronomy and cosmology.

Attitudes toward Heliocentrism

While neither Catholic nor Protestant churches had elevated geocentrism to the level of dogma, it rested on the seemingly unshakable foundations of received philosophical principles, scriptural corroboration, and plain common sense. The virtual sanctification of geocentric cosmology meant that the sun-centered (or heliocentric) planetary theory of Nicholas Copernicus (1473–1543) raised problems not only in theoretical astronomy but also in philosophy and theology. Galileo's efforts to convince the world of the truth of the Copernican theory thus took place at a time when the church sought to reaffirm its religious authority and when that authority seemed also to embrace questions in mathematical astronomy as well as theology. The publication of his *On the Two Chief World Systems* in 1632, sixteen years after Copernicus's work on the heliocentric system was placed on the Index of Prohibited Books, elicited a swift and punitive response from Rome. Sale of Galileo's book was immediately suspended, and Galileo was brought before the Inquisition. At the conclusion of his trial in 1633, heliocentrism was condemned as heretical, Galileo's works were placed on the Index, and Galileo was forced to recant his errors and sentenced to lifelong house arrest.

The startling vehemence of the Roman hierarchy in prosecuting Galileo has tended to mask the diversity of opinions found within the church itself. Dominican inquisitors argued that the Copernican theory cannot, in principle, be true because its claims rested on mathematical demonstrations. The problem with Galileo's mathematical argument in favor of heliocentrism was, therefore, not the insufficiency of evidence but the inherent limitations of his mode of reasoning. Some Jesuits, on the other hand, believed in the validity of mathematically based demonstrations but thought that Galileo's proofs were incomplete and, therefore, that he should not argue for the physical truth of heliocentrism. Still other clerics argued that Galileo had presented compelling arguments and that, as a result, a reinterpretation of certain passages of the Bible was necessary.

Diversity of clerical opinion notwithstanding, the Inquisition declared that heliocentrism was "philosophically absurd and false, and formally heretical." Despite the fact that this condemnation came from the highest levels of the Roman Curia, practical constraints limited both the scope and the execution of its decrees. Both the Roman Inquisition and the Index depended largely upon secular rulers to enforce their decrees in Catholic lands, and so their authority was limited to the obedient and like-minded. Protestants, of course, ignored them completely. France, for example, failed to promulgate the decrees of 1633 (though the faculty of the Sorbonne in Paris would issue its own condemnation of heliocentrism), and fewer than 10 percent of all surviving copies of Copernicus's book show signs of actually having been "corrected." The Spanish Inquisition operated independently of the one in Rome and neither endorsed Rome's injunctions against Copernicanism nor issued any of its own. Eventually, the vexed issue of heliocentrism achieved a belated, if incomplete, resolution in the mid–eighteenth century. In 1741, thirteen years after the discovery of the aberration of star light (a phenomenon understood to arise from the motion of the earth around the sun) and one year into his papacy, Pope Benedict XIV (b. 1675, p. 1740–58) effectively lifted the injunction against the heliocentric theory by having the Holy Office grant the first edition of *The Complete Works of Galileo* an imprimatur. In 1757, one year before the end of his papacy, Benedict ordered that all works espousing the heliocentric theory be removed from the Index.

Clerical Science, 1600–1800

Despite the church's disastrous condemnation of Copernicanism, it retained an important role as patron of a wide range of scientific activity. Members of Catholic religious orders, especially the Jesuits, continued to pursue research in observational and practical astronomy. Jesuit astronomers in Rome were the first to confirm Galileo's telescopic observations, while confreres in Germany discovered sunspots independently of Galileo, made important improvements in telescope design, and undertook extensive telescopic observations of the sun, comets, moon, and planets. In several Italian cities, Catholic cathedrals were used as solar observatories by clerics who had obtained permission to have holes drilled in the walls and brass meridian lines embedded in the floors so that the motion of the sun could be studied with precision. Moreover, the Jesuit, Benedictine, and Oratorian religious orders operated conventional observatories and collectively made important contributions not only to observational astronomy but also to meteorology, geography, and geodesy. Closely allied with astronomy was the teaching of mathematics, and, by 1700, the Jesuit order alone controlled more than one hundred chairs of mathematics, making it the single largest purveyor of mathematical education in Europe.

In addition to their contributions to observational astronomy and mathematics, churchmen were also active in the newly emerging experimental and empirical sciences. And while clerics tended to adopt a conservative stance in regard to interpretation, often seeking to preserve Aristotelian notions (e.g., arguing against the existence of the vacuum), they did so while insisting upon the importance of experiment in ascertaining the properties of physical reality and its validity as a means of testing theoretical claims. Because of their participation in overseas missions, Catholic clerics were well situated to engage in a wide range of empirical field sciences. In the period before the French Revolution, Dominican, Benedictine, Franciscan, and Jesuit missionaries together formed a loose but extensive network of amateur naturalists that literally spanned the globe. Their published reports of novel lands and peoples and their knowledge of indigenous herbal remedies added significantly to the fields of geography, natural history, botany, and medicine. Only in the eighteenth century, with the rise of large scientific societies and stable overseas trading companies, were networks of lay observers able to supplant missionary-naturalists.

Although clerical science continued to thrive well into the eighteenth century, the priest-scientist was brought to near extinction by 1800. Factors contributing to this demise were the Papal States' gradual loss of temporal wealth and political authority, the consolidation of state power under absolutist monarchies, and the pervasive anticlericalism of the Enlightenment. Perhaps the most severe blows were the suppression of the Jesuit order in 1773, which terminated what had been the richest scientific tradition within the post-tridentine church, and the sequestration of monastic properties in the wake of the French Revolution. Despite various papal initiatives in the nineteenth and twentieth centuries, the church has never recovered its prerevolutionary levels of scientific support and productivity.

Papal Patronage of Science

Modest recovery of the church's patronage of science came with the reestablishment in the mid–nineteenth century of two post-tridentine institutions, the Vatican Observatory and the Pontifical Academy of Sciences. Though founded in 1576 by Pope Gregory XIII (b. 1502, p. 1572–85) to facilitate the calendar reform that bears his name, the Vatican Observatory fell into disuse even before 1600. Reestablished in 1839, the Pontifical Observatory (as it was then called) flourished under the patronage of Pius IX (b. 1792, p. 1846–78) and the directorship of the capable Jesuit astronomer Pietro Angelo Secchi (1818–78). In 1879, the Italian government confiscated the Pontifical Observatory and began operating it as a state-run institution. In 1888, Pope Leo XIII (b. 1810, p. 1878–1903) reopened the Vatican Observatory; it has operated without interruption ever since, largely under the direction of Jesuit astronomers.

The Pontifical Academy of Sciences, like the Vatican Observatory, claims descent from a much-earlier institution, in this case the Academia Linceorum (Academy of the Lynx). Founded in 1603 by Prince Frederico Cesi (1585–1630), the Lincei flourished briefly in the 1610s and 1620s (when it could claim Galileo as its most illustrious member), but its activities came to a halt with the death of its founder in 1630. In 1847, Pius IX, invoking—and perhaps exploiting—the memory of the Lincei, founded the Pontificia Accademiae dei Nuovi Lincei as an official body of the Pontifical States. After a brief moment of reflected prestige under the astronomer Secchi, who served as the president from 1874 until his death in 1878, the New Lincei slowly slipped into invisibility. Under the initiative of Pius XI (b. 1857, p. 1922–39), the academy was reestablished in 1936 and rechristened the Pontifical Academy of Sciences. According to its charter, its membership was to be drawn from all nations and creeds, and its goal was "to honor pure science wherever it is found, assure its freedom and promote its researches."

Mendel, Teilhard, Lemaître

The renewal of the Vatican's direct patronage of scientific institutions since the mid–nineteenth century has been accompanied by a revival of scientific practice among Roman Catholic clerics. Although modest in comparison to former days, the modern tradition of clerical science has not been without its significant episodes. The work of three of its most prominent—though perhaps not most representative—members, Gregor Mendel (1822–84) in genetics, Pierre Teilhard de Chardin (1881–1955) in paleontology, and Georges Lemaître (1894–1966) in cosmology, suggests not only the disciplinary breadth of the modern tradition but also the church's direct, if sometimes strained, engagement with one of the central themes of modern science, namely the evolutionary worldview.

Gregor Mendel was born into a poor peasant family near Oldlau, Moravia (then part of the Austro-Hungarian Empire, now part of the Czech Republic); his only opportunity for an education was at the local school run by Piarist clerics. He went on to study at the university in Olomouc (Olmütz). After two years of extreme privation, he followed the recommendation of his physics professor and entered the Augustinian monastery of St. Thomas in Brno (Brünn)—though, as he himself admitted, "out of necessity and without feeling in himself a vocation for holy orders." The monastery, however, proved to be well suited to his quiet, studious ways and an ideal place for his work on plant hybridization.

As well known as Mendel's contributions to genetics have since become, what remains less well known is the fact that his plant-breeding experiments were part of an ongoing program of agricultural research within the monastery. For twenty years prior to Mendel's arrival, the monks of St. Thomas had engaged in plant-breeding experiments and had disseminated their results through teaching, publication, and participation in local agricultural and scientific societies. Mendel's initial work on plant heredity was conducted under Matthew Klácel (1802–82), director of the monastery's research gardens (an office that Mendel later held). Klácel's speculations on evolution, inspired in part by Hegelian philosophy (an intellectual allegiance that contributed to his eventual dismissal from the Augustinian order), deeply influenced Mendel's own work. During the period of Mendel's most important experimental work (c. 1853–68), scientific discussions within the monastery frequently touched upon the role of variation in the evolution of plants. Mendel had read the German translation of Charles Darwin's (1809–82) *Origin of Species* (1859) and fully accepted the theory of evolution by means of natural selection. Thus, his most important discoveries, the law of segregation (that the paired genes of body cells separate during the production of sex cells, or gametes) and the law of independent assortment (that the genes responsible for an organism's characteristics are inherited independently of each other), arose from a milieu of evolutionary speculation and a local monastic tradition of controlled experiments in plant hybridization. Although Mendel ceased his plant experiments in 1869 when he was elected abbot of St. Thomas and his laws of heredity were ignored for the next thirty years, his work has since become foundational for modern genetics and a central component of modern evolutionary theory.

The question of evolution, especially human evolution, was most controversial in the work of Pierre Teilhard de Chardin. Teilhard, who entered the Society of Jesus in 1898 at the age of seventeen, studied in France and England before completing his doctorate in paleontology at the Sorbonne in 1922. During his studies, he was deeply influenced by the speculative evolutionary philosophy of Henri Bergson (1859–1941). While teaching at the Catholic Institute in Paris in the early 1920s, Teilhard lectured on the theological doctrine of original sin within the framework of (directed) human evolution. His ideas drew severe complaints from conservative theologians, and Teilhard's Jesuit superiors forbade him to lecture on these topics. In April 1926, he was transferred (some say "exiled") to China. There he continued his work in paleontology and geology, making significant contributions to both fields. In 1929, for example, he was part of the team that discovered the celebrated "Peking man" (or *Sinanthropus,* later assigned to *Homo erectus*) near Chou-k'ou-tien.

Teilhard remained in China until shortly after the end of World War II. During his last years in Asia, he continued to develop his philosophical speculations regarding human evolution. His central idea was one of a thoroughgoing cosmic evolution that embraced both inorganic and organic matter, as well as all organisms and human consciousness. According to Teilhard, evolution unfolds along an axis of increasing organizational complexity, including several levels of "consciousness." Hominid evolution—or the "hominization of matter," as he called it—marks the emergence of the noösphere on Earth (a "sphere" of thinking matter analogous to the biosphere of living matter) and points toward the next stage, "planetization," before culminating in the final stage of the complete self-consciousness of creation, which he called the "Omega Point." Grand in conception, often poetic in its expression, and mystical in tone, Teilhard's writings moved him into a new and untested territory situated between evolutionary theory and Catholic theology.

Once he was back in Paris, Teilhard found it difficult to present his ideas for public discussion, the resistance initially coming largely from within his own order. Not only did he fail to find a publisher for this work, but his Jesuit superiors forbade him in 1947 to write on philosophical topics and denied him permission to assume the prestigious chair in paleontology at the Collège de France when it was offered to him in 1949. The maneuvers on the part of Teilhard's Jesuit superiors to block or limit the public exposure of his ideas were undoubtedly bound up with developments in Rome—though not always in ways that help explain their decisions. In 1948, the Pontifical Biblical Commission reaffirmed earlier declarations of 1909 regarding Genesis and human evolution but also claimed that these pronouncements were "in no way a hindrance to further truly scientific examination of the problems [of human evolution]." This implicit loosening of the strictures of 1909 should have, at the very least, encouraged the placement of Teilhard in the most prestigious chair in paleontology in all of France, but it did not.

In 1950, Pope Pius XII (b. 1876, p. 1939–58) issued the encyclical *Humani generis,* in which questions of evolution and theology took center stage. The encyclical opened with a condemnation of both pantheism and philosophical materialism and went on to declare the philosophies of "evolutionism" and historicism suspect because of their complicity in "relativistic conceptions of Catholic dogma." The encyclical stated that "the evolution of the human body from preexisting and living matter . . . is not yet a certain conclusion from the facts and that revelation demands moderation and caution." Hesitations notwithstanding, the encyclical went on to offer—for the first time—a restrained acceptance of evolution: "The teaching authority of the Church does not forbid that, in conformity with the present stage of human sciences and sacred theology, research and discussion on the part of men experienced in both fields take place with regard to the doctrine of evolution." While such words could have been read as encouragement for just the sort of public discussion Teilhard was hoping to pursue in Paris, his Jesuit superiors thought otherwise. In 1951, his order transferred (again, some say exiled) him to New York, thus removing him entirely from the French intellectual scene for a second time. In New York, Teilhard continued his paleontological research and his philosophical writings, though none of his nonscientific work was published before his death in 1955.

With the publication of *Le Phénemène humain* (*The Phenomenon of Man*) in 1955 and several other of his philosophical works, the controversy intensified. In 1957, Rome sought to remove his published works from the shelves of Catholic libraries and bookshops, and in 1962 (as his works continued to gain in popularity), Pope John XXIII (b. 1881, p. 1958–63) issued a *monitum,* or warning to readers against the uncertainties of Teilhard's theology. Neither step, however, slowed the international enthusiasm for "Teilhardism." Since the *monitum* of 1962, Rome has placed no further restrictions on his works, and he has become one of the most widely read and discussed Catholic intellectuals of the twentieth century.

Despite the controversies surrounding Teilhard's speculations concerning human evolution, Georges Lemaître's theoretical work on cosmic evolution met with immediate approbation. Four years after his ordination as a priest and in the same year that he completed his second doctoral thesis (1927), Lemaître published a short paper in which he laid out the basic framework of big bang cosmology. By combining the mathematical formalism of Albert Einstein's (1879–1955) theory of general relativity with Edwin Hubble's (1889–1953) empirical evidence indicating a general outward motion of distant galaxies, Lemaître postulated a dynamic model of an expanding universe of finite age. He was the first to understand that the recessional velocities of galaxies observed at present meant that, at some time in the distant past (Lemaître's initial estimate was between 20 billion and 60 billion years ago), all of the matter in the universe must have been confined to a sphere of small volume and enormous density. Lemaître postulated further that this "primeval atom" would break apart through spontaneous radioactive decay, and, as the fragments dispersed, lower densities would allow the formation of conventional atoms and, eventually, of stars, planets, and galaxies. Although Lemaître's model has since been modified in several of its details, his was the first rigorously scientific theory of the origin and evolution of the cosmos, and his assumptions regarding a physically definable beginning point in space-time and cosmic expansion still form the basis of all modern theories in big bang cosmology.

Lemaître's recognition within the church came swiftly and from the highest levels. In 1936, he was elected as a lifelong member of the newly reorganized Pontifical Academy of Sciences. The first international symposium sponsored by the academy was to be on "The Problem of the Age of the Universe" (scheduled for December 1939, it was canceled because of the outbreak of World War II). Later symposia were on such topics as stellar evolution, cosmic radiation (organized by Lemaître), the nuclei of galaxies, and the relationship between cosmology and fundamental physics. Lemaître played an active role in the academy throughout his life and served as its president from 1960 until his death in 1966. At his request, the academy in 1961 began awarding annually the Pius XI Gold Medal to outstanding young researchers in the natural sciences.

Catholicism and Modern Science

The apparent contradiction in the church's responses to Teilhard and Lemaître may be explained, in part, in terms of the particular brand of philosophy of science it has chosen to adopt, a philosophy perhaps best summarized under the notions of "autonomy" and "separation." Almost every pope since Pius XI has taken pains to reaffirm the autonomy of science. This autonomy, they have argued, is guaranteed on the one side by adherence to the methods of science and on the other by a theological view grounded in St. Augustine (354–430), who taught that Scripture was not to be read as a textbook on nature but as a guidebook to salvation. Moreover, they have repeatedly invoked the traditional Catholic doctrine of the "two truths" (i.e., natural or scientific knowledge can never contradict revealed or supernatural knowledge, since both issue from the same source) to maintain the separation between the domain of science and the domain of religion.

Lemaître drew upon these very principles in his discussions of his own work. In a lecture before the Solvay Conference in 1958, he stated that, "as far as I can see, [the primeval atom hypothesis] remains entirely outside any metaphysical or religious question. It leaves the materialist free to deny any transcendental Being [while] for the believer, it removes any attempt at familiarity with God." And despite the self-evident resonance between Lemaître's cosmogony and the story of Creation as related in Genesis, he never pursued such a connection in his technical or philosophical writings. Teilhard, on the other hand, consciously sought to blend together in his philosophical works scientific evidence of human evolution with the theological issues of original sin and salvation. His attempted synthesis thus brought him into conflict with the principle of separation between science and theology—a principle that had enabled the modern church to distance itself from the mistakes of the Galileo affair. At the same time, Teilhard chose to write in a domain (speculative philosophy) unprotected by the claims of autonomy and ungoverned by the methods and norms of scientific investigation.

The principles of autonomy and separation have also been invoked retrospectively in the case of Galileo. In 1981, Pope John Paul II (b. 1920, p. 1978–) appointed a commission of historians, theologians, and scientists to reexamine the trial and condemnation of Galileo and to "rethink the whole question" of the relationship between science and religion. After reviewing the commission's finding, John Paul announced in 1992—some three hundred sixty years after the fact—that the church had, indeed, erred in its condemnation of heliocentrism and its censure of Galileo. Furthermore, he pointed to the lessons to be learned from that affair:

> The error of the theologians of the time . . . was to think that our understanding of the physical world's structure was in some way imposed by the literal sense of Sacred Scripture. . . . In fact, the Bible does not concern itself with the details of the physical world. . . . There exist two realms of knowledge, one that has its source in revelation and one that reason can discover by its own power. . . . The methodologies proper to each make it possible to bring out different aspects of reality (John Paul II 1992, 373).

More recently still, John Paul has directly confronted the question of the relationship between Roman Catholic doctrine and human evolution—a question that has the potential of becoming as vexed as the question of heliocentrism in the seventeenth century. Evidently not wishing to repeat the mistakes of the past, the pope has made what have been seen as additional gestures of reconciliation and accommodation. In his welcoming address to participants in a symposium on "Evolution and the Origins of Life" sponsored by the Pontifical Academy of Sciences in October 1996 (the sixtieth anniversary of its refoundation), the pope forthrightly acknowledged the compelling advances that had been made in evolutionary theory:

> Today, almost half a century after the publication of the encyclical [*Humani generis*, 1950], new knowledge has led to the recognition of the theory of evolution as more than a hypothesis. It is indeed remarkable that this theory has been progressively accepted by researchers, following a series of discoveries in various fields of knowledge. The convergence, neither sought nor fabricated, of the results of the work that was conducted independently is in itself a significant argument in favor of this theory (John Paul II 1996).

Choosing neither to relinquish the matter of human evolution to scientists ("the Church's magisterium is directly concerned with the question of evolution, for it involves the conception of man") nor to abandon the long-held belief in the fundamental compatibility between science and theology ("truth cannot contradict truth"), the pope sought to reaffirm the church's authority "within the framework of her own competence" by pronouncing upon the allowable philosophical interpretations of human evolution. Thus, as Pius XII had done before, John Paul II reiterated that "theories of evolution which . . . consider the [human] spirit as emerging from the forces of living matter or as a mere epiphenomenon of this matter, are incompatible with the truth about man." Despite such insistence upon interpretative restrictions, the pope's remarks were in general more scientifically informed, more nuanced in regard to the relationship between science and philosophy (as well as between philosophy and theology), and more conciliatory in tone than the encyclical from 1950. (It must be kept in mind, however, that John Paul's address was to a lay audience and, thus, did not carry the same ecclesiastical authority as an encyclical.)

Conclusion

What this broader perspective on the relationship between Roman Catholicism and science reveals is scarcely the unrelieved high drama of confrontation implied by the conflict thesis. Rather, it is a story characterized by long periods of support for certain branches of science and indifference toward others, punctuated by occasional instances of controversy (chiefly heliocentrism in the seventeenth century and evolutionary theory in the twentieth). While the church's responses to controversial scientific innovations have been marked by a cautious conservatism, they have been monolithic neither across time nor even across a given generation. The complex and historically contingent relationship between Roman Catholicism and science since the Council of Trent cannot, therefore, be easily reduced to a single, all-embracing thesis.

BIBLIOGRAPHY

Ashworth, William B. "Catholicism and Early Modern Science." In *God and Nature: Historical Essays on the Encounter between Christianity and Science*, ed. David C. Lindberg and Ronald L. Numbers. Berkeley: University of California Press, 1986, 136–66.

Bowler, Peter J. *The Mendelian Revolution: The Emergence of Hereditarian Concepts in Modern Science and Society.* Baltimore: Johns Hopkins University Press, 1989.

Buckley, Michael J., S.J. *At the Origins of Modern Atheism.* New Haven: Yale University Press, 1987.

Delfgaauw, Bernard. *Evolution: The Theory of Teilhard de Chardin.* New York: Harper and Row, 1969.

Fantoli, Annibale. *Galileo: For Copernicanism and for the Church.* Trans. George V. Coyne, S.J. Vatican City: Vatican Observatory Press, 1994.

Feldhay, Rivka. *Galileo and the Church: Political Inquisition or Critical Dialogue?* Cambridge: Cambridge University Press, 1995.

Finocchiaro, Maurice A., ed. and trans. *The Galileo Affair: A Documentary History.* Berkeley: University of California Press, 1989.

Glick, Thomas F. "Teilhard de Chardin, Pierre." In *Dictionary of Scientific Biography*, ed. Charles Gillispie. New York: Scribners, 1976, 13:274–77.

Godart, O., and M. Heller. *Cosmology of Lemaître.* Tucson, Ariz.: Pachart, 1985.

Gould, Stephen Jay. "Nonoverlapping [sic] Magisteria." *Natural History* 3 (1997): 16–22, 60–62.

Harris, Steven. "Confession-Building, Long-Distance Networks, and the Organization of Jesuit Science." *Early Science and Medicine* 1 (1996): 287–318.

Heilbron, John. *Elements of Early Modern Physics.* Berkeley: University of California Press, 1982.

———. "Science in the Church." *Science in Context* 3 (1989): 9–28.

———. *The Sun in the Church: Cathedrals as Solar Observatories.* Cambridge: Harvard University Press, 1999.

John Paul II. "Lessons of the Galileo Case." *Origins* (CNS Documentary Service) 22, no. 22 (1992): 371–73.

———. "Truth Cannot Contradict Truth." Address to the Pontifical Academy of Sciences, October 22, 1996.

Kragh, Helge. "The Beginning of the World: Georges Lemaître and the Expanding Universe." *Centaurus* 32 (1987): 114–39.

Kruta, V., and V. Orel. "Mendel, Gregor." *Dictionary of Scientific Biography,* ed. Charles Gillispie. New York: Scribners, 1974, 9: 277–83.

Lemaître, Georges. "La culture catholique et les sciences positives." *Actes du VIe Congres Catholique de Malines* 5 (1936): 65–70.

———. *The Primeval Atom.* New York: Van Nostrand, 1950.

Marini-Bettòlo, G. B. *The Activity of the Pontifical Academy of Sciences, 1936–1986.* Vatican City: Pontificiae Academiae Scientiarum Scripta Varia, 1987.

Olby, Robert C. *Origins of Mendelism.* 2d ed. Chicago: University of Chicago Press, 1985.

Pius XII. "*Humani Generis:* Encyclical Letter Concerning Some False Opinions Which Threaten to Undermine the Foundations of Catholic Doctrine," August 12, 1950.

Poupard, Cardinal Paul. "Galileo: Report on Papal Commission Findings." *Origins* (CNS Documentary Service) 22, no. 22 (1992): 374–75.

Shea, William R. "Galileo and the Church." In *God and Nature: Historical Essays on the Encounter between Christianity and Science,* ed. David C. Lindberg and Ronald L. Numbers. Berkeley: University of California Press, 1986, 114–35.

Teilhard de Chardin, Pierre. *The Phenomenon of Man.* New York: Harper, 1959.

20 Evangelicalism and Fundamentalism

Mark A. Noll

Evangelicals and fundamentalists are for the most part religious descendants of British and American Protestant movements that have always taken science seriously. In the early eighteenth century, when older forms of religious authority were beginning to give way, evangelicals turned to "heart religion," to their own readings of Scripture, and to scientific procedures as new supports for their faith. There developed as a result a large evangelical investment in natural theology and proofs for God. For some time these means seemed to work well in coordinating religious beliefs and scientific concerns, but from the mid–nineteenth century the settled harmonies of earlier days began to give way. Increasingly, scientists seemed not to care about keeping God in the picture. More and more evangelicals (and their fundamentalist successors) thought that some of those who spoke in the name of science were abusing both science and religion. The result has been an often tangled twentieth-century history of coalitions, confrontation, cooperation, and combat. The particular concerns of fundamentalists and evangelicals parallel the concerns of many other religious groups in other places and at other times, but in America and wherever scientific authority is valued, these concerns have had an especially volatile relationship with scientific enterprises.

Mark A. Noll is McManis Professor of Christian Thought at Wheaton College, Wheaton, Illinois. He received his Ph.D. in American religious history from Vanderbilt University. He is the author of works on North American religious history and has edited (with David N. Livingstone) *Charles Hodge's What is Darwinism? and Other Writings on Science and Religion* (Grand Rapids: Baker, 1994); (with David N. Livingstone and D. G. Hart) *Evangelicals and Science in Historical Perspective* (New York: Oxford University Press, 1999); and (with David N. Livingstone) *B. B. Warfield on Evolution, Science, and Scripture: Selected Writings* (Grand Rapids: Baker, 2000).

EACH OF THE TERMS defining this essay—*evangelicalism, fundamentalism*, and *science*—is ambiguous. Yet, however plastic their definitions, it is clear that, since the mid–eighteenth century, the parts of the Anglo–North American Protestant world designated evangelical or fundamentalist have been deeply engaged with the practice of science. Even more, they have been deeply involved in political and cultural contests over the role of science in public life. Because evangelicalism itself was a product of the early modern consciousness that arose in part from an exalted respect for "science," it should not be surprising that evangelical traditions have nearly everywhere and always been preoccupied with scientific questions. This essay addresses (1) problems of definition, (2) the evangelical reliance on science, (3) the record of evangelical scientists, (4) attempts at narrowly evangelical science, and (5) evangelical concern for the larger meanings of science.

Definition

While *evangelical* has many legitimate meanings, it is used here to describe a family of Protestant traditions descended from the English Reformation, which espouses a basic set of religious convictions described by D. W. Bebbington as "conversionism, the belief that lives need to be changed; activism, the expression of the gospel in effort; biblicism, a particular regard for the Bible; and what may be called crucicentrism, a stress on the sacrifice of Christ on the cross" (Bebbington 1989, 2–3). The evangelical awakening, which began in the 1730s and affected most regions of Great Britain, Ireland, and the North American colonies, was part of a European-wide turn toward pietism that placed new emphasis on heartfelt religion and encouraged new skepticism about inherited, traditional religious authority. The spellbinding preaching of the British itinerant George Whitefield (1714–70), the pietistic theology of the Massachusetts minister Jonathan Edwards (1703–58), and, by the 1770s, dramatic growth among churches founded by John Wesley's (1703–91) Methodist missionaries made revivalism the defining heart of evangelicalism in the United States. By comparison with its British and Canadian counterparts, American evangelicalism has usually been more activistic, oriented to the immediate, and anticlerical.

The intellectual consequences of the evangelical movement have been ambivalent. On the one hand, since evangelicalism represented only a new set of emphases within historic Christianity, much historic Christian concern for reconciling faith and reason—for working out amicable connections between revelation from the Book of Nature and revelation from the Book of Scripture—

remained an important part of later evangelical movements. On the other hand, since evangelicalism promoted immediate experience over adherence to formal authorities, the individual over the collective, the Bible over tradition (even of the Protestant Reformation), and revival over less convulsive forms of Christian nurture, it has sometimes encouraged abandonment of traditional Christian thought and led to disputes with the learning of the larger world. Some evangelicals have, thus, promoted anti-intellectual attitudes. More germane to questions of science, some evangelicals have advanced conclusions about the natural world that they contend are taken from Scripture directly and so can be considered disinfected from the false science of the sinful world.

Although *fundamentalism* is now sometimes used in a generic sense for all conservative religious movements that resist the tides of modernity, the term arose to define a clearly demarcated segment of Protestant Christianity in the United States. According to one of its most perceptive students, George Marsden, fundamentalism became a distinct movement during and after World War I as a form of "militantly anti-modernist Protestant evangelicalism" (Marsden 1980, 4). Fundamentalism overlaps many other Protestant traditions, but its zealous defense of nineteenth-century revivalism and the ethics of nineteenth-century American piety separate fundamentalists (at least conceptually) from more generic Protestant evangelicalism, as well as from European immigrant pietism, the holiness movements emerging from Methodism, Pentecostalism, Calvinist or Lutheran confessionalism, Baptist traditionalism, and other denominational orthodoxies.

Strife during the 1920s among Baptists and Presbyterians in the northern United States marked the debut of a well-defined fundamentalist movement. These denominational conflicts pitted doctrinal conservatives agitated about larger changes in American society against denominational loyalists who, when it came to traditional doctrines, preferred peace to precision. When the inclusivists won these denominational battles, fundamentalists faded out of sight but not out of existence. Rather, they regrouped in powerful regional associations, publishing networks, preaching circuits, Bible schools, separate denominations, and independent churches. The tumults of the 1960s and following decades brought descendants of these cultural conservatives back to the public square.

Fundamentalist intellectual life was decisively influenced by dispensational premillennialism, a theological system first brought to America in the mid-nineteenth century by John Nelson Darby (1800–1882), an early leader of the

Plymouth Brethren. Dispensationalism interprets the Bible as literally as possible, and it has been preoccupied with the prophetic parts of Scripture. The heightened supernaturalism of dispensationalism also renders its adherents suspicious of exclusively natural explanations for the physical world.

Discriminating between *evangelicalism* and *fundamentalism* is difficult, since adherents of these movements, as well as outside observers, use the terms inconsistently. Most historians, however, usually treat Protestant fundamentalism as a subsection of evangelicalism while suggesting that many kinds of evangelicals should not be considered fundamentalists.

Science has always been an ambiguous, negotiated term in the history of evangelicalism and fundamentalism, since, in the domains of popular culture where evangelicalism and fundamentalism flourish, the term is used with multiple (often inconsistent) meanings. In an infinite variety of actual practices, evangelicals and fundamentalists embrace, disdain, ignore, or equivocate upon these meanings—*science* as a methodological commitment to observation, induction, rigorous principles of falsification, and a scorn for speculative hypotheses ("Scientists deal with knowledge of the world derived from testable empirical hypotheses"); *science* as shorthand for generalizations about the natural world (or the human person and human society) that are thought to have been established by experts ("Scientists have shown that the Grand Canyon was formed over millions of years"); and *science* as a principle of reasoning amounting to an autonomous source of social, moral, or even political authority ("Science holds our greatest hope for the future"). Flexibility in the use of the term *science* by evangelicals and fundamentalists, as well as by the general public, accounts for considerable intellectual confusion.

Finally, the connection between evangelicals and fundamentalists, on the one side, and science, on the other, is also beset with ambiguity. The subject can refer to practicing scientists who are evangelicals or fundamentalists (but where religious principles in the practice of science may not be distinct), to the stances of popular evangelical or fundamentalist leaders on scientific matters like evolution (but where engagement with actual research results may be next to nil), to forms of antiestablishment science promoted by ardent Bible-believers (but where other evangelicals or fundamentalists repudiate their conclusions as violating the true meaning of Scripture), or to many other possibilities. Ambiguity of definition, in sum, means that the following discussion can only sample the extraordinarily diverse facets of this protean subject.

Evangelical Reliance on Science

Evangelical commitments to the Book of Scripture and fundamentalist willingness to contest the authority of mainstream science loom large in general impressions of these groups. Yet, because evangelicalism came into existence, at least in part, as a result of its ability to exploit emphases in the increasingly scientific perspective of the eighteenth-century world, evangelicalism from the start made full use of scientific language, procedures, and warrants. Early leaders like John Wesley and George Whitefield shared much, at least formally, with the era's promoters of science—including an exploitation of sense experiences (to encourage what they called "experimental" Christianity) and an antitraditionalist reliance on empirical information. By the end of the eighteenth century, evangelicals in both Great Britain and the United States had also committed themselves fully to apologetical natural theology—the effort to demonstrate the truthfulness of Christianity by appealing in a scientific manner to facts of nature and the human personality.

In the United States, evangelical spokesmen enlisted scientific concepts to contend against the irreligion and disorder of the Revolutionary period. Led especially by the Scottish immigrant John Witherspoon (1723–94), president of the College of New Jersey (later Princeton University), American evangelicals tried to meet challenges from deism, radical democracy, and the disorderliness of the frontier with an appeal to universal standards of reason and science. In the 1790s and for several decades thereafter, evangelicals on both sides of the Atlantic recommended the natural theology of William Paley (1743–1805), even though they were often uneasy with Paley's utilitarian ethics and the ease with which he accounted for apparent waste and violence in nature.

Later, as evangelicals in America began to write their own apologetical textbooks, they drew ever more directly on methods of science. When Timothy Dwight (1752–1817) became president of Yale in 1795, he used arguments from natural theology to confront undergraduate doubts about the veracity of the Bible. Scientific arguments of one sort or another were a staple in the lengthy battles between the Unitarians and the trinitarians of New England. Widespread as the recourse to scientific demonstration was among the Congregationalists, it was the Presbyterians who excelled at what historian T. Dwight Bozeman has called a "Baconian" approach to the faith. In divinity, rigorous empiricism became the standard for justifying belief in God, revelation, and the Trinity. In the moral sciences, it marked out the royal road to ethical certainty. It also provided a key for using physical science itself as a demonstration of religious truths. In each case, the appeal was, as the successor of Witherspoon at Princeton, Samuel Stanhope Smith (1750–1819), put it, "to the evidence of facts, and to conclusions resulting from these facts which . . . every genuine discipline of nature will acknowledge to be legitimately drawn from her own fountain" (Smith 1787, 3). Among both Congregationalists and Presbyterians, the most theologically articulate evangelicals in the early republic, this approach predominated in rebuttals to Tom Paine's (1737–1809) *Age of Reason* (1794–96) in the 1790s and to other infidels thereafter. Their kind of "supernatural rationalism" was also useful for counteracting the impious use of science, by making possible the harmonization of the Bible first with astronomy and then with geology.

Revivalism, perhaps the least likely feature of antebellum evangelical life to reflect the influence of a scientific worldview, nonetheless took on a new shape because of that influence. Charles G. Finney (1792–1875), the greatest evangelist of the antebellum period and one of the most influential Americans of his generation, did not, by any means, speak for all evangelicals. But his vocabulary in a widely read book, *Lectures on Revivals of Religion* (1835), showed how useful scientific language had become. If God had established reliable laws in the natural world, so he had done in the spiritual world. To activate the proper causes for revivals was to produce the proper effect. In Finney's words: "The connection between the right use of means for a revival and a revival is as philosophically [i.e., scientifically] sure as between the right use of means to raise grain and a crop of wheat. I believe, in fact, it is more certain, and there are fewer instances of failure" (Finney 1960, 33). Because the world spiritual was analogous to the world natural, observable cause and effect must work in religion as well as in physics.

Nowhere did the language of evangelical Protestantism and the inductive ideals of modern science merge more thoroughly than in the American evangelical appropriation of the Bible. The orthodox Congregationalist Leonard Woods Jr. (1774–1854) wrote in 1822, for example, that the best method of Bible study was "that which is pursued in the science of physics," regulated "by the maxims of Bacon and Newton." Newtonian method, Woods said, "is as applicable in theology as in physics, although in theology we have an extra-aid, the revelation of the Bible. But in each science reasoning is the same—we inquire for facts and from them arrive at general truths" (quoted in Hovenkamp

1978, 3). Many others from North, South, East, and West said the same. The best-known statement of scientific biblicism appeared after the Civil War in Charles Hodge's (1797–1878) *Systematic Theology* (1872–73), but it was a position that he, with others, had been asserting for more than fifty years: "The Bible is to the theologian what nature is to the man of science. It is his storehouse of facts; and his method of ascertaining what the Bible teaches, is the same as that which the natural philosopher adopts to ascertain what nature teaches. . . . The duty of the Christian theologian is to ascertain, collect, and combine all the facts which God has revealed concerning himself and our relation to him. These facts are all in the Bible" (Hodge n.d., 1:10–11).

Such attitudes were by no means limited to the established denominations with reputations to protect. To cite just one of many possible examples, Alexander Campbell (1788–1866) led the Restorationist movement—which eventuated in the Disciples of Christ, the Churches of Christ, and the Christian Churches—in using scientific language as a principle of biblical interpretation. In self-conscious imitation of Francis Bacon (1561–1626), one of Campbell's successors, James S. Lamar, published in 1859 his *Organon of Scripture: Or, the Inductive Method of Biblical Interpretation,* in which deference to scientific thinking was unmistakable: "The Scriptures admit of being studied and expounded upon the principles of the inductive method; and . . . when thus interpreted they speak to us in a voice as certain and unmistakable as the language of nature heard in the experiments and observations of science" (quoted in Hughes and Allen 1988, 156).

Later in the nineteenth century, when new higher critical views of Scripture came to the United States from Europe, evangelicals resisted them by appealing directly to scientific principles, which they identified with inductive methods even as some of the new university science was assuming a more hypothesis-deductive approach. As they did so, an irony emerged, for America's new research universities, where higher critical views prevailed, also prided themselves on being scientific. In the 1870s and 1880s, graduate study on the European model began to be offered at older universities like Harvard and newer ones like Johns Hopkins. At such centers, objectivist science was exalted as the royal road to truth, and the new professional academics reacted scornfully to what was perceived as parochial, uninformed, and outmoded scholarship. All fields, including the study of the Bible, were to be unfettered for free inquiry. The sticking point with evangelicals was that university scholarship, in keeping with newer intellectual fashions, relied heavily upon evolutionary notions; ideas, dogmas, practices, and society all evolved over time, as did religious consciousness itself. Thus was battle joined not only on the meaning of the Bible but also on proper uses of science.

The inaugural public discussion of the new views occurred between Presbyterian conservatives and moderates from 1881 to 1883 in the pages of the *Presbyterian Review.* Both sides, as would almost all who followed in their train, tried, as if by instinct, to secure for themselves the high ground of scientific credibility. At stake was not just religion but the cultural authority that evangelical Protestants had exercised in American society. The moderates, led by Charles A. Briggs (1841–1913), were committed to "the principles of Scientific Induction." Since Old Testament studies had "been greatly enlarged by the advances in linguistic and historical science which marks our century," it was only proper to take this new evidence into account (Briggs 1881, 558). The conservatives were just as determined to enlist science on their side. William Henry Green (1825–1900), for example, chose not to examine W. Robertson Smith's (1846–94) "presumptions" that led him to adopt critical views of the Old Testament, but chose instead the way of induction: "We shall concern ourselves simply with duly certified facts" (Green 1882, 111).

Once the terms of the debate were set in this scientific form, the evangelicals defended their position tenaciously. In 1898, one of evangelist D. L. Moody's (1837–99) colleagues, R. A. Torrey (1856–1928)—who had studied geology at Yale—published a book entitled *What the Bible Teaches.* Its method was "rigidly inductive. . . . the methods of modern science are applied to Bible study—thorough analysis followed by careful synthesis." The result was "a careful, unbiased, systematic, thorough-going, *inductive* study and statement of Bible truth" (Torrey 1898, 1 [author's italics]).

Almost since their emergence as a distinct form of Protestantism, evangelicals adopted, promoted, and exploited the language of science as their own language. In the last three decades of the twentieth century, many evangelical and fundamentalist enterprises—including the widely used apologetic manuals of the popular evangelist Josh McDowell and the myriad presentations promoting creation science—have maintained this reliance on early modern scientific demonstration. The fact that such full and consistent efforts to exploit the prestige of early modern science have accompanied evangelical resistance to certain conclusions of modern scientific effort means that simple statements about evangelicals and science are always wrong.

Evangelical Scientists

The evangelical engagement with science includes also the professional scientific labors of self-confessed evangelicals. In both Britain and North America, evangelical scientists were especially prominent during the nineteenth century. After Darwinism and other potentially naturalistic explanations began to dominate professional science from the last third of that century, the presence of evangelicals and fundamentalists was not as obvious, but the numbers have always been greater than the stereotype of evangelical-scientific strife would suggest.

In Britain, a lengthy roster of evangelicals enjoyed considerable scientific repute for well over a century. Among these were a trio of evangelical Anglicans—Isaac Milner (1750–1820), Francis Wollaston (1762–1823), and William Farish (1759–1837)—who occupied in succession the Jacksonian Chair of Natural and Experimental Philosophy at Cambridge. Michael Faraday (1791–1867), the renowned pioneer of electromagnetism, was the member of a small evangelical sect, the Sandemanians or Glasites (after founder John Glas [1695–1773] and major promoter Robert Sandeman [1718–71]), who zealously practiced their unusual modification of traditional Calvinism. For later evangelicals, the Victoria Institute provided an ongoing base for efforts to use respectful science in harmony with, rather than in opposition to, faith.

In Scotland, the combination of Presbyterian seriousness about learning and the empirical bent of the Scottish Enlightenment produced several notable evangelical scientists. Sir David Brewster (1781–1868), after training for the ministry, became a specialist in optics, especially the polarization of light. Eventually, he served as principal of the University of Edinburgh. The Rev. John Fleming (1785–1857) was professor of natural history at King's College, Aberdeen, and the leading Scottish zoologist of his day. Hugh Miller (1802–56), a well-known geologist, opposed evolution but not the idea that the earth could be very old. His pioneering work included investigations of fossilized fish. Until his death in 1847, the leading Scottish minister of his age, Thomas Chalmers (1780–1847), not only supported his friends Brewster, Fleming, and Miller but also himself gave popular lectures on astronomy and offered other encouragements in scientific matters. The theological college of the Scottish Free Church, founded under Chalmers's leadership in 1843, maintained a chair of natural science whose incumbents included noteworthy theologian-scientists like John Duns (1820–1909). A final notable among British evangelicals, who was born in Ireland, was Sir George Stokes (1819–1903), professor of mathematics at Cambridge, who, for more than fifty years, was one of the most respected mathematicians and physicists of his day.

In North America, a similar roster of evangelicals gained scientific eminence. Joseph Henry (1797–1878), student of electromagnetism, diligent meteorologist, and first director of the Smithsonian Institution, was a longtime Presbyterian who, during his years as a professor at the College of New Jersey, regularly joined his friends at Princeton Theological Seminary to discuss issues at the intersection of theology and science. Asa Gray (1810–88), a botanist and taxonomist of extraordinary energy, became Charles Darwin's most active disciple in the United States but without giving up his beliefs, as an active Congregationalist, in historic Christianity or his efforts to convince Darwin that natural selection could be construed as a teleological system. James Dwight Dana (1813–95) eventually accepted a form of evolution in the last edition of his influential *Manual of Geology* (first published in 1862) but (with Gray) only in a teleological sense. The Canadian geologist and paleobotanist John William Dawson (1820–99) won his reputation through fieldwork in Nova Scotia, eventually became principal of McGill University, remained a dedicated Presbyterian, and participated actively in meetings and publications of the international Evangelical Alliance. George Frederick Wright (1838–1921) was a minister and a geologist who published important papers on the effects of glaciers on North American terrain and who encouraged Asa Gray to write essays promoting a Christianized form of Darwinism. Wright lived long enough to become disillusioned with developments in evolutionary theory, but he never lost his earlier confidence that science, properly carried out, would reinforce Christian theology, properly conceived. Just about the same could be said for several important evangelical geologist-educators of the nineteenth century, including Benjamin Silliman (1779–1864), Edward Hitchcock (1793–1864), Arnold Guyot (1807–84), and Alexander Winchell (1824–91).

In the twentieth century, professional scientists with evangelical convictions have found a home in Britain with the Victoria Institute (founded 1865) and in the United States with the American Scientific Affiliation (founded 1941). Both groups have received the unwelcome compliment of being criticized by the scientific establishment as too religious and by their fellow evangelicals as too naturalistic.

During the nineteenth century, most self-identified evangelical scientists looked upon their research as a way of confirming design in the universe. Twentieth-century evangelical scientists usually speak with greater restraint about the apologetical value of natural theology, but they join their predecessors in viewing scientific investigation as a way of glorifying God as Creator and Sustainer of the natural world. What remains to be investigated is the extent to which specifically evangelical beliefs or practices, as distinguished from more general Christian convictions shared with Roman Catholics and Protestants who are not evangelicals, have shaped the actual practices of their science.

Narrowly Evangelical Science

In the popular stereotype, evangelicals are better known as promoters of alternative scientific visions than as participants in the scientific mainstream, and with at least reasonably good cause. The modern proponents of what is variously called Flood geology, biblical creationism, or creation science are, in fact, carrying on an evangelical tradition that is almost as old as the tradition of evangelical professional science.

Among the first generations of evangelicals, for example, were some who found congenial the anti-Newtonian science of John Hutchinson. Hutchinson (1674–1737) developed his views of the material world in direct opposition to what he held to be the materialistic implications of Newton's gravitational mechanics. If in the Newtonian world objects could attract each other at a distance with no need for an intervening medium, Hutchinson concluded, Newton was setting up the material world as self-existent and, hence, in no need of God. From a painstakingly detailed study of the linguistic roots, without vowel points, of Old Testament Hebrew, Hutchinson thought he had discovered an alternative Bible-based science. The key was the identity of the roots for *glory* and *weight*, which led Hutchinson to see God actively maintaining the attraction of physical objects to each other through an invisible ether. Moreover, by analogous reasoning from the New Testament's full development of the Trinity, it was evident that a threefold reality of fire, air, and light offered a better explanation for the constituency of the material world than did modern atomism.

Hutchinson's ideas were promoted by several dons and fellows at Oxford and by several highly placed bishops in the Church of England, but, despite their appeal to the Bible as sole authority, they never received much allegiance from evangelicals. To be sure, in Britain, several early evangelicals, including

John Wesley, felt the tug of Hutchinsonianism, and William Romaine (1714–95), a leading evangelical Anglican preacher in London, held something like Hutchinsonian views. In America, there were similar indications of interest, including a respectful mention by Archibald Alexander (1772–1851) in 1812 during his inaugural sermon as first professor at Princeton Theological Seminary. Yet, Hutchinsonianism no more caught on among American evangelicals than it did among their British colleagues. The reason probably rests in the commitments that evangelicals had made to Baconian-Newtonian ideals and to a distaste for the high-church environments in which Hutchinsonianism flourished. The fact that the most visible Hutchinsonians in both Britain and North America were Tories, high-church Anglicans, and students of the Bible in Hebrew and Greek conveyed an elitist, authoritarian ethos entirely foreign to the populist, self-taught, and voluntaristic character of the evangelical movement.

Other forms of Bible-only science gained somewhat more allegiance among evangelicals during the course of the nineteenth century. In Britain, a school of "scriptural geology," advocated by a book with that title by George Bugg (c. 1769–1851) in 1826, gained some public credibility early in the century. Bible-only approaches to science were advanced unsystematically during the 1820s and 1830s by Edward Irving (1792–1834), a leading figure of the Catholic Apostolic Church and promoter of an intensely supernaturalistic, romantic evangelicalism. Irving and his associates tended to devalue the results of natural investigation and to exalt their own interpretations of Scripture as a source of knowledge opposed to other forms of human learning. The result was a heightened supernaturalism affecting doctrines of the Bible, the Second Coming of Christ, and the special presence of the Holy Spirit, as well as heightened supernaturalism concerning the operation of the physical world.

Irving's biblicism was far different from that promoted by Philip Gosse (1810–88), a naturalist of wide experience in Canada, the United States, and Jamaica, as well as England. Gosse was a well-respected student of marine invertebrates who came to oppose what he thought were the antibiblical implications of evolutionary theory. In his response, given fullest airing in his *Omphalos* (1857), he tried to retain both a literal interpretation of early Genesis and his own life's work as an observer of nature. To gain this end, Gosse proposed that evidence for the ancient age of the earth might be the result of God's deliberate creation of the world with the marks of apparent age.

Significantly, these varieties of Bible-only or Bible-dependent science enjoyed only modest acceptance among evangelicals. In the United States, a few evangelicals accepted Gosse's views on the apparent age of the earth, but they tended to be clerics like the Southern Presbyterian theologian Robert L. Dabney (1820–98) rather than practicing scientists like Gosse. Evangelicals were much more likely to seek accommodating adjustments between biblical authority and new scientific findings. More prominent were efforts to finesse earlier and simpler allegiance to the early chapters of Genesis. By the start of the twentieth century, the most popular of these accommodations were the "gap theory" (in which a vast expanse of time was postulated between God's original creation of the world and the creative acts specified in Genesis 1:3 and following) and the "day-age" theory (in which the days of Creation in Genesis 1 were interpreted as standing for lengthy geological eras). At least into the twentieth century, even in debates over evolutionary theories—which began well before Charles Darwin (1809–82) published his *Origin of Species* in 1859 and which always involved much more than Darwin's own notions of development through natural selection—evangelicals were as likely to propose accommodations between biblical revelation and scientific conclusions as they were to set the Bible against science.

So long as evangelicals took a substantial part in mainstream professional research, contrarian views of science never enjoyed more than local popularity. The success among evangelicals of Flood geology or creation science began slowly in the 1920s, precisely when tensions had emerged between evangelicals and fundamentalists, on the one hand, and proponents of university-certified specialized scientific knowledge, on the other. Unlike Hutchinsonianism and, to a certain extent, earlier forms of Bible-derived science, Flood geology or creation science has been able, especially since the 1960s, to exploit alienation from the centers of learning and to make its case in democratic, populist, and voluntarist forms that accentuate, rather than contradict, major themes in the evangelical tradition. The long history of evangelical engagements with science, however, suggests that the antagonisms promoted by creation science owe at least as much to recent developments as to historic patterns among either evangelicals or scientists.

Larger Meanings of Science

Public debates over evolution and creation science highlight the fact that evangelical engagement with science has regularly focused on grand metaphysical implications rather than on minute particulars.

In the first century after the evangelical awakenings of the 1730s and 1740s, evangelicals were ardent promoters of the age's new science but very often for extrinsic interests. They valued the language and some of the procedures of science, not so much to increase understanding of the physical world as to refurbish natural theology for the purpose of apologetics. American evangelicals used scientific reasoning straightforwardly to defend the traditional Christian faith in an era when their countrymen were setting aside other props that once had supported the faith—respect for history, deference to inherited authorities, and a willingness to follow tradition itself. A few evangelicals in the eighteenth and nineteenth centuries occasionally complained that too much authority was being given to natural theology at the expense of simple preaching or simple trust in Scripture. But more common were attitudes like those of Thomas Chalmers, who lectured and published widely on themes from natural theology but who regularly paused to show the limited value of those arguments. In a work published in 1836, Chalmers wrote: "It is well to evince, not the success only, but the shortcomings of Natural Theology; and thus to make palpable at the same time both her helplessness and her usefulness—helpless if trusted to as a guide or an informer on the way to heaven; but most useful if, under a sense of her felt deficiency, we seek for a place of enlargement and are led onward to the higher manifestations of Christianity" (Chalmers 1844, xiv).

Evangelical apologists in the United States were somewhat more inclined to wager higher stakes on the results of natural theology, which may be one of the reasons that later clashes between fundamentalists and modernists (involving great strife over the question of who was using the proper form of scientific procedure) were sharper in the United States than in Canada or Britain, where natural theology had been promoted with Chalmers's spirit.

Modern contentions over evolution, fomented by fundamentalists and some evangelicals, have regularly begun as debates over scientific results, procedures, and verifications. But, almost invariably, they have rapidly moved on to arguments over issues only remotely related to what practicing scientists do in their laboratories or in the field. From the defenders of modern scientific procedures have come protests about professional expertise, qualifications, and decorum. From the fundamentalists and evangelicals have come protests about the decline of Western morality. In moving so rapidly to great moral questions, evangelicals have only followed a longstanding tradition, which had been expressed with great clarity by the evangelical populist William Jennings Bryan (1860–1925). For Bryan, it was necessary to oppose evolution not because it imperiled traditional interpretations of Genesis 1 or sabotaged empirical investigations but because evolution was a threat to a treasured social ideal. As Bryan put it in 1925, the year of his appearance at the Scopes trial, human evolution is "an insult to reason and shocks the heart. That doctrine is as deadly as leprosy; . . . it would, if generally adopted, destroy all sense of responsibility and menace the morals of the world" (Bryan 1925, 51). In making this assertion, Bryan upheld a long evangelical tradition that subsumed the narrowly research-oriented aspects of science to its broad social implications.

The irony of the evangelical engagement with science is that, while evangelicalism emerged as a potent religious force in part by exploiting the prestige of science that was so important for Anglo–North American culture in the eighteenth and nineteenth centuries, descendants of this earlier evangelicalism, especially in fundamentalist forms, now view recent forms of science as a grave threat to what Christians value most. That irony, however, is also eloquent testimony to the depth and persistence of evangelical engagement with science, an engagement that has always been more complicated than either the champions of or the detractors from evangelicalism have been willing to concede.

BIBLIOGRAPHY

Bebbington, David W. *Evangelicalism in Modern Britain: A History from the 1730s to the 1980s.* London: Unwin Hyman, 1989.

———. "Science and Evangelical Theology in Britain from Wesley to Orr." In *Evangelicals and Science in Historical Perspective,* ed. David N. Livingstone, D. G. Hart, and Mark A. Noll. New York: Oxford University Press, 1999, 120–41.

Bozeman, Theodore Dwight. *Protestants in an Age of Science: The Baconian Ideal and Antebellum American Religious Thought.* Chapel Hill: University of North Carolina Press, 1977.

Briggs, Charles A. "Critical Theories of Sacred Scripture." *Presbyterian Review* 2 (July 1881): 550–79.

Bryan, William Jennings. *The Last Message of William Jennings Bryan.* New York: Fleming H. Revell, 1925.

Chalmers, Thomas. *On Natural Theology.* Edinburgh, 1836. Reprint. New York: Robert Carter, 1844.

Conser, Walter H., Jr. *God and the Natural World: Religion and Science in Antebellum America.* Columbia: University of South Carolina Press, 1993.

Davis, Edward B. "A Whale of a Tale: Fundamentalist Fish Stories." *Perspectives on Science and Christian Faith* 43 (December 1991): 224–37.

Finney, Charles G. *Lectures on Revivals of Religion,* ed. W. G. McLoughlin. 1835. Reprint. Cambridge: Harvard University Press, 1960.

Green, William Henry. "Professor W. Robertson Smith on the Pentateuch." *Presbyterian Review* 2 (January 1882): 108–56.

Gundlach, Bradley John. "The Evolution Question at Princeton, 1845–1929." Ph.D. diss., University of Rochester, 1995.

Hodge, Charles. *Systematic Theology,* 3 vols. 1872–73. Reprint. Grand Rapids, Mich.: Eerdmans, n.d.

Hovenkamp, Herbert. *Science and Religion in America, 1800–1860.* Philadelphia: University of Pennsylvania Press, 1978.

Hughes, Richard L., and C. Leonard Allen. *Illusions of Innocence: Protestant Primitivism in America, 1630–1875.* Chicago: University of Chicago Press, 1988.

Larson, Edward J. *Summer for the Gods: The Scopes Trial and America's Continuing Debate over Science and Religion.* New York: Basic Books, 1997.

Livingstone, David N. *Darwin's Forgotten Defenders: The Encounter between Evangelical Theology and Evolutionary Thought.* Grand Rapids, Mich.: Eerdmans, 1987.

———. "Darwinism and Calvinism: The Belfast-Princeton Connection." *Isis* 83 (1992): 408–28.

Livingstone, David N., and Mark A. Noll. "B. B. Warfield (1851–1921): A Biblical Inerrantist as Evolutionist." *Isis* 91 (2000): 283–304.

Marsden, George M. *Fundamentalism and American Culture.* New York: Oxford University Press, 1980.

Moore, James R. *The Post-Darwinian Controversies.* New York: Cambridge University Press, 1979.

———. "Interpreting the New Creationism." *Michigan Quarterly Review* 22 (1983): 321–34.

———. *The Darwin Legend.* Grand Rapids, Mich.: Baker, 1994.

Noll, Mark A. "Science, Theology, and Society: From Cotton Mather to William Jennings Bryan." In *Evangelicals and Science in Historical Perspective,* ed. David N. Livingstone, D. G. Hart, and Mark A. Noll. New York: Oxford University Press, 1999, 99–119.

Numbers, Ronald L. "George Frederick Wright: From Christian Darwinist to Fundamentalist." *Isis* 79 (1988): 624–45.

———. *The Creationists.* New York: Knopf, 1992.

———. *Darwinism Comes to America.* Cambridge: Harvard University Press, 1998.

Perspectives on Science and Christian Faith 44 (March 1992): 1–24. Articles on the fiftieth anniversary of the American Scientific Affiliation.

Poe, Harry L., and Jimmy H. Davis. *Science and Faith: An Evangelical Dialogue.* Nashville: Broadman and Holman, 2000.

Roberts, Jon H. *Darwinism and the Divine in America: Protestant Intellectuals and Organic Evolution, 1859–1900.* Madison: University of Wisconsin Press, 1988.

Smith, Samuel Stanhope. *An Essay on the Causes of the Variety of Complexion and Figure in the Human Species.* Philadelphia: Robert Aitken, 1787.

Torrey, R. A. *What the Bible Teaches.* Chicago: Fleming H. Revell, 1898.

Young, Davis A. *The Biblical Flood: A Case Study of the Church's Response to Extrabiblical Evidence.* Grand Rapids, Mich.: Eerdmans, 1995.

21 Creationism since 1859

Ronald L. Numbers

Creationism is the belief that the universe was created by God in the manner described in Genesis 1. Strict creationists believe that the days of Genesis ought to be taken literally as seven 24-hour days, while progressive creationists interpret the days of Creation figuratively as long periods of time. Even before the publication of Charles Darwin's *Origin of Species* (1859), many Christians had come to believe in the great antiquity of the earth. Following Darwin, some were willing to accept one or another form of evolution as well. As a result of the Scopes trial in 1925, however, an antievolution movement grew rapidly in the United States and gained large numbers of adherents for its strict creationist views, particularly among conservative Christians. In 1961 the publication of *The Genesis Flood* by John C. Whitcomb Jr. and Henry M. Morris sparked a creationist revival of what was termed *scientific creationism.* Its adherents sought the adoption by public schools of a curriculum that taught scientific creationism as an alternative scientific explanation to biological evolution.

Ronald L. Numbers is Hilldale and William Coleman Professor of the History of Science and Medicine at the University of Wisconsin, Madison. He holds a Ph.D. from the University of California at Berkeley. His numerous publications include *The Creationists: The Evolution of Scientific Creationism* (New York: Knopf, 1992) and *Darwinism Comes to America* (Cambridge: Harvard University Press, 1998). This essay is abridged from "The Creationists," in *God and Nature: Historical Essays on the Encounter between Christianity and Science,* edited by David C. Lindberg and Ronald L. Numbers (Berkeley: University of California Press, 1986), 391–423, Complete documentation, including citations to sources quoted, can be found in the original essay.

SCARCELY TWENTY YEARS after the publication of Charles Darwin's (1809–82) *Origin of Species* in 1859, special creationists could name only two working naturalists in North America, John William Dawson (1820–

99) of Montreal and Arnold Guyot (1807–84) of Princeton, who had not succumbed to some theory of organic evolution. The situation in Great Britain looked equally bleak for creationists, and on both sides of the Atlantic liberal churchmen were beginning to follow their scientific colleagues into the evolutionist camp. By the closing years of the nineteenth century, evolution was infiltrating even the ranks of the evangelicals, and, in the opinion of many observers, belief in special creation seemed destined to go the way of the dinosaur. But, contrary to the hopes of liberals and the fears of conservatives, creationism did not become extinct. The majority of late-nineteenth-century Americans remained true to a traditional reading of Genesis, and as late as 1991 a public-opinion poll revealed that 47 percent of Americans, and 25 percent of college graduates, continued to believe that "God created man pretty much in his present form at one time within the last 10,000 years."

Such surveys failed, however, to disclose the great diversity of opinion among professing creationists. Risking oversimplification, we can divide creationists into two main camps: *strict creationists,* who interpret the days of Genesis literally, and *progressive creationists,* who construe the Mosaic days to be immense periods of time. But, even within these camps, substantial differences exist. Among strict creationists, for example, some believe that God created all terrestrial life—past and present—less than ten thousand years ago, while others postulate one or more creations prior to the seven days of Genesis. Similarly, some progressive creationists believe in numerous creative acts, while others limit God's intervention to the creation of life and perhaps the human soul. Since this last species of creationism is practically indistinguishable from theistic evolutionism, this essay focuses on the strict creationists and the more conservative of the progressive creationists, particularly the small number who have claimed scientific expertise. Drawing on their writings, it traces the development of creationism from the Darwinian debates in the late nineteenth century to the battles for equal time in the late twentieth. During this period, the leading apologists for special Creation shifted from an openly biblical defense of their views to one based increasingly on science. At the same time, they grew less tolerant of notions of an old earth and symbolic days of Creation, common among creationists early in the twentieth century, and more doctrinaire in their insistence on a recent Creation in six literal days and on a universal flood.

The Darwinian Debates

The general acceptance of organic evolution by the intellectual elite of the late Victorian era has often obscured the fact that the majority of Americans remained loyal to the doctrine of special Creation. In addition to the masses who said nothing, there were many people who vocally rejected kinship with the apes, and there were other, more reflective, persons who concurred with the Princeton theologian Charles Hodge (1797–1878) that Darwinism was atheism. Among the most intransigent foes of organic evolution were the premillennialists, whose predictions of Christ's imminent return depended on a literal reading of the Scriptures. Because of their conviction that one error in the Bible invalidated the entire book, they had little patience with scientists who, as described by the evangelist Dwight L. Moody (1837–99), "dug up old carcasses . . . to make them testify against God."

Such an attitude did not, however, prevent many biblical literalists from agreeing with geologists that the earth was far older than six thousand years. They did so by identifying two separate creations in the first chapter of Genesis: the first, "in the beginning," perhaps millions of years ago, and the second, in six actual days, approximately four thousand years before the birth of Christ. According to this so-called gap theory, most fossils were relics of the first Creation, destroyed by God prior to the Adamic restoration. In 1909, the *Scofield Reference Bible,* the most authoritative biblical guide in fundamentalist circles, sanctioned this view.

Scientists such as Guyot and Dawson, the last of the reputable nineteenth-century creationists in North America, went still further to accommodate science by interpreting the days of Genesis as ages and by correlating them with successive epochs in the natural history of the world. Although they believed in special creative acts, especially of the first humans, they tended to minimize the number of supernatural interventions and to maximize the operation of natural law. During the late nineteenth century, their theory of progressive Creation circulated widely in the colleges and seminaries of America.

The Antievolution Movement

The early Darwinian debates remained confined largely to scholarly circles and often focused on issues pertaining to natural theology; thus, those who objected to evolution primarily on biblical grounds saw little reason to participate. But when the debate spilled over into the public arena during the 1880s and 1890s, creationists grew alarmed. "When these vague speculations, scattered to the four winds by the million-tongued press, are caught up by ignorant and untrained men," declared one premillennialist in 1889, "it is time for earnest Christian men to call a halt."

The questionable scientific status of Darwinism undoubtedly encouraged such critics to speak up. Although the overwhelming majority of scientists after 1880 accepted a long earth history and some form of organic evolution, many in the late nineteenth century expressed serious reservations about the ability of Darwin's particular theory of natural selection to account for the origin of species. Their published criticisms of Darwinism led creationists mistakenly to conclude that scientists were in the midst of discarding evolution. The appearance of books with such titles as *The Collapse of Evolution* and *At the Death Bed of Darwinism* bolstered this belief and convinced antievolutionists that liberal Christians had capitulated to evolution too quickly. In view of this turn of events, it seemed likely that those who had "abandoned the stronghold of faith out of sheer fright will soon be found scurrying back to the old and impregnable citadel, when they learn that 'the enemy is in full retreat.'"

Early in 1922, William Jennings Bryan (1860–1925), Presbyterian layman and thrice-defeated Democratic candidate for the presidency of the United States, heard of an effort in Kentucky to ban the teaching of evolution in public schools. "The movement will sweep the country," he predicted hopefully, "and we will drive Darwinism from our schools." His prophecy proved overly optimistic, but, before the end of the decade, more than twenty state legislatures debated antievolution laws, and three—Tennessee, Mississippi, and Arkansas—banned the teaching of evolution in public schools. Oklahoma prohibited the adoption of evolutionary textbooks, while Florida condemned the teaching of Darwinism. At times, the controversy became so tumultuous that it looked to some as though "America might go mad." Many persons shared responsibility for these events, but none more than Bryan. His entry into the fray had a catalytic effect and gave antievolutionists what they needed most, "a spokesman with a national reputation, immense prestige, and a loyal following."

Who joined Bryan's crusade? As recent studies have shown, they came from all walks of life and from every region of the country. They lived in New York City, Chicago, and Los Angeles, as well as in small towns and in the country.

Few possessed advanced degrees, but many were not without education. Nevertheless, Bryan undeniably found his staunchest supporters and won his greatest victories in the conservative and still largely rural South, described hyperbolically by one fundamentalist journal as "the last stronghold of orthodoxy on the North American continent," a region where the "masses of the people in all denominations 'believe the Bible from lid to lid.'"

Leadership of the antievolution movement came not from the organized churches of America but from individuals such as Bryan and interdenominational organizations such as the World's Christian Fundamentals Association, a predominantly premillennialist body founded in 1919 by William Bell Riley (1861–1947), pastor of the First Baptist Church in Minneapolis. Riley became active as an antievolutionist after discovering, to his apparent surprise, that evolutionists were teaching their views at the University of Minnesota. The early twentieth century witnessed an unprecedented expansion of public education—enrollment in public high schools nearly doubled between 1920 and 1930—and fundamentalists such as Riley and Bryan wanted to make sure that students attending these institutions would not lose their faith. Thus, they resolved to drive every evolutionist from the public-school payroll. Those who lost their jobs as a result deserved little sympathy, for, as one rabble-rousing creationist put it, the German soldiers who killed Belgian and French children with poisoned candy were angels compared with the teachers and textbook writers who corrupted the souls of children and thereby sentenced them to eternal death.

The antievolutionists liked to wrap themselves in the authority of science, but, unfortunately for them, they could claim few legitimate scientists of their own: a couple of self-made men of science, one or two physicians, and a handful of teachers who, as one evolutionist described them, were "trying to hold down, not a chair, but a whole settee, of 'Natural Science' in some little institution." Of this group, the most influential were Harry Rimmer (1890–1952) and George McCready Price (1870–1963).

Rimmer, a Presbyterian minister and self-styled "research scientist," had obtained his limited exposure to science during a term or two at San Francisco's Hahnemann Medical College. After his brief stint in medical school, he attended Whittier College and the Bible Institute of Los Angeles for a year each before entering full-time evangelistic work. About 1919 he settled in Los Angeles, where he set up a small laboratory at the rear of his house to conduct experiments in embryology and related sciences. Within a year or two, he established the Research Science Bureau "to prove through findings in biology, paleontology, and anthropology that science and the literal Bible were not contradictory." The bureau staff—that is, Rimmer—apparently used income from the sale of memberships to finance anthropological fieldtrips in the western United States. By the late 1920s, the bureau lay dormant, and Rimmer signed on with Riley's World's Christian Fundamentals Association as a field secretary. Besides engaging in research, Rimmer delivered thousands of lectures, primarily to student groups, on the scientific accuracy of the Bible. Posing as a scientist, he attacked Darwinism and poked fun at the credulity of evolutionists. He also enjoyed success as a debater.

George McCready Price, a self-trained Seventh-Day Adventist geologist, was less skilled at debating than Rimmer but more influential scientifically. As a young man, Price attended an Adventist college in Michigan for two years and later completed a teacher-training course at the provincial normal school in his native New Brunswick. The turn of the twentieth century found him serving as principal of a small high school in an isolated part of eastern Canada, where one of his few companions was a local physician. During their many conversations, the doctor almost converted his fundamentalist friend to evolution, but each time Price wavered he was saved by prayer and by reading the works of the Seventh-Day Adventist prophet Ellen G. White (1827–1915), who claimed divine inspiration for her view that Noah's Flood accounted for the fossil record on which evolutionists based their theory. As a result of these experiences, Price vowed to devote his life to promoting creationism of the strictest kind.

By 1906 he was working as a handyman at an Adventist sanitarium in southern California. That year he published a slim volume entitled *Illogical Geology: The Weakest Point in the Evolution Theory,* in which he brashly offered one thousand dollars "to any one who will, in the face of the facts here presented, show me how to prove that one kind of fossil is older than another." He never had to pay. According to Price's argument, Darwinism rested "logically and historically on the succession of life idea as taught by geology," and "if this succession of life is not an actual scientific fact, then Darwinism . . . is a most gigantic hoax."

During the next fifteen years, Price occupied scientific settees in several Seventh-Day Adventist schools and authored six more books attacking evolution, particularly its geological foundation. Although not unknown outside his own church before the early 1920s, he did not attract national attention until then. Shortly after Bryan declared war on evolution, Price published *The New Geology* (1923), the most systematic and comprehensive of his many books. Uninhibited by false modesty, he presented his "great law of conformable stratigraphic sequences . . . by all odds the most important law ever formulated with reference to the order in which the strata occur." This law stated that "any kind of fossiliferous beds whatever, 'young' or 'old,' may be found occurring conformably on any other fossiliferous beds, 'older' or 'younger.'" To Price, so-called deceptive conformities (where strata seem to be missing) and thrust faults (where the strata are apparently in the wrong order) proved that there was no natural order to the fossil-bearing rocks, all of which he attributed to the Genesis Flood. Despite criticism and ridicule from the scientific establishment—and the fact that his theory contradicted both the day-age and gap interpretations of Genesis—Price's reputation among fundamentalists rose dramatically. By the mid-1920s, the editor of *Science* could accurately describe him as "the principal scientific authority of the Fundamentalists."

In the spring of 1925, John Thomas Scopes (1900–70), a high-school teacher in Dayton, Tennessee, confessed to having violated the state's recently passed law banning the teaching of human evolution in public schools. His subsequent trial focused international attention on the antievolution crusade and brought Bryan to Dayton to assist the prosecution. Although the court in Dayton found Scopes guilty as charged, creationists found little cause for rejoicing. Some members of the press had not treated them kindly, and the taxing ordeal no doubt contributed to Bryan's death a few days after the end of the trial. Nevertheless, the antievolutionists continued their crusade, winning victories in Mississippi in 1926 and in Arkansas two years later. By the end of the decade, however, their legislative campaign had lost its steam.

Contrary to appearances, the creationists did not give up; they simply changed tactics. Instead of lobbying state legislatures, they shifted their attack to local communities, where they engaged in what one critic described as "the emasculation of textbooks, the 'purging' of libraries, and above all the continued hounding of teachers." Their new approach attracted less attention but paid off handsomely, as school boards, textbook publishers, and teachers in both urban and rural areas, North and South, bowed to their pressure. Darwinism virtually disappeared from high-school texts, and for years many American teachers feared being identified as evolutionists. Instead of attempting to convert the world to their way of thinking, the creationists increasingly turned their energies inward, organizing their own societies and editing their own journals.

The Creationist Revival

In 1964, one historian predicted that "a renaissance of the [creationist] movement is most unlikely." And so it seemed. But even as these words were penned, a major revival was under way, led by a Texas engineer, Henry M. Morris (1918–). Raised a nominal Southern Baptist, and as such a believer in Creation, Morris as a youth had drifted unthinkingly into evolutionism and religious indifference. A thorough study of the Bible following graduation from college convinced him of its absolute truth and prompted him to reevaluate his belief in evolution. After an intense period of soul-searching, he concluded that Creation had taken place in six literal days because the Bible clearly said so and "God doesn't lie." In the late 1950s, he began collaborating with a young theologian, John C. Whitcomb Jr. (1924–), of the Grace Brethren denomination, on a defense of Price's Flood geology. By the time they finished their project, Morris had earned a Ph.D. in hydraulic engineering from the University of Minnesota and was chairing the Civil Engineering Department at Virginia Polytechnic Institute; Whitcomb, a Princeton alumnus, was teaching Old Testament studies at Grace Theological Seminary in Indiana.

In 1961, they brought out *The Genesis Flood*, the most impressive contribution to strict creationism since the publication of Price's *New Geology* in 1923. In many respects, their book appeared to be simply "a reissue of G. M. Price's views, brought up to date," as one reader described it. Beginning with a testimony to their belief in "the verbal inerrancy of Scripture," Whitcomb and Morris went on to argue for a recent Creation of the entire universe, a Fall that triggered the second law of thermodynamics, and a worldwide Flood that in one year laid down most of the geological strata. Given this history, they argued, "the last refuge of the case for evolution immediately vanishes away, and the record of the rocks becomes a tremendous witness . . . to the holiness and justice and power of the living God of Creation!"

Despite the book's lack of conceptual novelty, it provoked intense debate among evangelicals. Progressive creationists denounced it as a travesty on geology that threatened to set back the cause of Christian science a generation, while strict creationists praised it for making biblical catastrophism intellectually respectable. Its appeal, suggested one critic, lay primarily in the fact that, unlike previous creationist works, it "looked legitimate as a scientific contribution," accompanied as it was by footnotes and other scholarly appurtenances. In responding to their detractors, Whitcomb and Morris repeatedly refused to be drawn into a scientific debate, arguing that "the real issue is not the correctness of the interpretation of various details of the geological data, but simply what God has revealed in His Word concerning these matters."

Whatever its merits, *The Genesis Flood* unquestionably "brought about a stunning renaissance of flood geology," symbolized by the establishment in 1963 of the Creation Research Society. Shortly before the publication of his book, Morris had sent the manuscript to Walter E. Lammerts (1904–96), a Missouri Synod Lutheran with a doctorate in genetics from the University of California. As an undergraduate at Berkeley, Lammerts had discovered Price's *New Geology*, and, during the early 1940s while teaching at UCLA, he had worked with Price in a local creationist society. After the mid-1940s, however, his interest in creationism had flagged—until awakened by reading the Whitcomb and Morris manuscript. Disgusted by some evangelicals' flirtation with evolution, he organized in the early 1960s a correspondence network with Morris and eight other strict creationists, dubbed the "team of ten." In 1963, seven of the ten met with a few other like-minded scientists at the home of a team member in Midland, Michigan, to form the Creation Research Society (CRS). Of the ten founding members, five possessed doctorates in biology; a sixth had earned a Ph.D. degree in biochemistry; and a seventh held a master's degree in biology.

At the end of its first decade, the society claimed 450 regular members, plus 1,600 sustaining members, who failed to meet the scientific qualifications. Eschewing politics, the CRS devoted itself almost exclusively to education and research, funded "at very little expense, and . . . with no expenditure of public money." CRS-related projects included expeditions to search for Noah's ark, studies of fossil human footprints and pollen grains found out of the predicted evolutionary order, experiments on radiation-produced mutations in plants, and theoretical studies in physics demonstrating a recent origin of the earth. A number of members collaborated in preparing a biology textbook based on creationist principles. In view of the previous history of creation science, it was an auspicious beginning.

The creationist revival of the 1960s attracted little public attention until late in the decade, when fundamentalists became aroused about the federally funded Biological Sciences Curriculum Study texts, which featured evolution, and the California State Board of Education voted to require public-school textbooks to include Creation along with evolution. This decision resulted in large part from the efforts of two southern California women, Nell J. Segraves (1922–) and Jean E. Sumrall (1927–). In 1961, Segraves learned of the U.S. Supreme Court's ruling in the Madalyn Murray (1919–late 1990s?) case, protecting atheist students from required prayers in public schools. Murray's ability to shield her child from religious exposure suggested to Segraves that creationist parents like herself "were entitled to protect our children from the influence of beliefs that would be offensive to our religious beliefs." It was this line of argument that finally persuaded the Board of Education to grant creationists equal rights.

Flushed with victory, in 1970 Segraves and her son Kelly (1942–) joined an effort to organize a Creation Science Research Center (CSRC), affiliated with Christian Heritage College in San Diego, to prepare creationist literature suitable for adoption in public schools. Associated with them in this enterprise was Henry Morris, who resigned his position at Virginia Polytechnic Institute to help establish a center for creation research. Because of differences in personalities and objectives, in 1972 the Segraveses left the college, taking the CSRC with them; Morris thereupon set up a new research division at the college, the Institute of Creation Research (ICR), which, he announced with obvious relief, would be "controlled and operated by scientists" and would engage in research and education, not political action. During the 1970s, Morris added five scientists to his staff and, funded largely by small gifts and royalties from institute publications, turned the ICR into the world's leading center for the propagation of strict creationism.

The 1970s witnessed another major shift in creationist tactics. Instead of trying to outlaw evolution, as they had done in the 1920s, antievolutionists now fought to give Creation equal time. And instead of appealing to the authority of the Bible, as Morris and Whitcomb had done as recently as 1961, they consciously downplayed the Genesis story in favor of what they called "scientific creationism." By 1974, Morris was recommending that creationists ask public schools to teach "only the scientific aspects of creationism," which, in practice, meant leaving out all references to the six days of Genesis and Noah's ark and focusing instead on evidence for a recent worldwide catastrophe and on arguments against evolution. Thus, the product remained virtually the same; only the packaging changed. The ICR textbook *Scientific Creationism* (1974), for example, came in two editions: one for public schools, containing no references to the Bible, and another for use in Christian schools that included a chapter on "Creation According to Scripture."

Creationists professed to see no reason why their Flood-geology model should not be allowed to compete on an equal scientific basis with the evolution model. In selling this two-model approach to school boards, creationists pressed their scientific claims. This tactic proved extremely effective, at least initially. Two state legislatures, in Arkansas and Louisiana, and various school boards adopted the two-model approach, and an informal poll of American school-board members in 1980 showed that only 25 percent favored teaching nothing but evolution. In 1982, however, a federal judge declared the Arkansas law, requiring a "balanced treatment" of Creation and evolution, to be unconstitutional, a decision endorsed by the U.S. Supreme Court five years later.

The influence of the creationist revival sparked by Whitcomb and Morris was immense. Not least, it elevated the strict creationism of Price and Morris to a position of virtual orthodoxy among fundamentalists, and it endowed creationism with a measure of scientific respectability unknown since the deaths of Guyot and Dawson. Unlike the antievolution crusade of the 1920s, which remained confined mainly to North America, the revival of the 1960s rapidly spread overseas as American creationists and their books circled the globe. Partly as a result of stimulation from America, including the publication of a British edition of *The Genesis Flood* in 1969, the lethargic Evolution Protest Movement, founded in Great Britain in the 1930s, was revitalized, and two new creationist organizations, the Newton Scientific Association and the Biblical Creation Society, sprang into existence in Britain. On the Continent, the Dutch assumed the lead in promoting creationism, encouraged by the translation of books on Flood geology and by visits from ICR scientists. Similar developments occurred elsewhere in Europe, as well as in Australia, New Zealand, Asia, and South America. By 1980, Morris's books alone had been translated into Chinese, Czech, Dutch, French, German, Japanese, Korean, Portuguese, Russian, and Spanish. Strict creationism had become an international phenomenon.

BIBLIOGRAPHY

Eve, Raymond A., and Francis B. Harrold. *The Creationist Movement in Modern America.* Boston: Twayne, 1991.

Gatewood, Willard B., Jr. *Preachers, Pedagogues, and Politicians: The Evolution Controversy in North Carolina, 1920–1927.* Chapel Hill: University of North Carolina Press, 1966.

———, ed. *Controversy in the Twenties: Fundamentalism, Modernism, and Evolution.* Nashville: Vanderbilt University Press, 1969.

Larson, Edward J. *Trial and Error: The American Controversy over Creation and Evolution.* Updated ed. New York: Oxford University Press, 1989.

———. *Summer for the Gods: The Scopes Trial and America's Continuing Debate over Science and Religion.* New York: Harvard University Press, 1998.

Marsden, George M. *Understanding Fundamentalism and Evangelicalism.* Grand Rapids, Mich.: Eerdmans, 1991.

McIver, Tom. *Anti-Evolution: A Reader's Guide to Writings before and after Darwin.* Baltimore: Johns Hopkins University Press, 1992.

Moore, James R. *The Post-Darwinian Controversies: A Study of the Protestant Struggle to Come to Terms with Darwin in Great Britain and America, 1870–1900.* Cambridge: Cambridge University Press, 1979.

———. "The Creationist Cosmos of Protestant Fundamentalism." In *Fundamentalisms and Society: Reclaiming the Sciences, the Family, and Education,* ed. Martin E. Marty and R. Scott Appleby. Chicago: University of Chicago Press, 1993, 42–72.

Morris, Henry M. *A History of Modern Creationism.* San Diego: Master Book Publishers, 1984.

Nelkin, Dorothy. *The Creation Controversy: Science or Scripture in the Schools?* New York: Norton, 1982.

Numbers, Ronald L. *The Creationists: The Evolution of Scientific Creationism.* New York: Knopf, 1992.

———, ed. *Creationism in Twentieth-Century America: A Ten-Volume Anthology of Documents, 1903–1961.* New York: Garland, 1995.

———. *Darwinism Comes to America.* Cambridge: Harvard University Press, 1998.

Roberts, Jon H. *Darwinism and the Divine in America: Protestant Intellectuals and Organic Evolution, 1858–1900.* Madison: University of Wisconsin Press, 1988.

Szasz, Ferenc Morton. *The Divided Mind of Protestant America, 1880–1930.* Tuscaloosa: University of Alabama Press, 1982.

Toumey, Christopher P. *God's Own Scientists: Creationists in a Secular World.* New Brunswick: Rutgers University Press, 1994.

Trollinger, William Vance, Jr. *God's Empire: William Bell Riley and Midwestern Fundamentalism.* Madison: University of Wisconsin Press, 1990.

Webb, George E. *The Evolution Controversy in America.* Lexington: University Press of Kentucky, 1994.

22 The Scopes Trial

Edward J. Larson

During the summer of 1925, the state of Tennessee indicted, tried, and convicted John T. Scopes for teaching the Darwinian theory of human evolution to students at Rhea County High School in Dayton, Tennessee. The episode made headlines in newspapers throughout the United States and Western Europe, and it has justly remained famous ever since. The trial was not a hostile proceeding, however. It was initiated by persons and groups opposed to a then-new Tennessee statute outlawing the teaching of the theory of human evolution. Scopes volunteered to stand trial as a means to test the statute. It was a made-for-the-media event that at the time and ever since has dramatized the tension between the authority of science and that of religion in modern America.

Edward J. Larson is the Richard B. Russell Professor of History and Law and chair of the History Department at the University of Georgia. He received the 1998 Pulitzer Prize in history for *Summer for the Gods: The Scopes Trial and America's Continuing Debate over Science and Religion* (New York: Basic Books, 1977). His latest book is *Evolution's Workshop: God and Science on the Galapagos Islands* (New York: Basic Books, 2001).

THE SCOPES TRIAL remains the best-known encounter between science and religion to take place in the United States. It occurred in 1925, soon after the Tennessee state legislature passed a statute forbidding public-school teachers from instructing students in the theory of human evolution. The law was the first major success of an intense national campaign by Protestant fundamentalists against the teaching of organic evolution in public schools. The so-called antievolution crusade, which had begun in earnest three years earlier, aroused fierce opposition from many American scientists, educators, and civil libertarians. After passage of the Tennessee antievolution statute, the growing public controversy soon focused on the small Tennessee town of Dayton, where a local science teacher named John T. Scopes (1900–1970) accepted the invitation of the American Civil Liberties Union (ACLU) to challenge the new law in court. The media promptly proclaimed it "the trial of the century" as this young teacher (backed by the nation's scientific, educational, and cultural establishments) stood against the forces of fundamentalist religious lawmaking.

Setting the Stage

The trial itself was more a media circus than a serious criminal prosecution. It began when Dayton civil engineer George W. Rappleyea, a vocal opponent of the antievolution law, learned of the ACLU's offer to assist any Tennessee teacher willing to challenge the new restriction in court. He immediately saw a chance to strike the statute and promote his community. Such a trial would put Dayton on the map, Rappleyea explained to other civic leaders. They agreed and asked John Scopes to stand trial.

Except for the fact that he did not teach biology, Scopes was an ideal defendant for the test case. Twenty-four years old at the time, Scopes had just completed his first year of teaching general science and coaching football at Dayton's public high school. A popular newcomer to town, he had little to lose from challenging the law—unlike the biology teacher, who had a family and also served as the high-school principal. Scopes looked the part of an earnest young teacher, complete with horn-rimmed glasses and a boyish face that made him appear academic but not threatening. Naturally shy, cooperative, and well liked, he would not alienate parents or taxpayers with soapbox speeches on evolution or give the appearance of a radical or ungrateful public employee. Yet his friends knew that Scopes disapproved of the new law and accepted an evolutionary view of human origins. Further, Scopes's father, an immigrant railroad mechanic and labor organizer, was a militant socialist and religious agnostic. John Scopes inclined toward his father's views about government and religion but in an easygoing way. When asked by the local school board president and the school superintendent if he would stand trial for teaching evolution, Scopes readily consented, even though he doubted that he had ever violated the law.

Bryan versus Darrow

Ever since Charles Darwin had published his theory of evolution in 1859, some conservative Christians had objected to the materialistic implications of its naturalistic explanation for the origins of man. Many of them maintained that God especially created the first humans, as the Bible suggested in Genesis, and rejected the notion that people had evolved from "brutes." Early in the twentieth century, these objections intensified with the spread of fundamentalism as a reaction by some traditional American Protestants to increased religious liberalism within their mainline denominations, especially because many of their opponents embraced an evolutionary view of religion. Fighting for the fundamentals of biblical orthodoxy, many fundamentalist leaders denounced evolutionary thinking in both science and religion.

American politician and orator William Jennings Bryan (1860–1925), a former Democratic presidential nominee with decidedly conservative religious beliefs, added his voice to this chorus opposing evolutionary teaching after World War I, as he came to see Darwinian theories of survival of the fittest (known as *social Darwinism* when applied to human society) behind excessive militarism, imperialism, and laissez-faire capitalism—the three greatest sins in Bryan's political theology. With his Progressive political instinct of seeking legislative solutions to social problems, Bryan called for the enactment of state restrictions on teaching the Darwinian theory of human evolution in public schools. Indeed, the antievolution movement had all the trappings of a political campaign as Bryan spoke, wrote, and lobbied for such laws across America during the early 1920s. Following a near miss in Kentucky in 1922, antievolution bills appeared in state legislatures throughout the country—with Bryan behind them all—culminating in the passage of the Tennessee law in 1925. When it became apparent that the ACLU was seeking to discredit that new law through the trial of John Scopes, Bryan volunteered to assist the prosecutors with the object of explaining why such a restriction was needed.

Bryan's pending appearance in Dayton drew in Clarence Darrow (1857–1938). By the twenties, Darrow stood out as the most famous trial lawyer in America. He had gained fame as the country's premier defender of labor organizers and political radicals but also was known for his militant opposition to religious influences in public life, particularly to biblically inspired legal restrictions on personal freedom. In this cause, he was a pioneer. His opposition to religious lawmaking stemmed from his belief that revealed religion, especially Christianity, divided people into warring sects, caused them to be judgmental, and was an irrational basis for action in a modern scientific age. In popular speeches, books, and articles, Darrow sought to expose biblical literalism as both irrational and dangerous. He welcomed the hullabaloo surrounding the antievolution crusade. It rekindled interest in his attacks on the Bible, which once had appeared hopelessly out of date in light of modern developments in mainline Christian thought but now seemed to regain relevance with the rise of fundamentalism. When Bryan offered to prosecute Scopes, Darrow volunteered to defend him. Darrow immediately became the brightest light in the already luminous defense team assembled by the ACLU to challenge Tennessee's antievolution law.

With Bryan and Darrow on board, Dayton civic leaders could only marvel at the success of their publicity scheme. They feted both men with banquets in their honor and housed them in two of Dayton's finest private homes.

National Interest

The prospect of these two renowned orators—Bryan and Darrow—actually litigating the profound issues of science versus religion and academic freedom versus popular control over public education turned the trial into a media sensation at the time and into the stuff of legend thereafter. News of the trial dominated the headlines during the weeks leading up to it and pushed nearly everything else off American front pages throughout the eight-day event. Two hundred reporters covered the story in Dayton, including some of the country's best correspondents, who represented many of the major newspapers and magazines. Thousands of miles of telegraph wires were hung to transmit every word spoken in court, and pioneering live radio broadcasts carried the oratory to the listening public. Newsreel cameras recorded the encounter, with the film flown directly to major Northern cities for projection in movie houses. The media billed it as "the trial of the century" before it even began, and it lived up to its billing.

For those involved, the encounter resembled a popular debate before a national audience more than it did a criminal prosecution tried before local jurors, a fact that explains the decision of the judge to permit radio microphones, newsreel cameras, and telegraph key operators in the courtroom. The courtroom arguments and speeches by both sides addressed the nation rather than the jurors (who missed most of the oratory anyway because it had so little to do with the facts of the case that it was delivered with the jury excused). The defense divided its arguments among its three principal attorneys. The prominent New York attorney Arthur Garfield Hays raised the standard ACLU arguments that Tennessee's antievolution statute violated the individual rights of teachers. Former Assistant Secretary of State Dudley Field Malone, a liberal Catholic, argued that the scientific theory of evolution did not necessarily conflict with an open-minded reading of Genesis. Darrow, for his part, concentrated on debunking fundamentalist reliance on revealed Scripture as a source of knowledge about nature that was suitable for setting educational standards. Their common goal, as Hays stated at the time, was to make it "possible that laws of this kind will hereafter meet the opposition of an aroused public opinion."

The prosecution countered with a half-dozen local attorneys led by the state's able prosecutor and future U.S. senator, Tom Stewart, plus Bryan and his son, William Jennings Jr., a California lawyer. In court, they focused on proving that Scopes broke the law and objected to any attempt to litigate the merits of that statute. The public, acting through elected legislators, should control the content of public education, they maintained. The elder Bryan, who had not practiced law for three decades, stayed uncharacteristically quiet in court and saved his oratory for lecturing the assembled press and public outside the courtroom about the vices of teaching evolution and the virtues of majority rule. He also prepared a thunderous three-hour-long address on these points that he planned to deliver as the prosecution's closing argument. As the actual trial played itself out, however, Darrow managed to frustrate Bryan's plan by waiving his own close because, under Tennessee practice, the defense controlled whether there would be closing arguments.

The Trial Unfolds

First came jury selection. Darrow typically stressed this part of a trial as being critical for the defense and often spent weeks going through hundreds of veniremen before settling on twelve jurors who just might be open to his arguments and acquit his typically notorious defendant. Darrow had a different objective at the Scopes trial, however. He wanted to convict the statute rather than acquit the defendant, and only judges could do this. Jurors simply applied the law to the facts of the case. Darrow could have won an acquittal by arguing that Scopes (who, after all, was not even a biology teacher) never violated the statute, but that would have left the statute intact. Instead, the defense sought either to have the trial judge strike the statute, which was all but beyond his role, or to have Scopes convicted and then appeal to a higher court, which could review the statute. Hence, Darrow quickly accepted even the simplest of jurors, including several who had never heard of the theory of evolution and one who could not read. This, he hoped, would dramatize the folly of nonscientists, be they jurors or legislators, sitting in judgment on the teaching of a scientific idea such as evolution.

No sooner was the jury selected than it was excused from the courtroom—for days—as the parties wrangled over defense motions to strike the statute as unconstitutional. Although these arguments occasionally soared into dramatic pleas from the defense for academic freedom and from the prosecution for majority rule, they generally skirted the underlying issues of science versus religion. Strictly following Tennessee precedent, the trial judge denied the motions anyway. Such issues were reserved for the state supreme court to resolve on appeal, he ruled.

The prosecution then presented uncontested testimony by students and school officials that Scopes had taught evolution. After the prosecution's brief presentation, the defense offered the testimony of fifteen national experts in science and religion, all prepared to defend the theory of evolution as a valid scientific theory that could be taught without public harm. The prosecution immediately objected to such testimony as irrelevant to the issue of whether Scopes had violated the law. The law should not be on trial, prosecutors argued, only the defendant. After days of debate, the judge agreed with the prosecution and the trial appeared to have ended without ever directly taking up the promised issues of science and religion in public education.

Bryan's Testimony

Frustrated by his failure to discredit the antievolution law through the testimony of scientists and liberal theologians, Darrow sought the same result by inviting Bryan to take the witness stand and face questions about it. Although he could have declined, Bryan accepted Darrow's challenge. Up to this point, lead prosecutor Tom Stewart had masterfully limited the proceedings and, with help from a friendly judge, confined his wily opponents. But Stewart could not control his impetuous co-counsel, especially because the judge seemed eager to hear Bryan defend the faith. "They did not come here to try this case," Bryan explained early in his testimony. "They came here to try revealed religion. I am here to defend it, and they can ask me any questions they please." Darrow did just that.

Thinking the trial all but over, except for the much-awaited closing oratory, and hearing that cracks had appeared in the ceiling below the overcrowded, second-floor courtroom, the judge had moved the afternoon session outside onto the courthouse lawn. The crowd swelled as word of the encounter spread. From the five hundred persons who evacuated the courtroom, the number rose to an estimated three thousand people spread over the lawn—nearly twice the town's normal population. The participants appeared on a crude wooden platform erected for the proceedings, looking much like Punch and Judy puppets performing at an outdoor festival. Enterprising youngsters passed through the crowd hawking refreshments. Darrow posed the well-worn questions of the village skeptic: Did Jonah live inside a whale for three days? How could Joshua lengthen the day by making the sun (rather than the earth) stand still? Where did Cain get his wife? In a narrow sense, as Stewart persistently complained, Darrow's questions had nothing to do with the case because they never inquired about human evolution. In a broad sense, as Hays repeatedly countered, they had everything to do with it because they challenged biblical literalism. Best of all for Darrow, no good answers to the questions existed.

Darrow questioned Bryan as a hostile witness, peppering him with queries and giving him little chance to explain. At times, the encounter resembled a firing line:

> "You claim that everything in the Bible should be literally interpreted?"
>
> "I believe everything in the Bible should be accepted as it is given there; some of the Bible is given illustratively. . . ."
>
> "But when you read that . . . the whale swallowed Jonah . . . how do you literally interpret that?"
>
> ". . . I believe in a God who can make a whale and can make a man and make both of them do what he pleases. . . ."
>
> "But do you believe he made them—that he made such a fish and it was big enough to swallow Jonah?"
>
> "Yes sir. Let me add: One miracle is just as easy to believe as another."
>
> "It is for me . . . just as hard."
>
> "It is hard to believe for you, but easy for me. . . . When you get beyond what man can do, you get within the realm of miracles; and it is just as easy to believe the miracle of Jonah as any other miracle in the Bible."

Such affirmations undercut the appeal of fundamentalism. On the stump, Bryan effectively championed the cause of biblical faith by addressing the great questions of life: The special creation of humans in God's image gave purpose to every person, and the bodily resurrection of Christ gave hope for eternal life to believers. But Darrow did not inquire about these grand miracles. For many Americans, laudable simple faith became laughable crude belief when applied to Jonah's whale, Noah's flood, and Adam's rib. Yet Bryan acknowledged accepting each of these biblical miracles on faith and professed that all miracles were equally easy to believe.

Bryan fared little better when he tried to rationalize two of the biblical passages raised by Darrow. In an apparent concession to modern astronomy, Bryan suggested that God extended the day for Joshua by stopping the earth rather than the sun. Similarly, in line with nineteenth-century evangelical scholarship, Bryan affirmed his understanding that the Genesis days of creation represented periods of time, leading to the following exchange, with Darrow asking the questions:

> "Have you any idea of the length of these periods?"
>
> "No; I don't."
>
> "Do you think the sun was made on the fourth day?"
>
> "Yes."
>
> "And they had evening and morning without the sun?"
>
> "I am simply saying it is a period."
>
> "They had evening and morning for four periods without the sun, do you think?"
>
> "I believe in creation as there told, and if I am not able to explain it I will accept it."

The earth could be 600,000,000 years old, Bryan admitted. Though he had not ventured far beyond the bounds of biblical literalism, the defense made the most of it. "Bryan had conceded that he interpreted the Bible," Hays gloated. "He must have agreed that others have the same right."

Stewart tried to end the two-hour-long interrogation at least a dozen times, but Bryan refused to stop. "I am simply trying to protect the word of God against the greatest atheist or agnostic in the United States," he shouted, pounding his fists in rage. "I want the papers to know I am not afraid to get on the stand in front of him and let him do his worst." The crowd cheered this outburst and every counterthrust he attempted. Darrow received little applause but inflicted the most jabs. "The only purpose Mr. Darrow has is to slur the Bible, but I will answer his questions," Bryan exclaimed at the end. "I object to your statement," Darrow shouted back, both men now standing and shaking their fists at each other. "I am examining your fool ideas that no intelligent Christian on earth believes." The judge finally had heard enough and abruptly adjourned court for the day. He never let the interrogation resume.

Legacy

Although partisans on both sides claimed the advantage, at the time most neutral observers viewed the trial as a draw, and none saw it as decisive. America's adversarial legal system tends to drive parties apart rather than to reconcile them. That was certainly the result in this case. Despite Bryan's stumbling on the witness stand (which his supporters attributed to his notorious interrogator's wiles), both sides effectively communicated their message from Dayton—maybe not well enough to win converts but at least sufficiently to energize those already predisposed toward their viewpoint. Owing largely to the media's portrayal of Darrow's effective cross-examination of Bryan, later made even more cutting in the popular play and film *Inherit the Wind,* millions of Americans thereafter ridiculed religious opposition to the theory of evolution. Yet the widespread coverage that was given Bryan's impassioned objections to the theory made opposition to it all but an article of faith among American fundamentalists. When Bryan died a week later in Dayton, fundamentalists acquired a martyr to their cause.

Consequently, the pace of antievolutionist activism actually quickened after the trial, but it encountered heightened popular resistance. Two other states promptly passed antievolution statutes modeled on the Tennessee law and several others imposed lesser restrictions. An anticipated victory in the Minnesota legislature turned into a demoralizing defeat, however. When one Rhode Island legislator introduced such a proposal in 1927, his bemused colleagues referred it to the Committee on Fish and Game, where it died without a hearing or a vote. The divided and divisive response on this issue continues to the present, as some school officials impose restrictions on teaching evolution while others ridicule them. Underlying this rift, surveys of public opinion consistently reveal that Americans remain nearly evenly split between those accepting the scientific theory of human evolution and those believing that God specially created Adam and Eve within the past ten thousand years.

With time and countless retellings, the Scopes trial has become part of the fabric of American culture. For some, it grew to symbolize the threat to scientific freedom and progress posed not simply by antievolutionism but also by religiously motivated lawmaking generally. For others, it suggested a growing hostility to religious faith within the scientific community and modern American society. Each side drew its own lessons from the Scopes trial and viewed future developments in light of those lessons.

BIBLIOGRAPHY

Conkin, Paul K. *When All the Gods Trembled: Darwin, Scopes, and American Intellectuals.* Lanham, Md.: Rowman and Littlefield, 1998.

Darrow, Clarence. *The Story of My Life.* New York: Grosset, 1932.

Ginger, Ray. *Six Days or Forever? Tennessee v. John Thomas Scopes.* London: Oxford University Press, 1958.

Larson, Edward J. *Summer for the Gods: The Scopes Trial and America's Continuing Debate over Science and Religion.* New York: Basic Books, 1997.

———. *Trial and Error: The American Controversy over Creation and Evolution.* New York: Oxford University Press, 1989.

Lawrence, Jerome, and Robert E. Lee. *Inherit the Wind.* New York: Bantam, 1960.

Levine, Lawrence W. *Defender of the Faith: William Jennings Bryan, the Last Decade, 1915–1925.* New York: Oxford University Press, 1965.

Marsden, George M. *The Soul of the American University: From Protestant Establishment to Established Non-belief.* New York: Oxford University Press, 1994.

Numbers, Ronald L. *Darwinism Comes to America.* Cambridge: Harvard University Press, 1998.

———. *The Creationists: The Evolution of Scientific Creationism.* New York: Knopf, 1992.

Tierney, Kevin. *Darrow: A Biography.* New York: Croswell, 1979.

Tompkins, Jerry D., ed. *D-Days at Dayton: Reflections on the Scopes Trial.* Baton Rouge: Louisiana State University Press, 1965.

Walker, Samuel. *In Defense of American Liberties: A History of the ACLU.* New York: Oxford University Press, 1990.

World's Most Famous Court Case: Tennessee Evolution Case. Dayton: Bryan College, 1990.

Part VI :: The Theological Implications of Modern Science

23 Physics

Richard Olson

This essay provides a broad overview of interactions between physics (or *natural philosophy,* the term formerly used for physical science) and religion from the early eighteenth century to the end of the twentieth. In the eighteenth century Newtonian natural theology sought proofs of God's existence and attributes in the structure and operations of the physical universe. In the nineteenth century, in response to Kantian claims that no religious propositions could be inferred from natural knowledge, a more limited natural theology sought merely to establish the compatibility between traditional Christian doctrines and the workings of the physical universe. At the same time some natural philosophers sought to develop heterodox scientistic religions. In the twentieth century, scientists such as Albert Einstein, David Bohm, and Freeman Dyson and theologians such as Wolfhart Pannenberg sought to explore the religious implications of relativity theory and quantum mechanics.

Richard Olson received his Ph.D. in the history of science from Harvard University in 1967. He is currently professor of history and Willard W. Keith Fellow in Humanities at Harvey Mudd College and adjunct professor of history at Claremont Graduate University. His publications include two volumes (soon to be three) of *Science Deified and Science Defied* (Berkeley. University of California Press, 1982 and 1990).

PHYSICS (*physique* [French], *physicae* [Latin], or *physik* [German]) became widely used in its modern sense (i.e., excluding the life sciences, geology, and chemistry) during the second half of the eighteenth century. As late as 1879, however, the major English-language textbook that covered what we call physics was Sir William Thomson (Lord Kelvin [1824–1907]) and Peter Guthrie Tait's (1831–1901) *Treatise on Natural Philosophy,* and university courses in Britain and America were still labeled courses in natural philosophy. Hence, this discussion of the religious elements in, and the implications of, physics begins about the middle of the eighteenth century and counts as physicists many figures who identified themselves as natural philosophers.

The periodization of physics is also unusual, and its labeling is inconsistent with general philosophical and literary usage. Topics treated before the middle of the last decade of the nineteenth century—mechanics, optics, heat, electricity and magnetism, hydrostatics and hydrodynamics—coupled with the theories and procedures that existed before 1897 to treat them are said to be parts of *classical* physics, whereas a group of topics that emerged after about 1895, including natural radioactivity, quantum physics (subatomic, atomic, molecular, plasma, and solid-state), as well as special and general relativity, is said to make up *modern* physics. Modern physics not only challenged the physical intuitions associated with classical physics, it also seemed to many to suggest very different religious implications. Hence, the following discussion is separated into classical and modern periods, dividing the two in the first decade of the twentieth century.

Classical Physics and Religion

During the first half of the eighteenth century, the most important and characteristic interactions between religion and natural philosophy occurred in connection with Newtonian natural theology, several features of which are important to recognize to set the stage for the religious impact of post-Newtonian developments in physics. First, Newtonian natural theology emphasized the need for some kind of active, nonmaterial agent, either God or something added by God to matter, to account for gravitational attraction. Either emphasis continued to support the matter / spirit dualism that had emerged as central to Cartesian and corpuscular philosophy during the seventeenth century. The former had special appeal for those who sought to find scientific support for belief in a God who remained continuously active within the natural universe. Second, Isaac Newton (1642–1727) insisted that the massively improbable structure of the solar system supported the argument that it had to be the product of a designer God rather than of mere chance. Finally, Newtonian natural theology acknowledged (indeed, it insisted upon) the need for God's infrequent, but unquestionably miraculous, interference with natural processes, for, without such miraculous interventions, it seemed clear from calculations based on Newton's *Principia* (1687) that instabilities in the solar system would have caused it to collapse within the duration of historical time. All of these features of the physical universe were understood to provide proofs of the existence of God.

Beginning in the middle decades of the eighteenth century, continuing developments in classical physics undermined the Newtonian position and substantially modified the way in which physics and religion were understood to be connected. One of the simplest and most dramatic effects emerged out of the development of celestial mechanics by Louis Lagrange (1736–1813) and Pierre Simon Laplace (1749–1827). In his *Exposition du systeme du monde* (*Explanation of a World System* [1797]), Laplace was able to demonstrate that the approximations Newton had used in dealing with the motions of Saturn and Jupiter had been the cause of the apparent nonperiodic element in their motions and that a more thorough solution of the problem failed to predict a collapse of the solar system. Hence, the Newtonian argument for the necessity of God's miraculous intervention in the world was vitiated. Perhaps even more important in the long run, Laplace proposed his nebular hypothesis, which offered a purely physical account of the structure of the solar system, which Newton had used to justify a belief in God. Laplace's physics came to symbolize the position of most French scientists, who argued that physics no longer offered any support for the traditional notion of God and that its implications favored pure materialism.

A quite different development of Newtonian ideas was initiated in *Theoria philosophiae naturalis* (*Theory of Natural Philosophy* [1758]) by the Serbian Jesuit Rodger Joseph Boscovich (1711–87), who also tended toward a version of materialism, but one quite unlike that of Laplace. Boscovich demonstrated that all versions of the mechanical philosophy that depended on the transfer of motion by the impact of perfectly hard particles involved a set of foundational assumptions that were logically inconsistent with each other. The problems of the mechanistic hypothesis could be avoided if one admitted that our notion of matter is drawn from our experiences of repulsive and attractive forces. Indeed, Boscovich went on to argue, particles of matter are best understood as unextended point centers of patterned forces that extend through space. Near the point center these forces approach infinite repulsion, while at great distances they approach the gravitational force of attraction. Between, they oscillate between attractive and repulsive regions, accounting for such phenomena as chemical affinities, the different phases of matter, and electrical and magnetic attractions and repulsions.

By decoupling the definition of matter from its traditional grounding in extension and by focusing on the constitutive active powers of particles of matter to attract and repel other entities, Boscovich undermined both the traditional grounds for dualistic ontologies and the grounds for arguing that the activity of matter must be a direct manifestation of God or something added to passive matter by God. Though Boscovich's ideas seemed to many to advocate atheism or deism by challenging the need for an immanent God, they were appropriated by Joseph Priestley (1733–1804) in his *Disquisitions Relating to Matter and Spirit* (1777) in support of what he viewed as a necessary reform of Christianity. According to Priestley, an outstanding self-taught natural philosopher and the founder of British Unitarianism, the belief in a dualism between matter and spirit derived from a contamination of primitive Christianity by Greek, especially Platonist, philosophy. Moreover, it led to such perversions of true Christianity as the doctrines of the Trinity and of the immortality of the soul (which seemed to obviate the doctrine of the resurrection). It also led to hopeless philosophical difficulties like the problem of how the soul and the body could interact if one was material and the other spiritual. By breaking down this dualism, Boscovich's physics not only avoided the confusions associated with Cartesian dualism, it also pointed the way toward a recovery of what Unitarians believed was the original meaning of Christianity, including its central doctrine of the resurrection, which now became the complete reconstruction of the mortal person by God. Boscovich's revisions of Newtonian natural philosophy also suggested to Priestley another central notion: that progress in science was God's way of gradually eliminating error and prejudice, of ending usurped authority in religion and politics, and of leading to the ultimate triumph of Christianity.

In Germany, the development of Newtonian philosophy in ways similar to those of Boscovich led in a very different direction through the highly influential works of Immanuel Kant (1724–1804). In his *Metaphysische Anfangsgrunde der Naturwissenschaft* (*Metaphysical Elements of Natural Science* [1787]), Kant also argued that attractive and repulsive forces are the essence of matter and that they are, therefore, not something added by a spiritual entity. Indeed, argued Kant, no argument derived from nature could prove anything about the existence or the nature of God. For Kant, religious issues were fundamentally moral issues, and the "oughts" of morality could not be derived from the "is" of natural philosophy. This Kantian separation of science from religion had a major influence on German scientists and theologians, minimizing their interest in natural theology. Moreover, Kantian ideas entered British natural theology through the writings of William Whewell (1794–1866) and Scots such as Thomas Chalmers (1780–1847). In Britain, however, although Kantian arguments changed the character of some claims of sophisticated natural theologians, they did not undermine the traditionally close linkages between science and religion. Beginning with Whewell's *Astronomy and General Physics Considered with Reference to Natural Theology* (1833), British natural theology virtually abandoned its traditional attempts to derive the duties of Christians from the natural world, as well as its claims that natural theology actually provided proof of God's existence. Henceforth, most British works in natural theology, such as *The Unseen Universe; or, Physical Speculations on a Future State* (1875) by physicists Peter Guthrie Tait (1831–1901) and Balfour Stewart (1828–87), admitted that they could not prove God's existence. Instead, they limited themselves to demonstrating the compatibility between the structure of the physical universe and the claims of traditional Christianity.

Among the many arguments developed by Stewart and Tait, one is particularly interesting for the way in which it anticipates issues subsequently raised by theologians concerned with chaos theory in the late twentieth century. Stewart and Tait raised the theological question of how we can reconcile our experience of agency with the determinism of classical physics. In dealing with this issue, Stewart drew on his own work, analyzing the energetics of what he called "delicate," or "instable," mechanical systems—systems that we would now label "chaotic." Such systems, he argued, may be so sensitive that, if they are at all complex, the effects of any tiny change in force may be incalculable. Living beings are delicate in this way and, as a consequence, for practical purposes, they act in indeterminate ways (Stewart and Tait 1889, 185).

If nineteenth-century physics could be appropriated in the service of Christianity, it is also the case that it could be appropriated for completely different and unorthodox "religious" purposes. Some materialists, such as Ludwig Büchner (1824–99), whose *Force and Matter; or, Principles of the Natural Order of the Universe* (1855) went through seventeen German editions and twenty-two other language editions by 1920, simply claimed to be able to demonstrate that physics undermined all support for any kind of theism. A more interesting position was illustrated in the writings of the German Nobel Laureate, natural philosopher, and physical chemist Wilhelm Ostwald (1853–1932). The founder of "energetics" posited the identification of matter and energy in a single "Monist" universe. Ostwald became convinced that scientific knowledge could replace religion as a foundation for morality and happiness, ultimately arguing that science *is* the god of the modern world. As early as 1905, he offered a formula for happiness:

$$G = (E + W)(E - W),$$

where G is the amount of happiness that an individual feels; E is the quantity of energy expended in activities that one wills to do; and W is the energy expended on activities done against one's will. Later, as chair of the German Monist League, whose conventions drew up to four thousand persons, Ostwald wrote more than two hundred scientific Sunday sermons promoting his substitute religion, offering self-hypnosis as a substitute for prayer to a higher authority, and designing naturalistic holidays to replace those of the Christian churches.

Modern Physics

After the professionalization of natural science during the nineteenth century, though it was common for physicists to seek support in their science for their religious belief or lack of belief in a transcendent God, few reported turning to scientific study primarily out of religious motivations, as had been the case during the seventeenth and early eighteenth centuries. One remarkable exception was Albert Einstein (1879–1955), for whom self-reported religious reasons played a major role in both the motivation of his own scientific work and his interpretations of the scientific work of others. Since his views, or views very much like his, have strongly influenced the attitudes of many important theoretical physicists and cosmologists well into the late twentieth century, they deserve special attention.

Einstein's religion was in no sense based on the notion of the personal God of orthodox Judaism, who demanded obedience and punished disobedience. "I cannot conceive of a God who rewards and punishes his creatures," he wrote. "Neither can I, nor would I, want to conceive of an individual that survives his physical death; let feeble souls, from fear or absurd egoism, cherish such thoughts" (Einstein 1952, 11). After a brief period of Jewish orthodoxy before he was twelve, Einstein adopted a commitment to what he later identified as Baruch Spinoza's (1632–77) entirely impersonal and entirely rational God: "A firm belief, a belief bound up with deep feelings, in a superior mind that reveals itself in the world of experience, represents my conception of God" (Paul 1982, 56). Einstein's firm conviction in the impersonal, objective aspect of God led him to the unshakable belief that the universe had a real existence, independent of all observers, and that it had to be totally causal and deterministic.

Moreover, because God was completely rational, Einstein was convinced throughout his life that a complete understanding of the natural world must ultimately be accessible to the human intellect. These commitments led him to oppose both positivist assertions that science could be nothing but the systematized record of our sensations and all acausal and statistical interpretations of quantum mechanics. Indeed, because quantum mechanics failed uniquely to stipulate the state of physical systems between observations, Einstein believed throughout his life that it must be fundamentally incomplete and that it could eventually be subsumed within a more comprehensive theory that would unify his own work on general relativity and all topics dealt with by quantum mechanics.

The search for some kind of grand unified theory, or theory of everything, based on a conviction that physics must ultimately be not only consistent with our sensory experiences but also logically inevitable and capable of accounting for everything, including the reason for the origin of the universe, continues among physicists at the beginning of the twenty-first century. Theoreticians such as Steven Weinberg (1933–) and Stephen Hawking (1942–) allude directly to Einstein as their inspiration and persist in arguing that their work is allowing them to see into the mind of God. Experimentalists have more recently begun to take up this point of view, as evidenced by Leon Lederman's *The God Particle: If the Universe Is the Answer, What Is the Question?* (1993).

During the first half of the twentieth century, attempts to account for the character of physical phenomena on the astronomical scale depended heavily on Einstein's general theory of relativity, which posited that space was "warped" in the presence of gravitating bodies. This theory, which was to be confirmed in 1919 by the observed bending of light by the sun, offered two possible implications regarding the history of the universe. One possibility was that the universe existed in a steady state, so that, though it appeared to be expanding, its density remained constant because of the continuous formation of matter in "empty" space. Alternatively, it was possible that the universe originated in a "big bang" at some point in the past. The first of these two solutions would, on the face of it, clearly have undermined traditional theological arguments for a creation of the universe.

In the early 1960s, empirical evidence of a residual heat radiation indicated that the big bang theory was correct, leading to a period in which a few astrophysicists and theologians (including Robert Jastrow [1925–]) optimistically suggested that the big bang theory provided new support for the creation of the universe at a point in time by a transcendent God. In 1988, however, this optimism was dealt a substantial blow by Stephen Hawking, who was able to show the possibility of a cosmology based on a fusion of general relativity and quantum mechanics, in which all observations to date could be accounted for in a finite universe that has neither spatial nor temporal bounds. As Hawking took special care to point out, in such a universe "there would be no singularities at which the laws of science broke down and no edge of space-time at which one would have to appeal to God or some new law to set the boundary conditions for space time. . . . The universe would be self-contained and not affected by anything outside itself" (Hawking 1988, 135). This argument does not disprove any Creation story, nor does it have any bearing on notions of divinity that are not transcendent, such as those associated with some variants of process theologies. But it does decouple large-scale physical phenomena from traditional supports for notions of a transcendent God.

In one of the most intriguing ironies associated with recent appropriations of physical arguments and analogies by theologians, Wolfhart Pannenberg (1928–) has argued that field theories in modern physics provide support for belief in God's continuing activity, or "effective presence" within the universe, as well as for the priority of the whole over any of its parts that plays a part in all discussions of apparent evil. The irony here is that modern field theories since Michael Faraday (1791–1867) have been developments out of Boscovich's eighteenth-century arguments that were taken as antagonistic to the need for God's ongoing activities in nature. According to Pannenberg, on the other hand, the tendencies of field theories to undermine the importance of traditional notions of matter and to replace them with space-filling immaterial forces suggests the analogous notion that the cosmic activity of the divine Spirit is like a field of force (Pannenberg 1988, 12).

It is generally agreed that quantum mechanics, the central features of which were articulated almost simultaneously in 1926 by Werner Heisenberg (1901–76) and Erwin Schroedinger (1887–1961) in different but logically equivalent forms, have had far more radical philosophical and theological implications than has relativity theory. As early as 1900, Max Planck (1858–1947) had shown that a correct formula for the distribution of energy in the spectrum emitted by a heated black body can be derived from the second law of thermodynamics if energy is emitted by an oscillating charged particle only in multiples of its frequency of oscillation, the proportionality constant, h, being equal to 6.6 × 10^ – 27 erg. seconds. In 1905, Einstein showed that the so-called photoelectric effect could be accounted for if the energy carried by a photon of light was h times the frequency. A few years later, Niels Bohr (1885–1962) was able to account for the spectrum of light emitted by hydrogen by assuming that electrons circled a positively charged nucleus without continuously radiating. When they did radiate, it was in a kind of instantaneous spasm produced when the electron dropped from one allowable energy level to another, and the allowable energy levels were governed by Planck's constant. These and numerous phenomena that could not be understood classically all found explanations in the general theory of quantum mechanics.

Heisenberg was among the first to explore some of the counterintuitive features in his 1927 paper "On the Intuitive Contents of Quantum-Theoretic Kinematics and Mechanics." In this paper he focused on what he called the principle of indeterminacy or uncertainty. Within quantum theory, there are pairs of variables, q and p, called conjugate variables such that

$$qp - pq = h/2\pi i,$$

where i is the square root of minus one. Heisenberg showed that this relationship could be given a physical interpretation if one considered the experimental uncertainties in measuring p and q variables. The mathematical relationship between p and q implied that the product of the uncertainties in their simultaneous measurements was always equal to or greater than Planck's constant divided by 2π. Even with theoretically perfect instruments, one could not simultaneously measure the value of conjugate variables with arbitrary precision. Moreover, since position and momentum are conjugate variables, it follows that one could never know perfectly the position and the momentum of even one particle, let alone those of all the particles of the universe, which is what Laplace had articulated as the condition that had to be met for a predictable deterministic universe. Indeed, if Heisenberg were correct, and if it is also the case that God created the universe, then it operates in such a way that not even God can predict its precise course in advance, raising theological issues regarding both the omniscience and the omnipotence of God.

If one considers Schroedinger's formulation of quantum mechanics, the uncertainty relationships are capable of a more extended and extremely interesting interpretation. In Schroedinger's system, solutions to certain equations are produced that have the form of classical wavelike functions. The product of two such functions gives the probability that, if a measurement of some variable is made, the variable will have that value. According to Heisenberg, Bohr, and Schroedinger, the Schroedinger wave functions represent the "state" of a quantum system. If we consider solutions for position, the "particle" whose position is to be measured is literally everywhere that the wave function has magnitude until a measurement is made. At that instant, the wave function collapses and the particle is found at a particular place. Many measurements of identical systems would lead to a distribution of results whose frequencies would reflect the square of the wave function at the places indicated.

This interpretation of quantum mechanics highlights several startling implications. First, it emphasizes that the uncertainty relations of Heisenberg reflect on indeterminacy or a casuality that is more than a reflection of human ignorance. This is true because two measurements of identical systems are almost certain to give different results. The fact that two different consequences follow from the same laws and initial conditions violates traditional understandings of determinism (of course, at another level, the wave functions are determined; it is only the results of our observations that are not). If the universe is not deterministic, then a number of possibilities exist, including that of freedom and responsibility. Second, Schroedinger's interpretation challenges the notion, so insisted upon by Einstein, of the objectivity of the physical universe. Bohr insists that there is a strong sense in which the physical world literally does not exist in any classical way, except when it is being measured. Furthermore, since the conditions for observing are part of the formal conditions for solving Schroedinger's equation, the results of any measurement depend not only on what is being measured but also on how the observer is interacting with it.

In 1935, Einstein, along with Boris Podolsky (1896–1966) and Nathan Rosen (1909–1995), published an article that highlighted another odd consequence of quantum mechanics and challenged the claim that quantum mechanics could ever be a complete description of physical reality. They invited consideration of the following thought experiment. They assumed that a particle composed of two protons and with a net zero spin splits into two protons with opposite spin. The two protons are allowed to travel a substantial distance in opposite directions. The two protons have equal probabilities of having right- or left-hand spins before the spin of either is measured, but the two spins must be in opposite directions. Now, suppose that the spin of one is measured to be left-handed. Instantaneously, the spin of the other will become right-handed, implying that information is passed between the two particles faster than the speed of light, which is presumed to be impossible. Clearly, according to Einstein, Podolsky, and Rosen, something is going on that is not contained within quantum mechanics itself. During the 1950s, David Bohm (1917–92) suggested a causal, nonlocal interpretation of quantum mechanics, involving hidden variables, consistent with the Einstein–Podolsky–Rosen (EPR) expectations, but it drew little attention. Since 1982, when a group of French physicists managed to carry out a near variant of the EPR experiment, the most widely held view seems to be that quantum mechanics is, indeed, complete but that it implies that reality is nonlocal, so that how one instrument operates can, in fact, influence distant events. These results have led to a revival of interest in Bohm's work on the part of both physicists and theologians.

The upshot of quantum mechanics has been to reopen a large number of questions that had seemed closed, including the renewed possibility of God's simultaneous instantaneous knowledge and activity everywhere in the universe. One of the more intriguing readings of the implications of quantum mechanics includes a revised version of the old notion of the design of the universe by an intelligent agent, coupled with an argument for free will in humans. In 1987, British-American theoretical physicist Freeman Dyson (1923–) argued that quantum entities, such as electrons, are active, choice-making agents and that experiments force them to make particular choices from the many options open to them. At a second level, the brains of animals "appear to be devices for the amplification of . . . the quantum choices made by the molecules inside our heads. . . . Now comes the argument from design. There is evidence from particular features of the laws of nature that the universe as a whole is hospitable to the growth of mind [defined as the capacity for choice]. . . . Therefore it is reasonable to believe in the existence of a third level of mind, a mental component of the universe. If we believe in this mental component and call it God, then we can say we are small pieces of God's mental apparatus" (Dyson 1987, 60ff).

Once again, we see the key feature of the use of natural philosophy or physics in religious discourse since the beginning of the nineteenth century. Virtually no one—except those who make science into a religion—has argued for nearly two hundred years that religious propositions can be proved through physical arguments. Instead, physicists have shown a remarkable interest and aptitude in demonstrating that physical laws are consistent with, and even suggestive of, a wide variety of theological consequences.

BIBLIOGRAPHY

Davies, Paul C. W. *God and the New Physics.* New York: Simon and Schuster, 1983.

Dyson, Freeman J. "Science and Religion." In *Religion, Science, and the Search for Wisdom,* ed. David M. Byers. Washington, D.C.: Bishops' Committee on Human Values, National Conference of Catholic Bishops, 1987.

Einstein, Albert. *Ideas and Opinions.* New York: Crown, 1952.

Fagg, Lawrence W. *Electromagnetism and the Sacred: At the Frontier of Spirit and Matter.* New York: Continuum, 1999.

Gascoigne, John. "From Bentley to the Victorians: The Rise and Fall of British Newtonian Natural Theology." *Science in Context* 2 (1988): 219–56.

Hahn, Roger. "Laplace and the Mechanistic Universe." In *God and Nature: Historical Essays on the Encounter between Christianity and Science,* ed. David C. Lindberg and Ronald L. Numbers. Berkeley: University of California Press, 1986, 256–76.

Hakfoort, Caspar. "Science Deified: Wilhelm Ostwald's Energeticist World-View and the History of Scientism." *Annals of Science* 49 (1992): 525–44.

Hawking, Stephen. *A Brief History of Time: From the Big Bang to Black Holes.* New York: Bantam, 1988.

Hiebert, Erwin N. "Modern Physics and Christian Faith." In *God and Nature: Historical Essays on the Encounter between Christianity and Science,* ed. David C. Lindberg and Ronald L. Numbers. Berkeley: University of California Press, 1986, 424–47.

Jammer, Max. *Einstein and Religion: Physics and Theology.* Princeton: Princeton University Press, 1999.

Lederman, Leon, with Dick Teresi. *The God Particle: If the Universe Is the Answer, What Is the Question?* New York: Houghton Mifflin, 1993.

Margenau, Henry, ed. *Integrative Principles of Modern Thought.* New York: Gordon and Breach, 1972.

McAvoy, John, and J. E. McGuire. "God and Nature: Priestley's Way of Rational Dissent." *Historical Studies in the Physical Sciences* 6 (1975): 325–404.

Müller-Marcus, Siegfried. *Wen Sterne rufe: Gespräch mit Lenin.* Wiesbaden: Credo, 1960.

———. *Der Gott der Physiker.* Basel: Birkhauser, 1986.

Odum, Herbert H. "The Estrangement of Celestial Mechanics and Religion." *Journal of the History of Ideas* 27 (1966): 533–58.

Pannenberg, Wolfhart. "The Doctrine of Creation and Modern Science." *Zygon* 23 (1988): 12.

———. *Systematische Theologie.* Vols. 2 and 3. Göttingen: Vandenhoeck and Ruprecht, 1991 and 1993.

———. *Toward a Theology of Nature: Essays on Science and Faith.* Louisville: Westminster, 1993.

Paul, Iain. *Science, Theology, and Einstein.* New York: Oxford University Press, 1982.

Penrose, Roger. *The Emperor's New Mind: Concerning Computers, Minds, and the Laws of Physics.* Oxford: Oxford University Press, 1989.

Polkinghorne, John. *The Faith of a Physicist.* Princeton: Princeton University Press, 1994.

Russell, Robert John, William Stoeger, and George Coyne, eds. *Physics, Philosophy, and Theology: A Common Quest for Understanding.* Vatican City: Vatican Observatory, 1988.

Russell, Robert John, N. Murphy, and A. Peacock. *Chaos and Complexity: Scientific Perspectives on Divine Action.* Vatican City: Vatican Observatory, 1995.

Sharpe, Kevin J. *David Bohm's World: New Physics and New Religion.* Lewisburg, Pa.: Bucknell University Press, 1993.

Stewart, Balfour, and Peter Guthrie Tait. *The Unseen Universe; or, Physical Speculations on a Future State.* 1875. Reprint. London: Macmillan, 1889.

Tipler, Frank. *The Physics of Immortality.* New York: Doubleday, 1994.

Weinberg, Steven. *Dreams of a Final Theory: The Search for the Fundamental Laws of Nature.* New York: Pantheon Books, 1992.

Wertheim, Margaret. *Pythagoras' Trousers: God, Physics, and the Gender Wars.* New York: Random House, 1995.

Whewell, William. *Astronomy and General Physics Considered with Reference to Natural Theology.* London: Pickering, 1833.

Worthing, Mark William. *God, Creation, and Contemporary Physics.* Minneapolis: Fortress, 1996.

24 Twentieth-Century Cosmologies

Craig Sean McConnell

In the twentieth century the interactions between science and religion regarding the origin of the universe were quite varied. Throughout the century, some physical cosmologists encouraged discussion of the theological significance of their work. At midcentury the "steady state" theory was advanced, at least in part, in the hope of challenging theistic interpretations of "big bang" theories. The steady state theory was soon embraced by some Christians, who reframed it in theistic terms. In general, discussions of cosmic origins and evolution have been less rancorous than discussions of the origin of life and organic evolution, for at least three reasons. First, the astronomical evidence regarding the origin of the universe is sparser than the fossil record, and many people regarded physical cosmology as mere speculation. Second, cosmological theories of the twentieth century have become increasingly mathematical and inaccessible to nonspecialists. Third, many people have found the question of the origin and evolution of the universe to be less personal and more esoteric than the question of the origin and evolution of human life.

Craig Sean McConnell obtained his Ph.D. from the University of Wisconsin at Madison. He is assistant professor of liberal studies at California State University, Fullerton. He is currently working on a history of the big bang–steady state debate.

THE SCIENTIFIC STUDY of the structure and the origin of the universe has changed enormously in the twentieth century. At the turn of the century, the most contentious issue was the nature of nebulae—whether they were part of our galaxy or independent "island universes" outside it. By the 1980s, big bang cosmology had become a subdiscipline of astrophysics with hundreds of practitioners and a constant place in the public eye. The interaction between cosmology and religion also varied enormously in this period, ranging from disregard, to harmonious co-existence, to antagonistic opposition. The dynamics of this relationship have often been more dependent on the personalities of key figures in cosmology than on the intrinsic scientific content of the theories themselves.

In the nineteenth century, a careful distinction was made between the words *cosmogony* and *cosmology*, the former referring only to theories of the origins of the universe or solar system, the latter to theories of the structure and the evolution of the universe after its creation. In the twentieth century, these terms became conflated. In the first half of the century, *cosmogony* was used to refer to structure and evolution in addition to origins. By midcentury, the term *cosmology* was preferred, and, by the 1970s, *cosmogony* was an archaic word, unfamiliar to most. This essay traces the major developments in cosmogony, understood to be the study of both the origin and the structure of the universe, and the religious reaction to these theories. Adopting modern convention, these theories are referred to as *cosmologies*.

Observational and Theoretical Cosmology

For the first few decades of the twentieth century, two distinct lines of inquiry could properly be considered cosmology, but there was little religious reaction to either. The nature of nebulae was a major issue in observational astronomy that was debated in a public forum by Heber Curtis (1872–1942) and Harlow Shapley (1885–1972) in 1920. Shapley argued that nebulae were clouds of dust within our galaxy and that our galaxy was the extent of the observable universe. Curtis, echoing the work of John Herschel (1792–1871) and Jacobus Kapteyn (1851–1922), argued that nebulae were separate galaxies. The issue was settled when Edwin Hubble (1889–1953) and others observed separate stars in the Andromeda nebula. This confirmation of the extragalactic nature of nebulae greatly expanded astronomers' estimate of the size of the universe, but it had nothing to say about the origin of the universe and attracted little, if any, religious discussion.

The other line of cosmological inquiry was opened in 1917, when Albert Einstein (1879–1955) turned his general theory of relativity to the consideration of the whole universe. Considering the interaction between gravitation and space-time at the largest imaginable scale, Einstein wrote equations that governed the whole cosmos. These equations implied a space-time that was curved, so that it was "closed," like the surface of a sphere, though it was "unbounded," in that a line without end could be drawn on the surface. The equa tions were also unstable, which implied that the universe was either expanding or contracting. Einstein thought that the universe was stable, so he added a term to the equations that would make the equations stable as well.

This work opened the field of modern theoretical cosmology, though few entered it right away. The mathematics that was required to consider cosmological problems in relativistic terms was notoriously difficult and esoteric, so only those concerned with these problems were able to follow the arguments they contained. Einstein's equations were soon challenged by Willem de Sitter (1872–1934), who demonstrated in 1917 that other cosmologies were possible, and by Aleksandr Friedman (1888–1925), who wrote equations in 1922 for an expanding universe, but these models were considered by most physicists to be "mathematical exercises" of no physical import. By the 1920s, a few religious writers were challenging the tenets of relativity, particularly the writings of moral relativists, but few had the mathematical skills to investigate these early theoretical cosmologies. Most of the debate about science and religion in the period before 1930 took place between creationists and evolutionists.

The Expanding Universe

Relativistic cosmology and observational astronomy were brought together in the work of Edwin Hubble and Georges Lemaître (1894–1966). Hubble's study of nebulae led him to the observation that the nebulae are all rushing away from the earth and from each other at a tremendous rate. Though there was initial confusion about how to interpret this observation, by 1930 most astronomers agreed that it meant that the universe itself was expanding. This expansion of the universe could be reconciled with Friedman's work, and Lemaître did just that. Lemaître was dedicated to linking the work of Friedman to that of Hubble, bringing observational evidence to bear on theoretical cosmology. Lemaître proposed that the expansion of the universe could be traced back to a very dense state in the distant past, in which the particles of the whole universe existed as a huge atomic nucleus. He called this nucleus the "primeval atom" and claimed that the expansion of the universe was the aftermath of a process analogous to radioactive decay that took place on a cosmic scale.

Lemaître's original publication appeared in an obscure Belgian journal, but Arthur Eddington (1882–1944) brought it to a larger audience by having it republished in the *Monthly Notices of the Royal Astronomical Society* and by featuring it in his popular book *The Expanding Universe* (1933). Eddington, a Quaker, made overt attempts here to reconcile modern cosmology with religion, suggesting that science should include a spiritual as well as an intellectual appreciation of nature. James Jeans (1877–1946) did the same in his popular book *The Mysterious Universe* (1930). The Victoria Institute, a British association dedicated to reconciling science and religion, published a number of articles that were approving of the work of Eddington and Jeans. In particular, the distinction that was emerging between notions of an evolving universe, which might be consistent with theistic cosmology, and notions of evolving life, which seemed antagonistic to theistic descriptions of life, placated the members of the Victoria Institute.

In *God and the Universe* (1931), Chapman Cohen (1868–1954) took Eddington and Jeans to task for their conciliatory posture toward religion. This collection of essays, many of them previously published in the *Freethinker*, was sponsored by the Secular Society, a British antireligious organization. Cohen, who wanted to dismiss religion entirely, was annoyed at the presence of spiritual language in Eddington's *The Nature of the Physical World* (1928) and Jeans's *The Mysterious Universe* and was particularly irritated by the friendly reviews these books received from members of the clergy. Perhaps more influential was an attack by L. Susan Stebbing (1885–1943) in *Philosophy and the Physicists* (1937). Stebbing, who had more impressive academic credentials than Cohen, chided Eddington and Jeans for making unfounded emotional appeals to religion and for being bad philosophers as well. Subsequent popular works by both Eddington and Jeans contained fewer references to spiritual matters.

The Big Bang–Steady State Debate

Though Lemaître had hoped for an eventual synthesis of nuclear physics and theoretical cosmology, it was the work of George Gamow (1904–68) and his collaborators Ralph Alpher (1921–) and Robert Herman (1914–97) that brought the science of nuclear physics into modern cosmology. Gamow considered the heat that was required for nuclear reactions and the heat that he believed a primeval atom would contain in the first moments of its decay and proposed that the elements were formed in the first moments of this expansion. This work brought new attention to Lemaître's primeval atom and showed promise at first, but the theory suffered from several deficits. Gamow could not find a pathway of nuclear reactions to build elements of atomic number five. Worse, Hubble was refining his estimates for the age of the universe, and the figure was quite a bit shorter than the estimates for the age of the earth and the age of some stars.

A team of astronomers in England developed an entirely different cosmology in hope of setting these difficulties aside. According to their steady state theory, the universe is eternal. The expansion of the universe has been going on forever and is not evidence of any special moment of Creation. Thomas Gold (1920–) and Hermann Bondi (1919–) developed this theory from the philosophical principle they called the Perfect Cosmological Principle. Just as Copernicus (1473–1543) had claimed that the earth does not occupy any special place in the universe, Gold and Bondi argued that the present does not occupy any special time in the universe. To explain the apparent constancy of the density of matter, they had to propose that matter is continuously created in space. A small amount of hydrogen (approximately one atom in a space the size of a school assembly hall every one hundred thousand years) would be enough to balance the observed expansion of the universe and would be so rare an occurrence that physicists would likely never see such a creation occur. While Gold and Bondi presented this theory in philosophical terms, their collaborator Fred Hoyle (1915–) developed a relativistic field equation for the steady state theory.

The popular and religious reaction to the debate between these competing theories—the big bang, as Hoyle derisively referred to the primeval atom theories, and the steady state theory—was enormous. Public lectures and debates were staged, many of them broadcast on radio and television. Pope Pius XII (b. 1876, p. 1939–58), in a speech to the Pontifical Academy of Science in 1951, endorsed Lemaître's primeval atom, an action that amused Gamow, irritated Hoyle, and horrified Lemaître. Gamow cited Pius XII in a paper he published in *Physical Review* (1952), though he tried to distance himself from the connection between cosmology and biblical Creation in his popular text *The Creation of the Universe* (1952). Lemaître, who was an ordained priest as well as a physicist, advocated a "separate spheres" approach to the issues of science and religion. He thought that they operated on different epistemological foundations and had little of merit to offer each other.

Hoyle often claimed that the religious resonance between Genesis and the big bang made people believe in the latter irrationally. His *Nature of the Universe* (1950), ostensibly a defense of the steady state theory, ends with a long diatribe against religion in general and Christianity in particular. Hoyle revisited these antireligious themes often, in his introductory astronomy textbook *Frontiers of Astronomy* (1955), his autobiographical musing *Ten Faces of the Universe* (1976), and his essay *The Origin of the Universe and the Origin of Religion* (1993). Ironically, some religious writers preferred Hoyle's steady state theory on the grounds that the big bang seemed too deistic and that the hand of God was evident in a universe that was constantly balanced by the creation of new matter.

The Big Bang Paradigm

A series of observational discoveries in the late 1960s eventually settled the big bang–steady state debate in favor of the big bang, though the steady state theory was never fully abandoned by Bondi, Gold, or Hoyle. However, the debate raged on in fundamentalist Christian circles long after the issue was considered settled among astronomers and physicists. Indeed, the steady state theory attracted new adherents in the pages of the *Creation Research Society Quarterly*. This renewed interest in the steady state theory seems to have been largely motivated by the desire to discredit evolution by proxy—big bang cosmology and evolutionary biology were seen to be complementary theories, so a refutation of the evolutionary big bang might challenge biological evolutionary thought as well.

In an ironic bit of turnaround, Fred Hoyle's collaborator and colleague Chandra Wickramasinghe (1939–) was called as an expert witness for the creationists in the 1981 Arkansas creation-evolution trial, and Hoyle's work was enlisted in the defense of creationism. Hoyle and Wickramasinghe argued for the necessity of a cosmic Creator, based on calculations estimating the probability of life originating on the earth in the available time frame. Though Hoyle remained elusive about the exact nature of this Creator, Wickramasinghe associated the Creator directly with his Buddhist beliefs. Hoyle and Wickramasinghe also made claims about extraterrestrial origins of life and were largely marginalized in the scientific community. Philosopher of science Michael Ruse, testifying on behalf of the evolutionists, occasionally conflated big bang cosmology with evolution, but, for the most part, the trial was focused on the biological sciences.

In the 1970s and 1980s, some cosmologists returned to a more harmonious representation of the relationship between cosmology and religion. British cosmologists such as Paul Davies (1946– ; emigrated to Australia in 1990) in his

God and the New Physics (1983) have returned to speaking of cosmology and religion as addressing the same questions, while American cosmologists have largely ignored the question of the religious implications of cosmology. A noticeable exception is *The Physics of Immortality* (1994), in which Frank Tipler (1947–) claims to have reduced theology to a subdiscipline of physics. He uses cosmological arguments to demonstrate the existence of God and the certainty of an afterlife. The book sold well, though it received numerous skeptical reviews.

Consideration of cosmology by religious thinkers has typically been overshadowed by consideration of evolution. For many, cosmology is not as personally offensive as evolutionary biology. Others find cosmology so speculative that it is not worthy of protracted rebuttal. The esoteric mathematics of general relativity, the language of cosmology since 1917, makes it hard for most people to engage the arguments. Religious scientists concerned with reconciling science and religion have typically studied subjects in geology and evolution, and, in fact, many of the discussions of cosmology and theology drift back into discussions of evolution and theology.

BIBLIOGRAPHY

Bertotti, Bruno, Roberto Balbinot, Silvio Bergia, and Antonio Messina, eds. *Modern Cosmology in Retrospect.* Cambridge: Cambridge University Press, 1990.

Davies, Paul. *The Mind of God: The Scientific Basis for a Rational World.* New York: Touchstone, 1992.

Hetherington, Norriss S., ed. *Encyclopedia of Cosmology: Historical, Philosophical, and Scientific Foundations of Modern Cosmology.* New York: Garland, 1993.

Jaki, Stanley. *Science and Creation: From Eternal Cycles to an Oscillating Universe.* New York: Science History Publications, 1974.

———. *God and the Cosmologists.* Washington, D.C.: Regnery Gateway, 1989.

Kragh, Helge. *Cosmology and Controversy: The Historical Development of Two Theories of the Universe.* Princeton: Princeton University Press, 1996.

McMullin, Ernan. "Religion and Cosmology." In *Encyclopedia of Cosmology: Historical, Philosophical, and Scientific Foundations of Modern Cosmology,* ed. Norriss S. Hetherington. New York: Garland, 1993, 579–95.

Munitz, Milton. *Theories of the Universe.* New York: Free Press, 1965.

North, John. *The Measure of the Universe: A History of Modern Cosmology.* London: Oxford University Press, 1965.

———. *Astronomy and Cosmology.* New York: Norton, 1995.

Numbers, Ronald L. *The Creationists: The Evolution of Scientific Creationism.* New York: Knopf, 1992.

Singh, Jagjit. *Great Ideas and Theories of Modern Cosmology.* New York: Dover, 1970.

Smith, Robert. *The Expanding Universe: Astronomy's 'Great Debate,' 1900–1931.* Cambridge: Cambridge University Press, 1982.

Tilby, Angela. *Soul: God, Self, and the New Cosmology.* New York: Doubleday, 1992.

Yourgrau, Wolfgang, and Allen Breck, eds. *Cosmology, History, and Theology.* New York: Plenum, 1977.

25 Scientific Naturalism

Edward B. Davis and Robin Collins

There has recently been a renewed challenge to naturalism, especially to methodological naturalism, most visibly by a group of scholars who identify themselves as advocates of "intelligent design." Crucial to assessing this challenge is an understanding of naturalism in its various forms. Historically, it has been possible to interpret the regularity of nature either theistically or nontheistically. Although naturalism remains the dominant paradigm in modern scientific thought, major objections have been raised on several fronts and alternatives to the naturalistic paradigm have been proposed.

Edward B. Davis received his Ph.D. from Indiana University. He is professor of the history of science at Messiah College, Grantham, Pennsylvania. He is editor (with Michael Hunter) of *The Works of Robert Boyle* (14 vols., London: Pickering and Chatto, 1999–2000) and author of numerous articles on the historical relationship of Christianity and science. His current research focuses on the religious beliefs of modern American scientists.

Robin Collins holds a Ph.D. from Notre Dame University. He is associate professor of philosophy at Messiah College. He has written several essays on philosophical aspects of religion and science and is currently writing a book tentatively entitled *The Well-Tempered Universe: God, Fine Tuning, and the Laws of Nature.*

SCIENTIFIC NATURALISM is the conjunction of naturalism—the claim that nature is all that there is and, hence, that there is no supernatural order above nature—with the claim that all objects, processes, truths, and facts about nature fall within the scope of the scientific method. This ontological naturalism implies weaker forms of naturalism, such as the belief that humans are wholly a part of nature (anthropological naturalism), the belief that nothing can be known of any entities other than nature (epistemological naturalism), and the belief that science should explain phenomena only in terms of entities and properties that fall within the category of the natural, such as by natural laws acting either through known causes or by chance (methodological naturalism). Before the late nineteenth century, scientific naturalism was not the dominant way of understanding the world, nor is it in the early twenty-first century the only metaphysical position consistent with modern science. Technically, scientific naturalism is not the same thing as philosophical materialism, which is the belief that everything is ultimately material, but it is closely related, and today they are usually conflated. Traditional theists do not accept scientific naturalism, although they may agree with anthropological naturalism and / or methodological naturalism.

Naturalism before 1900

The first naturalists in the Western tradition were certain pre-Socratic philosophers who sought to explain all things as natural events rather than as the result of divine action. For example, Thales of Miletus (fl. c. 585 B.C.) attributed earthquakes to tremors in the water on which the disk of the earth floated. This was a naturalistic rendering of the older Greek view that the god of the sea, Poseidon, was responsible for causing them. Similarly, the author of the Hippocratic treatise *On the Sacred Disease* (which deals with epilepsy), written about 400 B.C., opens the work by rebuking those who attribute the cause of the disease to the gods. In his opinion, they are simply hiding their own ignorance of the real cause, like "quacks and charlatans." The atomist Empedocles (c. 492–432 B.C.) assigned the origin of all living things to a crude forerunner of evolution by survival of the fittest, an idea later developed by the Roman Epicurean poet Lucretius (c. 99–55 B.C.). The parts of animals would form by chance and then come together; only those combinations of parts that fit the right pattern were viable. Two other atomists of the fifth century B.C., Leucippus and Democritus, viewed the world as an infinite number of uncreated atoms moving eternally in an infinite, uncreated void, colliding by chance to form larger bodies. Epicurus (341–270 B.C.) extended this description to the gods themselves, holding that they were composed of atoms and situated in the spaces between the infinite number of universes that co-existed at any one time. In this way, atomists sought to combat superstition and bondage to irrational fears of capricious gods who needed to be appeased.

Although Plato (c. 427–347 B.C.) shared the atomists' opposition to Greek polytheism, he rejected their purely natural and nonteleological mode of explanation. In the dialogue *Timaeus,* Plato accounted for the origin of order in the world by means of a godlike figure, the Demiurge, who imposed form on undifferentiated matter. Plato asserted the impossibility of attaining real "knowledge" of the material world, about which one could have only "opinions." A true science of nature was possible only insofar as the human mind could grasp the essences of the eternal Forms or Ideas of things, especially the axioms of logic and mathematics. This was the basis of an important kind of idealism (not naturalism), according to which the rational soul has been "imprisoned" in the flesh, and true freedom is found in contemplation of the Forms.

Like his teacher Plato, Aristotle (384–322 B.C.) rejected the atomists' assumption that the world was a mindless chaos rather than an intelligently ordered cosmos. Unlike Plato, however, he made the Forms and the teleology deriving from them immanent within nature rather than the result of an external intelligence. In doing this, Aristotle naturalized Plato's account of knowledge by emphasizing that knowledge of the Forms arises out of studying matter itself. For Aristotle, scientific explanations required the identification of four causes to be complete. They included not only the secondary (or efficient) and material causes (corresponding to the mechanistic explanations of the atomists), but also the formal and final causes, containing the immediate plan and ultimate purpose of the thing or event. As Aristotle understood them, all four causes were entirely natural, but the final and formal causes would later be associated with the divine design for the creation, especially the organic creation. Aristotle also distinguished three levels of soul (vegetative, animal, and rational), understanding them, too, in natural terms, as principles of vitality and organization.

During the first millennium of Christianity, thinkers in the Latin West had limited access to Greek scientific literature, especially the works of Aristotle. In the absence of a sophisticated form of naturalism, Plato's idealism provided the philosophical framework for most Christian thought. The most influential early Christian author was undoubtedly Augustine (354–430), who followed Plato in granting little importance to the study of the material world. Throughout this early period, Christian thinkers typically thought of science as a "handmaiden" to theology, which was the "queen of the sciences": science might serve theology by assisting in understanding biblical references to nature, but it ought never challenge the sole authority of theology to define reality.

That situation changed dramatically with the recovery of a large body of Greek scientific and medical works previously unavailable in the Latin West, a process that began in the eleventh century and culminated in the twelfth and thirteenth centuries with the appearance in Europe of universities dominated by Aristotelian natural philosophy. Thomas Aquinas (c. 1225–74), a Dominican who taught theology at Paris, undertook an ambitious project to integrate Aristotle with Christian theology. Thomas was careful to limit the scope of reason and to reject those Aristotelian tenets that seemed most threatening to the faith, especially the claim that the world is external. He also modified Aristotle's view that the order in nature is rationally necessary, claiming that God is free to establish any particular order in the world, although once established, it would apply necessarily except for miracles and the free actions of humans. A fundamental problem for Thomas and other monotheistic scholars, such as the Jewish physician and philosopher Rambam (also called Moses Maimonides [1135–1204]), is still central to theology: how to relate God to nature in a manner that acknowledges his sovereignty and ongoing involvement in the world, while affirming the integrity of the world as a creation with a measure of independence. Ultimately, Thomas succeeded in creating a synthesis that would become the most widely accepted form of Christian philosophy until modern times, but his efforts were controversial at the time. Indeed, many ideas of Aristotle and certain Arabic commentators were formally banned from being taught at Paris, the last and largest ban coming in 1277, when 219 specific propositions were condemned as contrary to the faith. The thrust of this condemnation was boldly to assert God's absolute power to do things contrary to the tenets of Aristotelian naturalism.

Although the condemnation did not give birth to modern science—a claim that has sometimes been advanced—it did contribute to the appearance of a vigorous new supernaturalism in the form of medieval nominalism. According to its outstanding exponent, William of Ockham (c. 1280–c. 1349), God's absolute power can do anything short of a logical contradiction, so the laws of nature are not necessary truths. The order of nature is, therefore, contingent, with observed regularities reflecting God's faithfulness in upholding the creation as expressed through his ordinary power rather than through any rational necessity arising from the nature of things.

This belief in the ongoing supernatural activity of the Creator became even stronger in the sixteenth century with the spread of the doctrines of the Protestant Reformation about God's absolute sovereignty, which encouraged many early modern natural philosophers to downplay the independence of nature from God and to advocate an unambiguously empirical approach to scientific knowledge. As tools of the divine will, matter and its properties had to be understood from the phenomena, not from metaphysical first principles, giving empiricism a clear theological foundation. For example, Robert Boyle (1627–91), the most influential publicist of mechanistic science, held that the laws by which God governed matter were freely chosen, ruling out the possibility of an a priori science of nature.

Boyle's critique of the very ideas of nature and natural law in *A Free Enquiry into the Vulgarly Received Notion of Nature* (1686) underscores the general historical truth that the concept of natural law is metaphysically ambivalent: Either God or nature can be seen as the ultimate agent behind the laws we observe in operation. Thus, for supernaturalists like Boyle, Isaac Newton (1642–1727), and Marin Mersenne (1588–1648), miracles were defined as extraordinary acts of God outside the ordinary course of nature, but the ordinary course of nature itself was understood to be nothing other than the ordinary acts of God. Newton's belief that God occasionally adjusts the motions of the planets is often misunderstood as involving a "God of the gaps," in which God is conceived to be a "clockmaker" active only in extraordinary events that are inexplicable by natural law. Rather, Newton believed that all natural events were divinely caused, and he never endorsed the clock metaphor with which he is wrongly associated. If supernaturalists emphasized the miraculous character of the ordinary, Thomas Hobbes (1588–1679) took the opposite track, making God a material being and endowing matter with activity and thought. In another materialistic step, physicians such as William Harvey (1578–1657) and Francis Glisson (1597–1677) endowed matter with sensation and treated diseases as wholly natural disorders, in keeping with the tradition of a profession that was often thought by contemporary commentators to exude the faint odor of atheism.

Toward the end of the seventeenth century, the plausibility of miracles came under increasing attack. At the same time, critical approaches to Scripture began to take hold. In the next century, both of these currents tore into the biblical testimony concerning miracles. "Rational" religion, ultimately rooted in the crisis of religious authority that resulted from the Reformation, appealed widely to intellectuals of the eighteenth century.

By 1800, appeals to direct divine agency were increasingly rare in science, even in natural history. As the nineteenth century progressed, natural historians tended increasingly to admire and to imitate the lawlikeness of physics and to divorce their discipline from theology by finding ways to explain nature without miracles. An obvious example is the enormous influence of the nebular hypothesis of Pierre Simon Laplace (1749–1827), a naturalistic account of the origin of the solar system. This was consistent with a general trend among theologically inclined scientists, such as William Whewell (1794–1866) and Charles Babbage (1792–1871), to locate evidence of divine governance more in the regularity of nature (God as lawgiver) than in exceptions to it. Several attempts to give a naturalistic account of all living creatures culminated in Charles Darwin's (1809–82) *Origin of Species* (1859) and *The Descent of Man and Selection in Relation to Sex* (1871), the tenets of which are still regarded as essentially correct by most scientists today.

Twentieth-Century Naturalism

Darwin's theory spawned the widespread use of the concept of evolution to justify various social, political, and religious agendas, claiming for them a scientific basis. But, more important, it played a pivotal role in scientific naturalism's becoming the dominant worldview of the academy by the middle of the twentieth century. Indeed, in every discipline today, except in some schools of theology, a strict methodological naturalism is observed, and typically an ontological naturalism is presupposed by most of the practitioners of these disciplines. For example, a largely unspoken rule in both the sciences and the humanities is that, insofar as one attempts to explain human behavior or beliefs, they must be explained by natural causes, not by appealing to such things as an immaterial soul or a transcendent ethical or supernatural order, as previous thinkers had done.

Although Darwinism largely set the stage for the dominance of scientific naturalism, it was not until the twentieth century that serious attempts were made to work out and defend it as a comprehensive philosophy, especially with regard to ethics and our understanding of the human mind, two areas that have presented particularly difficult problems for naturalism. Two major approaches have been used by naturalists in response to these difficulties: reductionism and eliminativism. (Since the 1980s, however, a third alternative has gained some adherents in the philosophy of mind, in which a naturalism concerning the mind is affirmed, but the metaphysical nature of the mind and its metaphysical relation to the body are claimed to be largely intrinsically incomprehensible by us. We do not discuss this alternative here, however.) *Reductionism* affirms the validity of the area of discourse in question but interprets it as really being about the natural world, even though the discourse might appear to be about some other realm. In ethics, for example, this approach takes the form of attempting to reduce all terms of ethical appraisal—terms such as *good, bad, right,* and *wrong*—to statements about social customs or human happiness or to expressions of emotions or approval, all of which are subject to scientific investigation. So, for instance, under one commonly held version of ethical naturalism, to say something is wrong is merely to say that it is contrary to the customs of one's culture.

In philosophy of mind and related fields, such as cognitive science and artificial intelligence, reductionism takes the form of attempting to reduce everyday statements about the mind—statements such as that John is in pain or that Sally believes that Jim likes her—to statements about natural processes or interrelations between natural processes. For example, behaviorism, one of the first proposals along these lines, argues that statements about beliefs, feelings, and other mental states are ultimately merely statements about our physical behavior. Similarly, functionalism, a successor to behaviorism and the dominant view, claims that statements about human mental states are analogous to statements ascribing human mental characteristics to a computer running a piece of software. To say, for instance, that a computer is figuring out its next move in a chess game is to say something about a complex functional interrelationship between elements of the computer's hardware and the environment. Similarly, functionalists hold that statements about human mental states are really disguised statements about "functional" interrelations both between brain states themselves and between these states and the environment.

Most philosophers agree that reductionism has not succeeded in ethics and the philosophy of mind. This, among other things, has led many naturalists to adopt the second major approach, that of *eliminativism.* To be an eliminativist about a certain domain of discourse is to claim that the primary terms or concepts of the discourse fail to correspond to anything in the world, and, hence, statements using these terms are, strictly speaking, false. For example, ethical eliminativists, such as J. L. Mackie, claim that nothing is really morally right or morally wrong, or good or bad, since there are no properties such as rightness or wrongness existing out in the world. Similarly, eliminativists in neuroscience and philosophy of mind, such as Paul Churchland, claim that most mental items such as beliefs do not really exist, and, hence, statements referring to these items, such as that John *believes* that Sue loves him, are always false.

Eliminativism is still largely a minority position among naturalists, especially in the philosophy of mind, mostly because it is widely thought either to contradict what is obvious or to be self-refuting. For example, many philosophers, such as Lynn Rudder Baker, argue that the mind/body eliminativist's denial of the existence of beliefs—especially the more radical eliminativist denial that anyone is conscious—not only runs against what is absolutely obvious from experience but also is self-refuting since, if we do not have beliefs, then the eliminativists themselves cannot really believe in eliminativism.

Even though both reductionism and eliminativism have encountered serious problems, many naturalists reject nonnaturalism as a serious option, largely for the following reasons. First, they cite the success of the physical sciences in support of their position, claiming that, in case after case, science has been able to take events and processes that were once thought mysterious or ascribed to supernatural agencies (such as diseases, earthquakes, and psychological disorders) and offer well-supported naturalistic explanations of them. Hence, they argue, we have every reason to believe that mysteries such as human consciousness will one day be explained by physical science, even though we cannot at present see how. More implicit, naturalists often assume that the modes of explanation adopted in the physical sciences, with their requirement that explanations be given solely in terms of natural causes, are paradigmatic of what it is to explain or even to understand a phenomenon. Thus, for instance, Colin McGinn, a leading naturalist in the philosophy of mind, states that "naturalism about consciousness is not merely an option; it is a condition of understanding" (McGinn 1991, 47), a view echoed by other leading workers in the field such as Daniel Dennett.

Second, naturalists have associated nonnaturalism with superstition, antiscience, and traditional religious worldviews, all of which, they claim, are ultimately harmful to human progress. Thus, in the words of Bertrand Russell (1872–1970), many adopt materialism (today the most common form of naturalism) not so much from an independent conviction of its truth but "as a system of dogma set up to combat orthodox dogma" (Russell 1925, xi), particularly religious dogma and superstition. Indeed, according to John Searle, himself a prominent materialist, the "unstated assumption behind the current batch of [materialist] views is that they represent the only scientifically acceptable alternatives to the antiscientism that went with traditional dualism, the belief in the immortality of the soul, spiritualism, and so on" (Searle 1992, 3).

Third, in support of their position, naturalists have pointed to the purportedly intractable problem of relating a hypothesized nonnatural order—whether it be transcendent ethical principles, an immaterial soul, or God—to the natural order. For instance, since the time of René Descartes (1596–1650), naturalists have persistently criticized Cartesian dualism for supposedly being unable to explain how an immaterial mind could causally interact with the material brain.

Finally, naturalists have often blamed traditional Western forms of dualism, along with the attendant claim going back to Plato that the nonnatural order is superior, for the denigration and neglect of the natural order and, hence, the denigration of the body, the denigration of women (who have been traditionally associated with the body), and the oppression of non-Western cultures (especially cultures that have often been thought of as primitive, i.e., close to a state of nature). Similarly, thinkers such as Karl Marx (1818–83), Sigmund Freud (1856–1939), Bertrand Russell, and many modern secular humanists have blamed nonnaturalism, particularly in the form of traditional religious worldviews, for not only oppressing human freedom and creating various psychological and sociological disorders but also impeding human progress by diverting our attention to some putative supernatural order instead of helping us scientifically deal with our problems.

More than any other group, those who value religion (whether or not they are believers) have been particularly concerned with scientific naturalism and have offered a variety of responses both to it and to the arguments offered in its favor. One extreme response has been to reinterpret religious beliefs naturalistically. Gordon Kaufman, for example, claims that the word *God* should be interpreted as referring to those cosmic evolutionary forces that gave rise to our existence (Kaufman 1981, 54–6). Another extreme response in the opposite direction has been vigorously to reject any form of naturalism, such as the young-earth creationists have done with regard to human and animal origins. The majority of religious believers in the academy, however, have attempted to accommodate and incorporate as much as possible the generally accepted insights of naturalists, even as they retain what they believe is essential to their religion. Today, for example, most religious believers working in psychology and psychiatry would first attempt to explain a mental disorder naturalistically, instead of appealing to supernatural causes as in times past, and many religious people accept some type of theistic evolution. Similarly, many biblical scholars, even many of those who are quite theologically conservative, attempt to account for the origin, genre, and meaning of biblical texts largely by appealing to natural causes.

Of course, this adaptation takes different forms, largely depending on what is regarded as essential to a particular religion. For instance, some Christian philosophers (including some who are otherwise quite orthodox) have rejected the idea of an immaterial soul, claiming that it is not essential to Christianity, while on similar grounds many Christian biblical scholars have denied the occurrence of a bodily resurrection. On the other hand, many religious believers think that a fully supernatural understanding of the inspiration of their scriptures is essential to their religion: For instance, some Christians believe in the inerrancy of the Bible, while some Jews and almost all Muslims believe that every individual letter of their scriptures was dictated by God.

Along with the appropriation of the insights of scientific naturalism, religious believers and other nonnaturalists have also offered a systematic critique of scientific naturalism as a comprehensive philosophy. They have both critiqued the arguments offered in favor of it and pointed out its current lack of success in adequately accounting for such things as ethics, rationality, and the human mind, despite decades of effort, something that most naturalists concede. Moreover, nonnaturalists have begun to pursue more positive programs of their own ranging over a wide variety of issues. For example, philosopher Richard Swinburne has argued for a model of explanation based on reasons and actions of a personal agent (Swinburne 1979); physicist John Polkinghorne has attempted to develop models of God's providential control over creation that respect the internal integrity of the created order (Polkinghorne 1989); Nobel Prize–winning physiologist John Eccles (1903–) has attempted to develop a nonphysicalist account of the mind (Eccles and Popper 1981); and a wide range of authors are engaged in exploring the hypothesis of an intelligent designer as the ultimate explanation for the so-called fine-tuning of the basic structure of the universe and / or the evolution of life.

In addition, some theists, such as philosopher Alvin Plantinga and professor of law Phillip Johnson, have challenged the hegemony of methodological naturalism in all areas of the academy (Johnson 1995; Plantinga 1996). Instead, these thinkers have advocated the acceptance of a pluralism of research methodologies, each with its own assumptions. Under their proposal, for example, some groups of biologists would construct theories and research programs under the assumption that life is ultimately a result of divine design, whereas others (perhaps the majority) would continue to operate under methodological naturalism. Other theists, however, such as philosopher of science Ernan McMullin and physicist Howard van Till, accept the general validity of methodological naturalism for scientific inquiry, while they reject its extrapolation into a broader, ontological naturalism.

Finally, a highly diverse, largely popular movement has developed that seeks to find an alternative paradigm to mechanistic naturalism and that proposes ideas that tend to fall outside current mainstream Western forms of thought. Thinkers in this movement typically stress ideas such as holism, vitalism, the primacy of consciousness, and various ideas from Eastern philosophies and religions, which they often attempt to relate to certain interpretations of modern physics. Although this sort of movement is frequently called "New Age," such a label is inadequate, since it tends to leave the false impression that the movement forms some homogeneous whole, when actually it is simply a loosely related set of dissatisfactions with current reductionistic and dualistic worldviews. Moreover, this label implies that these ideas are new, when many of them are actually ancient. Finally, it is often unclear whether the conclusions of a specific thinker in this movement fall outside the confines of scientific naturalism or whether he or she is merely advocating a new, nonmechanistic form of it. Thinkers who postulate such things as telepathy or advocate the virtues of acupuncture, for instance, typically consider them to be part of the natural, scientifically explicable world, whereas authors who believe in the existence of ghosts, spirit guides, reincarnation, and the like usually think that they fall outside the confines of scientific naturalism.

Although scientific naturalism, particularly in its methodological and mechanistic varieties, dominates in the academy, it is increasingly being challenged both inside and outside academic circles. Whether naturalism in any variety can ultimately meet these challenges is uncertain. But one thing seems clear. As philosopher Thomas Nagel and others have stressed with regard to human consciousness, scientific naturalism in its current form will need to undergo radical conceptual revision to account for important features of the world and human experience.

BIBLIOGRAPHY

Baker, Lynn Rudder. *Saving Belief: A Critique of Physicalism.* Princeton: Princeton University Press, 1987.

Boyle, Robert. *A Free Enquiry into the Vulgarly Received Notion of Nature*, ed. Edward B. Davis and Michael Hunter. Cambridge: Cambridge University Press, 1996.

Brooke, John H. "Natural Law in the Natural Sciences: The Origins of Modern Atheism?" *Science and Christian Belief* 4 (1992): 83–103.

Cannon, Walter F. "The Problem of Miracles in the 1830s." *Victorian Studies* 4 (1960): 5–32.

Churchland, Paul M. *Matter and Consciousness*. Rev. ed. Cambridge: MIT Press, 1988.

Craig, William Lane, and J. P. Moreland, eds. *Naturalism: A Critical Analysis*. New York: Routledge, 2000.

Davis, Edward B. "Newton's Rejection of the 'Newtonian World View': The Role of Divine Will in Newton's Natural Philosophy." *Science and Christian Belief* 3 (1991): 103–17.

Dear, Peter. "Miracles, Experiments, and the Ordinary Course of Nature." *Isis* 81 (1990): 663–83.

Dembski, William, and James Kushiner. *Signs of Intelligence: Understanding Intelligent Design*. Grand Rapids, Mich.: Brazos Press, 2001.

Eccles, John, and Karl R. Popper. *The Self and Its Brain*. New York: Springer International, 1981.

Grant, Edward. "The Condemnation of 1277, God's Absolute Power, and Physical Thought in the Late Middle Ages." *Viator* 19 (1979): 211–44.

Henry, John. "The Matter of Souls: Medical Theory and Theology in Seventeenth-Century England." In *The Medical Revolution of the Seventeenth Century*, ed. Roger French and Andrew Wear. Cambridge: Cambridge University Press, 1989, 87–113.

Hooykaas, R. "Science and Theology in the Middle Ages." *Free University Quarterly* 3 (1954): 77–163.

Hutchison, Keith. "Supernaturalism and the Mechanical Philosophy." *History of Science* 21 (1983): 297–333.

Johnson, Philip. *Reason in the Balance: The Case against Naturalism in Science, Law, and Education*. Downers Grove, Ill.: InterVarsity, 1995.

Kaufman, Gordon D. *Constructing the Concept of God*. Philadelphia: Westminister, 1981.

Krikorian, Yervant H. *Naturalism and the Human Spirit*. New York: Columbia University Press, 1987.

Mackie, J. L. *Ethics: Inventing Right and Wrong*. New York: Penguin, 1983.

McGinn, Colin. *The Problem of Consciousness*. Cambridge, Mass.: Blackwell, 1991.

McMullin, Ernan. "Plantinga's Defense of Special Creation." *Christian Scholar's Review* 21 (1991): 55–79.

Miller, Kenneth. *Finding Darwin's God: A Scientist's Search for Common Ground between God and Evolution*. New York: HarperCollins, 1999.

Nagel, Thomas. "What Is It Like to Be a Bat?" *Philosophical Review* 83 (1974): 435–50.

Numbers, Ronald L. *Creation by Natural Law: Laplace's Nebular Hypothesis in American Thought*. Seattle: University of Washington Press, 1977.

Oakley, Francis. *Omnipotence, Covenant, and Order: An Excursion in the History of Ideas from Abelard to Leibniz*. Ithaca: Cornell University Press, 1984.

Plantinga, Alvin. "Methodological Naturalism?" In *Facets of Faith and Science*. Vol. 1: *His-*

toriography and Modes of Interaction, ed. J. van der Meer. Lanham, Md.: University Press of America, 1996, 177–221.

Polkinghorne, John. *Science and Providence: God's Interaction with the World.* London: SPCK, 1989.

Robinson, Howard. *Matter and Sense: A Critique of Contemporary Materialism.* New York: Cambridge University Press, 1982.

Russell, Bertrand. "Introduction: Materialism, Past and Present." In *The History of Materialism*, by Frederick Albert Lange. London: Routledge and Kegan Paul, 1925. Reprint. 1950, 1957.

Searle, John R. *The Rediscovery of the Mind.* Cambridge: MIT Press, 1992.

Swinburne, Richard. *The Existence of God.* Oxford: Clarendon, 1979.

26 The Design Argument

William A. Dembski

At the heart of the design argument is whether the order and complexity exhibited by the world are the result of an intelligent cause or a blind, undirected process. The design argument attempts to establish the first of these options. Historically, in looking for evidence of design, the argument has focused on the world as a whole, its laws, and structures within the world, notably life. The design argument has two recent incarnations. One employs the Anthropic Principle and focuses on the fine-tuning, or "just-so" aspects, of the physical universe required for human observers. The other constitutes a revival of design-theoretic reasoning in biology and is known as Intelligent Design.

William A. Dembski holds a Ph.D. in mathematics from the University of Chicago and a Ph.D. in philosophy from the University of Illinois at Chicago. He is an associate research professor at Baylor University and a senior fellow with Seattle's Discovery Institute. His most important books on the topic of design are *The Design Inference* (Cambridge: Cambridge University Press, 1998), which lays out the logic of inferring design, and *No Free Lunch* (Lanham, Md.: Rowman and Littlefield, 2001), which shows how that logic applies to biology.

Historical Development

The design argument reasons from features of the physical world that exhibit order or complexity to an intelligent cause responsible for those features. Just what features signal an intelligent cause, what the nature of that intelligent cause is (e.g., a personal agent or telic process), and how convincingly those features establish an intelligent cause remain subjects for debate and account for the variety of design arguments over the centuries. The design argument is also called the teleological argument.

The design argument must be distinguished from a prior metaphysical commitment to design. For instance, in the *Timaeus* Plato (427–347 B.C.) proposed a Demiurge (Craftsman) who fashioned the physical world but not because the physical world exhibits features that cannot be explained apart from the Demiurge. Plato knew the work of the Greek atomists, who needed no such explanatory device. Rather, within Plato's philosophy, the world of intelligible forms constituted the ultimate reality, of which the physical world was but a dim reflection. Plato, therefore, posited the Demiurge to transmit the design inherent in the world of forms to the physical world.

Often the design argument and a metaphysical commitment to design have operated in tandem. This has been especially true in the Christian tradition, in which the design argument is used to establish an intelligent cause, and a metaphysical commitment to God then identifies that intelligent cause with God. The design argument and a metaphysical commitment to design have also tended to be conflated within the Christian tradition, so that the design argument often appears to move directly from features of the physical world to the triune God of Christianity.

Full-fledged design arguments have been available since classical times. Both Aristotle's (384–322 B.C.) final causes and the Stoics' seminal reason were types of intelligent causation inferred at least in part from the apparent order and purposiveness of the physical world. For example, in *De natura deorum* (*On the Nature of the Gods*), Cicero (106–43 B.C.) writes: "When we see something moved by machinery, like an orrery or clock . . . , we do not doubt that these contrivances are the work of reason; when therefore we behold the whole compass of heaven moving with revolutions of marvelous velocity and executing with perfect regularity the annual changes of the seasons with absolute safety and security for all things, how can we doubt that all this is effected not merely by reason, but by a reason that is transcendent and divine?" (Cicero 1933, 217–19).

Throughout the Christian era, theologians have argued that nature exhibits features that nature itself cannot explain but that instead require an intelligence beyond nature. Church fathers like Minucius Felix (third century A.D.) and Gregory of Nazianzus (A.D. c. 329–89), medieval scholars like Moses Maimonides (1135–1204) and Thomas Aquinas (c. 1225–74), and common-sense realists like Thomas Reid (1710–96) and Charles Hodge (1797–1878) were all theologians who made design arguments, arguing from the data of nature to an intelligence that transcends nature. Thomas's fifth proof for the existence of God is perhaps the best known of these.

With the rise of modern science in the seventeenth century, design arguments took a mechanical turn. The mechanical philosophy that was prevalent at the birth of modern science viewed the world as an assemblage of material particles interacting by mechanical forces. Within this view, design was construed as an externally imposed form on preexisting inert matter. Paradoxically, the very clockwork universe that the early mechanical philosophers like Robert Boyle (1627–91) used to buttress design in nature was in the end probably more responsible than anything for undermining design in nature. Boyle (1686) advocated the mechanical philosophy because he saw it as refuting the immanent teleology of Aristotle and the Stoics, for whom design arose as a natural outworking of natural processes. For Boyle this was idolatry, identifying the source of creation not with God but with nature.

The mechanical philosophy offered a world operating by mechanical principles and processes that could not be confused with God's creative activity and yet allowed such a world to be structured in ways that clearly indicated the divine handiwork and, therefore, design. What's more, the British natural theologians always retained miracles as a mode of divine interaction that could bypass mechanical processes. Over the subsequent centuries, however, what remained was the mechanical philosophy and what dropped out was the need to invoke miracles or God as designer. Henceforth, purely mechanical processes could do all the design work for which Aristotle and the Stoics had required an immanent natural teleology and for which Boyle and the British natural theologians required God.

The British natural theologians of the seventeenth through the nineteenth centuries, starting with Robert Boyle and John Ray (1627–1705) and finding their culmination in William Paley (1743–1805), looked to biological systems for convincing evidence that a designer had acted in the physical world. Accordingly, they thought it incredible that organisms, with their astonishing complexity and superb adaptation of means to ends, could originate strictly through the blind forces of nature. William Paley's *Natural Theology* (1802) is largely a catalogue of biological systems he regarded as inexplicable apart from a superintending intelligence. Who was this designer posited by the British natural theologians? For many it was the traditional Christian God, but for others it was a deistic God, who had created the world but played no ongoing role in governing it.

Criticisms

Criticisms of the design argument have never been in short supply. In classical times, Democritus (c. 460–370 B.C.) and Lucretius (c. 99–55 B.C.) con ceived of the natural world as a whirl of particles in collision, which sometimes chanced to form stable configurations exhibiting order and complexity. David Hume (1711–76) referred to this critique of design as "the Epicurean Hypothesis": "A finite number of particles is only susceptible of finite transpositions: and it must happen, in an eternal duration, that every possible order or position must be tried an infinite number of times. This world, therefore, with all its events, even the most minute, has before been produced and destroyed, and will again be produced and destroyed, without any bounds and limitations. No one, who has a conception of the power of infinite, in comparison of finite, will ever scruple this determination" (Hume 1779 [reprint], 67). Modern variants of this critique are still with us in the form of inflationary cosmologies (Guth and Steinhardt 1989), many-worlds interpretations of quantum mechanics (Deutsch 1997), and certain formulations of the anthropic principle (Barrow and Tipler 1986).

Though Hume cited the Epicurean hypothesis, he never put great stock in it. In *Dialogues Concerning Natural Religion* (1779), Hume argued principally that the design argument fails as an argument from analogy and as an argument from induction. He also emphasized the problem of imperfect design or dysteleology. Though widely successful in discrediting the design argument, Hume's critique is no longer as convincing as it used to be. As Elliott Sober (1993, chap. 2) observes, Hume incorrectly analyzed the logic of the design argument, for the design argument is properly speaking neither an argument from analogy nor an argument from induction but an inference to the best explanation. Inference to the best explanation confirms hypotheses according to how well they explain the data under consideration. So staunch a Darwinist as Richard Dawkins (1987, 5–6) agrees that in Hume's day design was the best explanation for biological complexity.

Whereas Hume attempted a blanket refutation of the design argument, Immanuel Kant (1724–1804) limited its scope. According to Kant, "The utmost . . . that the [design] argument can prove is an *architect* of the world who is [constrained] by the adaptability of the material in which he works, not a *creator* of the world to whose idea everything is subject" (Kant 1787 [reprint], 522). Far from rejecting the design argument, Kant objected to overextending it. For Kant, the design argument legitimately establishes an "architect" (i.e., an intelligent cause whose contrivances are constrained by the materials that make up the world), but it can never establish a Creator who originates the very materials that the architect then fashions. Charles Darwin (1809–82) delivered the design argument its biggest blow. Darwin was ideally situated historically to do this. His *Origin of Species* (1859) fitted perfectly with an emerging positivistic conception of science that was loath to invoke intelligent causes and sought as far as possible to assimilate scientific explanation to natural law. Hence, even though Darwin's selection mechanism remained much in dispute throughout the second half of the nineteenth century, the mere fact that Darwin had proposed a plausible naturalistic mechanism to account for biological systems was enough to convince the Anglo-American world that some naturalistic story or other had to be true.

Even more than cosmology, biology had, under the influence of British natural theology, become the design argument's most effective stronghold. It was here more than anywhere else that design could assuredly be found. To threaten this stronghold was, therefore, to threaten the legitimacy of the design argument as a reputable intellectual enterprise. Richard Dawkins sums up the matter thus: "Darwin made it possible to be an intellectually fulfilled atheist" (Dawkins 1987, 6). God might still exist, but the physical world no longer required him to exist.

Nevertheless, design did not simply wither and die with the rise of Darwinism. Instead, its roots went deeper, ramifying into the physical laws that structure the universe. To many scholars of the late nineteenth and the twentieth century, thinking of design in terms of biological contrivance was no longer tenable or intellectually satisfying. The focus, therefore, shifted from finding specific instances of design within the universe to determining whether and in what way the universe as a whole was designed.

The Anthropic Principle

The Anthropic Principle underlies much of the contemporary discussion about the design of the universe. Astrophysicist Brandon Carter coined the term in 1970. In its original formulation, the Anthropic Principle merely states that the physical laws and fundamental constants that structure the universe must be compatible with human observers. Since human observers exist, the principle is obviously true.

The Anthropic Principle is relevant to design because the conditions that need to be satisfied for the universe to permit the existence of human observers are so specific that slight variations in these conditions would no longer be compatible with human observers. These conditions are usually defined as the laws and fundamental constants of physics. For instance, if the gravitational constant were slightly larger, stars would be too hot and burn too quickly for life to form. On the other hand, if the gravitational constant were slightly smaller, stars would be so cool as to preclude nuclear fusion and with it the production of the heavy elements necessary for life. In either case, the existence of human observers (and the development of human life) would have been physically impossible. Hence, the inhabited universe is anthropocentric: It was designed for human beings, whose emergence was not an accidental by-product of purposeless cosmic evolution.

The requirement that such precise conditions be satisfied for the existence of human observers seems itself to require explanation and has led to renewed design arguments by both theists and nontheists (e.g., Collins 1999; Davies 1992; Leslie 1989; Swinburne 1979). They argue that design is the best explanation for the fine-tuning of physical laws and constants.

Nonetheless, using design to explain cosmological fine-tuning (or Anthropic Coincidences, as they are also called) is controversial. The usual move for refuting such cosmological design arguments is to invoke a selection effect. Accordingly, cosmological fine-tuning is said not to require explanation because without it human observers would not exist to appreciate its absence. Like a lottery in which the winner is pleasantly surprised at being the winner, so human observers are pleasantly surprised to find themselves in a finely tuned universe. No design is required to explain the winning of the lottery, and likewise no design is required to explain human observers residing in a finely tuned universe.

Stated thus, the selection-effect antidesign argument is easily rebutted (Swinburne 1979, 138). What makes chance a viable alternative to design in the lottery analogy is the existence of other lottery players. The reason lottery winners are surprised at their good fortune is because most lottery players are losers. But there is only one universe—or is there? For a selection effect successfully to refute a design argument based on cosmological fine-tuning requires an ensemble of universes in which most universes are losers in the quest for human observers (Barrow and Tipler 1986). But this solution requires inflating one's ontology, with a consequent blurring of physics and metaphysics, which is itself problematic (Lindley 1993).

Intelligent Design

Although cosmological design arguments that locate design at the level of the whole universe remain popular, since the 1990s interest has also reverted to biological design arguments that locate design in actual biological systems. The shift here is from looking at the design of the universe as a whole to looking at particular instances of design within the universe and specifically in biology. This brand of design arguments falls under the rubric of Intelligent Design.

Proponents of Intelligent Design attempt to develop a scientific research program as a positive alternative to Darwinism and other naturalistic approaches to the origin and history of life. What has kept design outside the scientific mainstream during the past 140 years is the absence of precise methods for distinguishing intelligently caused objects from unintelligently caused ones. For design to be a fruitful scientific concept, scientists have to be sure they can reliably determine whether something is designed. Johannes Kepler (1571–1630) thought the craters on the moon were intelligently designed by moon dwellers. We now know that the craters were formed by blind natural processes. It is the fear of falsely attributing something to design only to have the attribution overturned later that has prevented design from entering science proper. With precise methods for discriminating intelligently from unintelligently caused objects, design theorists claim that they can avoid Kepler's mistake and reliably locate design in biological systems.

As a theory of biological origins and development, Intelligent Design's fundamental claim is that intelligent causes are necessary to explain the complex, information-rich structures of biology and that these causes are empirically detectable. To say that intelligent causes are empirically detectable is to say that there exist well-defined methods that, on the basis of observational features of the world, are capable of reliably distinguishing intelligent causes from undirected natural causes. Many special sciences have already developed such methods for drawing this distinction—notably forensic science, cryptography, archeology, and the Search for Extraterrestrial Intelligence (Ratzsch 2001).

Whenever these methods detect intelligent causation, the underlying entity they uncover is a type of information known alternately as *specified complexity* or *complex specified information* (Dembski 1998a, 2001). Think of the signal that convinced the radio astronomers in the film *Contact* that they had found an extraterrestrial intelligence. The signal was a long sequence of prime numbers. Because of its length, the signal was complex and could not be assimilated to any natural regularity. And yet, because of its arithmetic properties, it matched an objective, independently given pattern. The signal was thus both complex and specified. What's more, the combination of complexity and specification convincingly pointed those astronomers to an extraterrestrial intelligence. Design theorists contend that specified complexity is a reliable indicator of design, is instantiated in certain (though by no means all) biological structures, and lies beyond the capacity of blind natural causes to generate it.

To say that specified complexity lies beyond the capacity of blind natural causes to generate it is not to say that naturally occurring systems cannot exhibit specified complexity or that natural processes cannot serve as a conduit for specified complexity. Naturally occurring systems can exhibit specified complexity, and nature operating without intelligent direction can take preexisting specified complexity and shuffle it around. But that is not the point. The point is whether nature (conceived as a closed system of blind, unbroken natural causes) can *generate* specified complexity in the sense of originating it when previously there was none.

Take, for instance, a Dürer woodcut. It arose by mechanically impressing an inked woodblock on paper. The Dürer woodcut exhibits specified complexity. But the mechanical application of ink to paper via a woodblock does not account for that specified complexity in the woodcut. The specified complexity in the woodcut must be referred back to the specified complexity in the woodblock, which in turn must be referred back to the designing activity of Dürer himself (who in this case deliberately chiseled the woodblock). Specified complexity's causal chains end not with nature but with a designing intelligence.

Intelligent Design properly formulated is a theory of information. Within such a theory, complex specified information (or specified complexity) becomes a reliable indicator of intelligent causation as well as a proper object for scientific investigation. Intelligent Design thereby becomes a theory for detecting and measuring information, explaining its origin, and tracing its flow. It is therefore not the study of intelligent causes per se but of informational pathways induced by intelligent causes. As a result, Intelligent Design presupposes neither a Creator nor miracles. It is theologically minimalist. It detects intelligence without speculating about the nature of the intelligence.

Biochemist Michael Behe (1996) connects specified complexity to biological design. Behe defines a system as *irreducibly complex* if it consists of several interrelated parts so that removing even one part completely destroys the system's function. For Behe, irreducible complexity is a sure indicator of design. One irreducibly complex biochemical system that Behe considers is the bacterial flagellum. The flagellum is an acid-powered rotary motor with a whiplike tail that spins at 20,000 rpm and whose rotating motion enables a bacterium to navigate through its watery environment. Behe shows that the intricate machinery in this molecular motor—including a rotor, a stator, O-rings, bushings, and a drive shaft—requires the coordinated interaction of at least thirty complex proteins and that the absence of any one of these proteins would result in the complete loss of motor function. Behe argues that the Darwinian mechanism is in principle incapable of generating such irreducibly complex systems. In *No Free Lunch*, William Dembski (2001) attempts to demonstrate that Behe's notion of irreducible complexity is a special case of specified complexity. In particular, he argues that systems like the bacterial flagellum exhibit specified complexity and are therefore designed.

Design has had a turbulent intellectual history. The chief difficulty with design to date has consisted in discovering a conceptually powerful formulation of it that will fruitfully advance science. It is the empirical detectability of intelligent causes that promises to make Intelligent Design a full-fledged scientific theory and distinguishes it from the design arguments of philosophers and theologians, or what has traditionally been called "natural theology." According to the Intelligent Design thesis, the world contains events, objects, and structures that exhaust the explanatory resources of undirected natural causes and that can be adequately explained only by recourse to intelligent causes. Intelligent Design purports to demonstrate this rigorously. Hence, it takes a long-standing philosophical intuition and makes it into a scientific research program. This program depends on advances in probability theory, computer science, the concept of information, molecular biology, and the philosophy of science—to name but a few. It remains to be seen whether it can turn design into an effective conceptual tool for investigating and understanding the natural world.

BIBLIOGRAPHY

Barrow, John, and Frank Tipler. *The Anthropic Cosmological Principle.* Oxford: Oxford University Press, 1986.

Behe, Michael. *Darwin's Black Box.* New York: Free Press, 1996.

Boyle, Robert. *A Free Enquiry into the Vulgarly Received Notion of Nature.* 1686. Reprint. Ed. M. Hunter and E. B. Davis. Cambridge: Cambridge University Press, 1996.

Cicero, Marcus Tullius. *De Natura Deorum.* Trans. H. Rackham. Cambridge: Harvard University Press, 1933.

Collins, Robin. "A Scientific Argument for the Existence of God." In *Reason for the Hope Within,* ed. Michael Murray. Grand Rapids, Mich.: Eerdmans, 1999.

Darwin, Charles. *The Origin of Species.* 1859. Reprint. London: Penguin, 1985.

Davies, Paul. *The Mind of God: The Scientific Basis for a Rational World.* New York: Touchstone, 1992.

Dawkins, Richard. *The Blind Watchmaker.* New York: Norton, 1987.

Dembski, William. *The Design Inference: Eliminating Chance through Small Probabilities.* Cambridge: Cambridge University Press, 1998a.

———. *No Free Lunch: Why Specified Complexity Cannot Be Purchased without Intelligence.* Lanham, Md.: Rowman and Littlefield, 2001.

———, ed. *Mere Creation: Science, Faith, and Intelligent Design.* Downers Grove, Ill.: InterVarsity Press, 1998b.

Deutsch, David. *The Fabric of Reality: The Science of Parallel Universes and Its Implications.* New York: Penguin, 1997.

Guth, Alan, and Paul Steinhardt. "The Inflationary Universe." In *The New Physics,* ed. Paul Davies. Cambridge: Cambridge University Press, 1989, 34–60.

Hume, David. *Dialogues Concerning Natural Religion.* 1779. Reprint. Buffalo, N.Y.: Prometheus Books, 1989.

Hurlbutt, Robert H. *Hume, Newton, and the Design Argument.* Lincoln: University of Nebraska Press, 1965.

Jeffner, Anders. *Butler and Hume on Religion.* Stockholm: Diakonistyrelsens Bokforlag, 1966.

Kant, Immanuel. *Critique of Pure Reason,* 1787. Trans. N. K. Smith. New York: St. Martin's, 1929.

Leslie, John. *Universes.* London: Routledge, 1989.

Lindley, David. *The End of Physics: The Myth of a Unified Theory.* New York: Basic Books, 1993.

Mackie, J. L. *The Miracle of Theism: Arguments for and against the Existence of God.* Oxford: Clarendon Press, 1982.

Paley, William. *Natural Theology.* 1802. Reprint. Boston: Gould and Lincoln, 1852.

Ratzsch, Del. *Nature, Design, and Science: The Status of Design in Natural Science.* Albany: State University of New York Press, 2001.

Sober, Elliott. *Philosophy of Biology.* Boulder, Colo.: Westview, 1993.

Swinburne, Richard. *The Existence of God.* Oxford: Oxford University Press, 1979.

27 Ecology and the Environment

David N. Livingstone

The relationship between environmental thinking and Christian theology has been so historically complex that monocausal accounts—like the claim that Western Christianity is to blame for the modern ecological crisis—are inadequate. In the following essay some of these heterogeneous connections are explored through an examination of several key metaphors that have been called upon to express the relationship between God and nature, notably, the Divine Economist, Mother Nature, and the Celestial Mechanic. The greening of theology in the last thirty years completes the survey and confirms the continuing vitality of the conversation between environmentalism and theology.

David N. Livingstone is a professor of geography and intellectual history at the Queen's University of Belfast, from which he received his Ph.D. in 1982. He is the author of several books, including *Darwin's Forgotten Defenders* (Edinburgh: Scottish Universities Press and Grand Rapids, Mich.: Eerdmans, 1987); *Nathaniel Southgate Shaler and the Culture of American Science* (Tuscaloosa: University of Alabama Press, 1987); *The Geographical Tradition* (Oxford, U.K., and Cambridge, Mass.: Blackwell Publishers, 1993); and (with R. A. Wells) *Ulster-American Religion* (Notre Dame: University of Notre Dame Press, 1999). His most recent book is *Spaces of Science: Chapters in the Historical Geography of Scientific Knowledge* (Chicago: University of Chicago Press, in press).

Religion and the Metaphors of Nature

Although the term *ecology* was not coined until the nineteenth century—by Ernst Haeckel (1834–1919) in his *General Morphology* (1866)—it was fundamentally a substitute for the earlier and widespread designation the *economy of nature*. Haeckel himself spoke of ecology as "the theory of the economy of nature" while, more recently, Richard Hesse defined it as "the science of the 'domestic economy' of plants and animals" (Hesse et al. 1937, 6). This metaphorical association—thinking of nature as if it were a political economy—is particularly significant for religious reasons because early proponents of the "economy of nature" or the "polity of nature" typically cast the Creator in the role of divine economist.

Nowhere, perhaps, is this conceptual alignment more clearly revealed than in the work of the Swedish botanist Carolus Linnaeus (1707–78), whose taxonomic enthusiasm was fired by the profound conviction that he was unearthing the very order of God's Creation. Indeed, his 1749 essay "The Oeconomy of Nature" was intended to identify the hand of God in nature's order. In this system, all living things were bound together into a chain of interlocking links. To Linnaeus, God was the Supreme Economist, for the analogy was with a well-run household under the watchful eye of a beneficent housekeeper. Hence, the Linnaean system could, at once, confute atheism and justify the social order. So, too, could the political economy of nature expressed in the writings of the Anglican clergyman Gilbert White (1720–93). In *The Natural History of Selborne* (1789), he recorded the natural order of his little parish, insisting throughout that providence had contrived to make "Nature . . . a great economist" who pervasively displayed the wisdom of God (quoted in Worster 1977, 7–8).

In more or less secularized forms this economic metaphor continued to condition ecological thinking from the period of the Enlightenment right into the twentieth century. Late-eighteenth- and nineteenth-century biogeographers, for example, routinely spoke of "nations" of plants. Alexander von Humboldt (1769–1859) treated plant associations as if they were political economies, while Goethe (1749–1832) deployed the fiscal concept of "budget" in his depiction of the natural world as a perfect economy with "inviolate balances." Given such connections between political economy and the "economy of nature," it is no surprise to find figures like Thomas Ewbank (1792–1870) writing, in 1855, that the world's economy "was designed for a Factory" by the great Designer (Worster 1977, 53). In the early twentieth century, ideas about the appropriate functioning of human economies and social communities continued to condition the new science of ecology. Eugenius Warming (1841–1924) and Frederic Clements (1874–1945), for instance, believed that plant communities had what William Coleman called "a definite general economy" with a specific set of occupying life-forms (Coleman 1986).

A different, though related, metaphorical conception of nature has rather earlier roots and can be traced back at least to the Middle Ages. This was the idea of nature as an organism, a living being, and it was only with the coming of the mechanical universe of Galileo (1564–1642), Isaac Newton (1642–1727), and Francis Bacon (1561–1626) that the potency of this image began to lose its appeal. Indeed, some historians, like Carolyn Merchant, have claimed that the origin of modern environmental despoliation is to be found in the substitution of an inert, mechanistic model of nature for an earlier life-filled, organic vision. Moreover, because the organic analogy was typically construed in gendered terms—as female—a number of ecofeminists have urged that the image of the earth as a nurturing mother had a culturally constraining effect on human action. While organicist ways of thinking were progressively to diminish in the wake of the scientific revolution, they certainly did not disappear from Western consciousness. To the contrary. In the past century or so, organismic modes of thought have blossomed in the development of ecological thinking—and in its accompanying ideological preoccupations. Frank Fraser Darling, a leading conservation spokesman during the 1960s, for example, called the West to adopt "the philosophy of wholeness" or "the truth of Zoroastrianism . . . that we are all of one stuff, difference is only in degree, and God can be conceived as being in all and of all, the sublime and divine immanence" (quoted in Passmore 1974, 173). More recently, organicism has been further rejuvenated in the much publicized Gaia hypothesis (Gaia was the Greek earth goddess) advocated by James Lovelock (1919–), a scientist who worked for NASA (National Aeronautics and Space Administration) and Hewlett-Packard. Lovelock describes the global system as "the largest living organism" and "a complex entity involving the Earth's biosphere, atmosphere, oceans and soil; the totality constituting a feedback or cybernetic system which seeks an optimal physical and chemical environment for life on earth" (Lovelock 1979, 11).

Similarly implicated in the organic vision is the "deep ecology" movement championed by the Norwegian philosopher Arne Naess. Deep ecology is not human centered but celebrates the close partnership of all forms of life and insists on the equal right to live and blossom. Thus, the deep-ecology movement rejects the separation of humanity from the rest of the natural order and honors the intrinsic value of every form of life. While Lovelock claims scientific objectivity for his Gaia hypothesis and the deep ecologists call for social reform, others find in organicism inspiration for what is called the New Paganism and the restoration of worship of the Earth Goddess. In many ways, this turn of events can be seen as part of a New Age rejection of scientific rationalism and the Enlightenment and as the perpetuation of organic ways of thinking about the natural world that flourished during the medieval period.

An altogether different image of nature received impetus with the advent of the scientific revolution of the seventeenth century—that of the machine. The triumph of this mechanical vision was due, in large measure, to the search for inexorable laws governing the physical world. Through the writings of figures like René Descartes (1596–1650), Newton, and Robert Boyle (1627–91), the triumph of the mechanical system was secured. According to some historians, the new science, particularly as championed by Francis Bacon, ushered in a new ethic that sanctioned the despoliation of nature. Courtesy of the mechanical arts, nature was dominated and bound into service. This transformation from the organic to the mechanistic, moreover, was not effected in cerebral isolation from changing social conditions. Rather, it was intertwined with the lengthy and complex shift from manorial farm economics to market capitalism, with its marked ecological consequences.

If, indeed, the new science initiated profound environmental change, it was *within* the mechanical philosophy that principles of environmental management began to be enunciated. Concerned at wasteful land practices, John Evelyn (1620–1706), for example, who published in 1662 his famous *Silva: A Discourse of Forest Trees and the Propagation of Timber in His Majesty's Dominions,* responded to the alarming drop in timber supply by appealing for the institution of sound conservation practices.

As often as not, such conservation principles were built upon the assumption that the human species had been created to be God's viceroy on Earth, a perspective that received impetus from the intimate connections between the new science and the mushrooming of a natural theology designed to uncover the ways in which the orderliness of the world machine attested to the sovereignty and beneficence of its Celestial Mechanic. Within this scheme, humans were seen as having a responsibility to exercise stewardship over the natural world to ensure that the marks of its designer were not effaced. God was a wise conservationist, and people, made in his image, were to act as caretakers of his world.

This form of beneficent dominion surfaced in the writings of Sir Matthew Hale (1609–76) and William Derham (1657–1735). Hale, England's mid-seventeenth-century Lord Chief Justice, told his readers that humanity's stewardship role was for the purpose of curbing the fiercer animals, protecting the other species, and preserving plant life. As for Derham, his *Physico-Theology* (1713) outlined a range of ecologically sound principles that included population stability, ecological interdependence, and species adaptation. All of these were rooted in his conviction that the Creator's "Infinite Wisdom and Care condescends, even to the Service, and Wellbeing of the meanest, most weak, and helpless insensitive Parts of the Creation" (Derham 1727, 425).

This fundamentally *managerial* approach to environment, adapted as it was to the rationalizing tendencies of the new mechanical world order, aimed at long-term planning, the maximization of energy production, sustained yield, ecosystem control, and the application of science to policy formation. It would ultimately issue in modern cost-benefit analysis, the concept of sustainable development, and environmental-impact assessment.

The White Thesis and Its Critics

Despite the fact that it was in the period of the scientific revolution that the mainsprings of environmental managerialism are to be found, there are those who urge that it was the coming of the new science that played a crucial role in the emergence of the modern environmental dilemma. Chief among these critics was Lynn White Jr. (1907–87), whose famed diagnosis of "The Historical Roots of Our Ecologic Crisis" appeared in *Science* in 1967. His claim was that environmental devastation had its roots in the Western marriage between science and technology, a union whose intellectual origins predated the scientific revolution. During the Middle Ages, he argued, a profound dislocation in the understanding of "man and nature" had taken place. Instead of humanity's being thought of as *part* of nature, the human race was seen as having dominion *over* nature and, thus, as licensed to violate the physical environment. This attitudinal shift, when conjoined to new technology, wreaked ecological havoc. As for the origins of this exploitative turn of events, White asserted that it was the consequence of the triumph of Christianity over paganism. For Christianity, he insisted, held that nature existed for the benefit of man, who was made in the image of God. This "most anthropocentric religion," he went on, stood in stark contrast to earlier religious traditions in which every tree, spring, and stream had its own guardian spirit. Christianity, he concluded, fostered environmental indifference by eradicating pagan animism.

While it rapidly provoked a furious controversy and a suite of refutations, White's paper should be read in the context of Bert Hall's comment that "White was a believing Christian, and in his early publications he argued for the importance of medieval Christianity in our cultural makeup" (Hall 1988). Indeed, in a 1975 commencement address to the San Francisco Theological Seminary—

of which he was a trustee—White concluded that "the study and contemplation of nature are an essential part of the Christian life both because they are acts of praise, and also because they teach us how our fellow creatures praise God in their own ways" (White 1975, 11). Besides, White's diagnosis was a good deal more subtle than conventional résumés suggest. He was fully aware that Western Christianity encompassed a variety of traditions, some of which—notably that of St. Francis of Assisi (1181 / 2–1226), whom he proposed as the patron saint of ecologists—were more reverential toward the created order.

White was certainly not alone in finding religious sentiments at the headwaters of the environmental crisis. Max Nicholson, for fourteen years director-general of the Nature Conservancy in Great Britain, for instance, insisted that Christianity was ecologically culpable because of the doctrine of "man's unqualified right of dominance over nature" and called for the obliteration of "the complacent image of Man the Conqueror of Nature, and of Man Licensed by God to conduct himself as the earth's worst pest" (Nicholson 1970, 264). (It is perhaps significant that he more recently insisted that the "need for theological rethinking on man's place in nature is urgent" [Nicholson 1987, 195].) Arnold Toynbee located the origins of environmental improvidence in biblical monotheism and claimed that the only solution lay in resorting to the *Weltanschauung* of pantheism. Similar sentiments have been expressed by many other writers, but perhaps the most articulate defender of a revisionist version of the White thesis is John Passmore, who claimed that a combination of traditional belief in human dominion over creation and Stoic philosophy encouraged a morally unconstrained use of nonhuman nature.

Despite this chorus of support, White's analysis has not escaped criticism. Lewis Moncrief expressed misgivings about attempts to account for ecological insensitivity in terms of single-factor causes, arguing instead for the significance of a range of "cultural variables," of which two were especially prominent: democratization following in the wake of the French Revolution and, in the American context, the frontier experience. The absence of a public and a private environmental morality and the inability of social institutions to adjust to the ecological crisis Moncrief attributed to these factors. The geographer Yi-Fu Tuan approached the topic rather differently by examining environmental conditions in a number of Eastern regions. It turns out that, despite their ostensibly ecologically sensitive religious traditions, their *practices* were every bit as destructive as those in the West. Hence, the "official" line on attitudes toward environment (the quiescent, adaptive line) in Chinese religions, for example, is actually vitiated by behavior as mistreatments of nature abound through deforestation and erosion, rice terracing, and urbanization.

From yet another perspective, the historian Keith Thomas argues that White and his supporters overestimate religious motivation in human behavior. For Thomas, it was the coming of private property and a money economy that fostered the exploitation of the environment and the disenchantment of nature. In addition, he points to the contested character of the Judeo-Christian stance toward nature: Alongside the tradition sanctioning the human right to exploit nature's bounty was a persistent theology of human stewardship. This, too, is emphasized by Robin Attfield, who insists that the idea that everything exists to serve humanity is not the position of the Old Testament and that there is "much more evidence than is usually acknowledged for . . . beneficent Christian attitudes to the environment and to nonhuman nature" (Attfield 1983a, 369).

Historical Retrieval

Partly as a response to the charges of critics like White, a number of scholars have scrutinized the history of the Christian West to determine just what the legacy of Christianity's attitudes to nature has actually been. We have already seen that the principle of stewardship was promulgated during the scientific revolution by writers urging a restrained human use of nature. But both before and after this crucial moment in Western history, Christian voices urging environmental sensitivity were to be heard.

The case of St. Francis of Assisi, for example, is well known. Committed to a life of poverty and a gospel of repentance, he treated all living and inanimate objects as brothers and sisters and stressed the importance of communion with nature. Some, however, have thought that these very sentiments came too close to heresy and so have turned to other sources of environmental inspiration such as St. Benedict (c. 480–c. 543). The principles of stewardship that he espoused amounted to an early wise-use approach to nature. Indeed, it is for this reason that René Dubos believes that St. Benedict is much more relevant than St. Francis to human life in the modern world. Of course, Benedict did not emerge from a theological vacuum. There were ethical resources embedded even earlier in the patristic period upon which to call. In the *Hexaemeron,* Basil the Great (c. 330–79), one of the Cappadocian Fathers, for instance, displayed a profound interest in nature, as did his contemporary St. Ambrose (339–97) in his own writings; both sought to unveil the wisdom of the Creator in the balance and harmony of nature and to insist on the partnership between God and humanity in the task of improving the earth.

On the other side of the scientific revolution, during the eighteenth and early nineteenth centuries, theological efforts to erode an arrogant anthropocentrism began to surface. Worldwide geographical reconnaissance, expanding astronomical horizons, and an emerging sense of "deep time" all tended to diminish the significance of the human subject. But it also became more common *within* the Christian church to find those urging that all members of the Creation were entitled to be used with civility. Christian writers like John Flavel (a Presbyterian divine [c. 1630–91]), Thomas Taylor (a Seeker [1618–82]), Christopher Smart (a religious poet [1722–71]), and Augustus Montague Toplady (a Calvinist minister and hymn writer [1740–78]) variously showed that, in the Bible, animals were regarded as good in and of themselves and not just for their potential service to humanity. John Wesley (1703–91) instructed parents not to let their children cause needless harm to living things, such as snakes, worms, toads, or even flies. So powerful was this Christian impulse toward a new sensibility that Keith Thomas believes that the "intellectual origins of the campaign against unnecessary cruelty to animals . . . grew out of the (minority) Christian tradition that man should take care of God's creation" (Thomas 1984, 180).

This new sensibility manifested itself in two conceptually significant ways for the growth in ecological thinking. First, there was the enormous significance of the environmental knowledge—such as herbals and county natural histories—produced by dozens of parson-naturalists. Indeed, the natural-history pursuit in the English-speaking world was, by the middle decades of the eighteenth century, a combination of religious impulse, intellectual curiosity, and aesthetic pleasure. Second, Christian theology contributed enormously to an emerging sense of ecological interconnectedness. As Clarence Glacken amply demonstrated, the "real contribution of physico-theology . . . was that it saw living interrelationships in nature concretely. It documented them. It had already—before Darwin's 'web of life'—prepared men for the study of ecology" (Glacken 1967, 427).

The Greening of Theology

At least in part, the retrieval of some of these historical voices is a consequence of what might be called "the greening of theology" over the last quarter of the twentieth century. Joseph Sittler, for example, drew attention to the affirmation of creation in the church's liturgy and hymnody; Paul Santmire traced environmental motifs in the writings of Irenaeus (c. 130–c. 200), Augustine (354–430), Martin Luther (1483–1546), and John Calvin (1509–64); and, even more recently, James Nash recalled the ecological sensitivity of the desert fathers and the Celtic saints, among others. Historical revisionism, however, does not exhaust the contemporary interface between ecology and religion. Prior to the publication of White's diagnosis, Sittler had been developing a theology of the earth and urging that environmental malpractice was an affront to God, while Richard Baer had spoken of environmental misuse as a theological concern. Since then, numerous pronouncements on the environment have been forthcoming from a variety of theological traditions. Drawing inspiration from the process thinking of Alfred North Whitehead (1861–1947) and Charles Hartshorne (1897–), writers like Conrad Bonafazi and John Cobb have sought to cultivate an ecological conscience. Evangelical contributions have been forthcoming from writers like Francis Schaeffer, Rowland Moss, Lawrence Osborne, Loren Wilkinson, Calvin de Witt, and the Calvin Center for Christian Scholarship. More theologically radical is the Creation spirituality championed by the American Dominican priest Matthew Fox. Roman Catholic writers like Thomas Berry and Paul Collins have developed ecological theologies, and a variety of theological ecofeminists have urged that the struggle against the domination of women is intimately connected with other forms of domination, including the environment. As these recent writings reveal, the continuing vitality of the debate over the connections between religion and ecology and the production of ecotheologies shows little sign of diminishing.

BIBLIOGRAPHY

Attfield, Robin. "Christian Attitudes to Nature." *Journal of the History of Ideas* 44 (1983a): 369–86.

———. *The Ethics of Environmental Concern.* New York: Columbia University Press, 1983b.

Black, John S. *The Dominion of Man: The Search for Ecological Responsibility.* Edinburgh: Edinburgh University Press, 1970.

Browne, Janet. *The Secular Ark: Studies in the History of Biogeography.* New Haven: Yale University Press, 1983.

Coleman, William. "Evolution into Ecology? The Strategy of Warming's Ecological Plant Geography." *Journal of the History of Biology* 19 (1986): 181–96.

Derham, William. *Physico-Theology; or, A Demonstration of the Being and Attributes of God, from His Works of Creation.* 8th ed. London: 1727.

Glacken, Clarence. "The Origins of the Conservation Philosophy." *Journal of Soil and Water Conservation* 11 (1956): 63–66.

———. *Traces on the Rhodian Shore: Nature and Culture in Western Thought from Ancient Times to the End of the Eighteenth Century.* Berkeley: University of California Press, 1967.

Hall, Bert S. "Lynn White, Jr., 29 April 1907–30 March 1987." *Isis* 79 (1988): 478–81.

Hesse, R., W. C. Allee, and K. P. Schmidt. *Ecological Animal Geography: An Authorized, Rewritten Edition Based on* Tiergeographie auf Oekologischer Grundlage, *by Richard Hesse.* New York: Wiley, 1937.

Jackson, Myles W. "Natural and Artificial Budgets: Accounting for Goethe's Economy of Nature." *Science in Context* 7 (1994): 409–31.

Larson, James. "Not without a Plan: Geography and Natural History in the Late Eighteenth Century." *Journal of the History of Biology* 19 (1986): 447–88.

Lovelock, James. *Gaia: A New Look at Life on Earth.* Oxford: Oxford University Press, 1979.

Merchant, Carolyn. *The Death of Nature: Women, Ecology, and the Scientific Revolution.* New York: Harper and Row, 1980.

———. "Earthcare: Women and the Environmental Movement." *Environment* 23 (1981): 2–13.

Moncrief, Lewis W. "The Cultural Basis of Our Environmental Crisis." *Science* 170 (October 30, 1970): 508–12.

Naess, Arne. "The Shallow and the Deep, Long-Range Ecology Movement: A Summary." *Inquiry* 16 (1973): 95–100.

Nash, Roderick Frazier. *The Rights of Nature: A History of Environmental Ethics.* Madison: University of Wisconsin Press, 1989.

Nicholson, Max. *The Environmental Revolution: A Guide for the New Masters of the World.* London: Hodder and Stoughton, 1970.

———. *The New Environmental Age.* Cambridge: Cambridge University Press, 1987.

Passmore, John. *Man's Responsibility for Nature: Ecological Problems and Western Tradition.* London: Duckworth, 1974.

———. "Attitudes to Nature." In *Nature and Conduct,* ed. R. S. Peters. Royal Institute of Philosophy Lectures. Vol. 8. London: Royal Institute of Philosophy, 1975, 251–64.

Thomas, Keith. *Man and the Natural World: Changing Attitudes in England, 1500–1800.* London: Penguin, 1984.

Toynbee, Arnold. "The Religious Background of the Present Environmental Crisis." *International Journal of Environmental Studies* 3 (1972): 141–46.

Tuan, Yi-Fu. "Our Treatment of the Environment in Ideal and Actuality." *American Scientist* (May / June 1970): 246–49.
White, Lynn, Jr. "The Historical Roots of Our Ecologic Crisis." *Science* 155 (1967): 1203–7.
———. "Christians and Nature." *Pacific Theological Review* 7 (1975): 6–11.
Worster, Donald. *Nature's Economy: A History of Ecological Ideas*. Cambridge: Cambridge University Press, 1977.

Part VII : : Current Historiographical Issues

28 Gender

Sara Miles and John Henry

In Western thought both science and religion have sought to define women's nature, explain their conduct, and dictate their roles. Historically, scientific theories and theological dogmas have often provided rationales for the inferiority of women's nature, the bases for women's supposed emotional and irrational behaviors, and the justifications for circumscribing their actions, positions, and responsibilities, including their exclusion from scientific and religious activities. Questioning the assumptions underlying these theories and dogmas, several schools of twentieth-century scholarship began to look at the ways in which political and social commitments have influenced the presuppositions and conclusions of scientific and theological knowledge. Examining first the roles that some exceptional women played in scientific and religious endeavors, scholars then turned to analyzing claims about women. More recently, some historians, sociologists, and philosophers have argued that the scientific and theological fields themselves are androcentric, raising questions regarding whether either science or theology can say anything valid about women unless women are allowed to participate in and help to shape the disciplines that seek to understand them.

Sara Miles received her Ph.D. from the University of Chicago. She is currently vice president for institutional effectiveness and associate professor of history and biology at Eastern University, St. Davids, Pennsylvania. She is the author of several articles dealing with the relationships between science and religion, the latest analyzing the correspondence between Asa Gray and Charles Darwin on the subject of teleology and design in nature.

John Henry received his Ph.D. from the Open University in England. He is currently senior lecturer in the history of science at the University of Edinburgh, Scotland. Recent publications include *Moving Heaven and Earth: Copernicus and the Solar System* (Cambridge: Icon Books, 2001) and *The Scientific Revolution and the Origins of Modern Science*, 2d ed. (Basingstoke: Palgrave, 2002).

THE TERM *gender* is used in this essay in accordance with the way late-twentieth-century feminists borrowed the word, to differentiate those socially and politically variable meanings of *masculine* and *feminine* from the more fixed biological meanings. In this sense, the notion of gender is intended, as Donna Haraway (a leading feminist thinker) has suggested, "to contest the naturalization of sexual difference." Her point is that a wide range of supposed differences between the sexes have been invoked and exploited to support different attitudes toward, and treatment of, the sexes in various social and political contexts. To talk of gender differences in these contexts, rather than differences of sex, is to alert readers to the all too real possibility that such differences may have been socially constructed to serve particular interests.

Consideration of gender issues arising from, or occurring within, the natural sciences and its various forms of institutional organization was first explicitly signaled in an article called simply "Gender and Science" by Evelyn Fox Keller that appeared in 1978 (reprinted in Keller 1985, 75–94). Although it is possible to find earlier studies concerned with different aspects of relationships between women and the sciences, the subject has become a growth area in feminist scholarship since the publication of Keller's article. Essentially, there are three major aspects of gender and science that have attracted feminist attention: (1) the study of women by science, (2) the role of women in science, and (3) the "gendered" nature of science itself, which (it is alleged) traditionally excluded women and their experiences from any association with, or relevance to, scientific development. There is now a considerable literature in each of these areas. An increasing number of historical studies seek to show, on the one hand, the changing ways in which women, their sexuality, their anatomy, and their mentality have been viewed by almost exclusively male scientists and, on the other hand, the previously unacknowledged contributions that women themselves have made to scientific development. Meanwhile, there is an equally burgeoning area of feminist studies that addresses Keller's original concern that there exists a "pervasive association between masculine and objective, [and] more specifically between masculine and scientific." In what follows, each of these areas of feminist focus is considered in turn, but no attempt is made to give a comprehensive coverage of all of the issues. Our concern is to consider primarily how these different aspects of gender and science relate to religious or theological matters.

Women According to Science

Since ancient times, Western culture has viewed women as inferior to men, offering a justification for this view that has typically been religious, philosophical, or scientific in nature. In the *Republic* (5.25), Plato (c. 427–347 B.C.) accepted the theoretical possibility that women could be equal to men in abilities, differing only in reproductive functions. For Plato, it was important and just, therefore, to provide both sexes with a common education to allow individual differences to appear. Aristotle (384–322 B.C.), however, differed from his mentor and established the scientific basis for many of the standard arguments for women's inferiority. For Aristotle, biology was destiny. Defining a female as a "mutilated male," he developed a biological/philosophical theory that dichotomized traits hierarchically, with male traits being superior to female traits. Hence, men were hot, dry, active, rational, powerful, and spiritual, whereas women were cold, wet, passive, emotional, weak, and material, and no amount of education could overcome women's inherent inferiority. Aristotle's theory provided a scientific explanation for his society's views concerning women and men and justified its cultural rules and practices regarding the sexes.

Aristotle was to be immensely influential in the tradition of natural philosophy, particularly during the Western European Middle Ages, but there was a rival theory in the biological tradition, developed by the supremely influential Greek medical writer Galen (A.D. 129–c. 210). For both Aristotle and Galen, women were underdeveloped males, whose sexual organs remained inside their bodies instead of descending to form the penis, scrotum, and testicles. The ovaries received no name of their own until the seventeenth century, being referred to by medical writers as the female testicles, while the vagina was seen as homologous with the penis. But, while Aristotle believed that the female testicles must be useless on account of their lack of development, Galen insisted that they were fully functional. So while Aristotle was able to see women as mutilated or deformed males, Galen saw them as "perfect in their sex." This difference reflected the two thinkers' opposed views of procreation. For Aristotle, women were like the ground into which the sower plants his seed. They provided only the material from which the embryo was formed, while the man's sperm performed the act of shaping and organizing the matter into a human being. Galen, taking seriously the fact that children often resemble their mothers more than their fathers, believed that both partners contributed equally and that children were formed from a mixture of spermatic fluid from father and mother. (A corollary of this view was that women must achieve orgasm to conceive—another notion that ran counter to the influential Aristotelian view that women are passive in the sexual act.) It should be noted, however, that, while Galen regarded women as "perfect in their sex," there was no question that their sex was inferior to the male sex. Here Galen was in complete agreement with Aristotle.

While Hebrew attitudes toward women led to the same kinds of conclusions as did the Greek, they were justified on religious premises rather than the scientific and philosophical reasoning of Greek thought. Eve was tempted by the serpent and caused Adam to sin; as punishment, God placed women in a subordinate position, to be ruled over by men (Genesis 3). The author of the apocryphal Ecclesiasticus writes: "From a woman sin had its beginning, and because of her we all die" (25:24). Etymologically, the word *wife* in the Pentateuch often means "woman belonging to a man" (e.g., Genesis 2:24–25, 3:8, 17). A woman's mind and spirit were especially weak and susceptible to false teachings and deceptions. She therefore needed the protective authority of a male—father, brother, or husband. The Hebrew tradition nonetheless placed a positive value on many of the emotional characteristics viewed as feminine, such as compassion, love, and pity. Since the Jews believed that both men and women were created in the image of God and so reflect his rational, spiritual, and moral attributes, some of the feminine attributes were believed to belong to God as well. Hence, the prophet Isaiah taught that God will act in a "motherly" fashion as he comforts his people (Isaiah 66:13). The Greeks, by contrast, viewed such characteristics as weak because they were opposed to the rational attributes of the male.

The early Christian view of women in some ways challenged traditional attitudes toward women in both Hebrew and Greek thinking. The Apostle Paul (d. A.D. c. 67) wrote to the church at Galatia that, in Jesus Christ, the old divisions based on human understanding had been overcome and that "there is neither male nor female; for you are all one in Christ Jesus" (Gal. 3:28 New International Version). Early Christians were careful, however, not to upset traditional cultural norms. Women carried on active charitable work in local churches but did not participate in the ministry as elders (presbyters). With the growth of the monastic movement in the fourth century, women were active in forming convents. During the High Middle Ages (c. 1000–c. 1400), the influence of Greek ideas became more dominant after the establishment of Galenism in the medical faculties and Aristotelianism in the arts faculties of the medieval universities. The synthesis of Aristotelianism with Christian theology, initiated by Thomas Aquinas (c. 1225–74) and consolidated in subsequent university scholasticism, ensured that theories of the inferiority of women were fully endorsed by natural philosophy. The coupling of rationality with maleness and passion with femaleness was perhaps the most significant way that this Christian-Aristotelian synthesis influenced attitudes toward women. On the one hand, women's supposed inferior rational powers—and, hence, their inability to control their emotions—became the explanation for Eve's inability to withstand the serpent's wiles. For the Dominican authors of the highly influential fifteenth-century work on witchcraft, *Malleus maleficarum,* this rational inferiority explained why such a high proportion of witches were women. It made them more susceptible to the devil's deceptions and, combined with their unruly, passionate natures, prompted them to unnatural demonic alliances and to inappropriate emotional responses in human relationships. At the same time, the belief in the intellectual inferiority of women provided the basis for their exclusion from scientific and medical education and later even from those practices with which they had been traditionally involved, such as midwifery.

The period known as the scientific revolution did nothing to redress the balance. On the contrary, as a number of feminist scholars have pointed out, it saw a renewed emphasis on sexual metaphors of male dominance over the passive female. As the standard view of sexual politics was increasingly applied to Mother Nature, so the natural philosopher increasingly saw himself as ravishing and enslaving her. Francis Bacon (1561–1626), a leading spokesman for both the new empirical science and the usefulness of natural knowledge, said that nature must be captured and enslaved, and her secrets, like her inner chambers, penetrated. Concomitant with such views was an increased emphasis on natural philosophy, not merely as a way of understanding the physical world (as it had been previously) but as a means of controlling, manipulating, and exploiting it for the benefit of mankind. Feminist historians have also suggested that the mechanistic natural philosophy developed during the scientific revolution, and in many ways characteristic of it, was a masculine kind of natural philosophy that replaced the more feminine holistic, vitalistic, and magical worldviews that had preceded it. It should be noticed, however, that the justification for attributing gender to these differing approaches to nature is itself open to dispute. Magic, for example, was always an exploitative endeavor and was, for example, a major influence upon Francis Bacon's ideas about the reform of natural philosophy. Even so, feminist historical analyses of the gendered nature of the scientific revolution seem hard to deny. After all, if the magical tradition did, indeed, influence modern science, it did not survive the experience. As seventeenth-century natural philosophers took what they wanted from the magical worldview and turned it into the new philosophy, they vigorously denounced what was left of that tradition as superstition. Moreover, the branch of magic known as witchcraft became, during the period of the European witch craze, a major means of discrediting magic as blasphemous, heretical, or superstitious.

If the scientific revolution saw a renewed emphasis upon the biological and, therefore, sociopolitical inferiority of women, a further change in sexual politics was required during the eighteenth century. The replacement of absolutist political systems (with their belief in the divine right of kings and a rigid hierarchical organization) by social-contract theories of politics (which held that monarchy or other forms of government were based upon the delegation of political power by the people to the government to act on their behalf) gave rise to more egalitarian notions of social organization. Thomas Laqueur, Londa Schiebinger, and others have argued that, with the newly pervading political theory of liberal egalitarianism, it no longer seemed acceptable to maintain the old hierarchical positioning of men over women. Some thinkers, accordingly, argued for equality of the sexes, but, for most, this notion was unacceptable. Hence, the authority of science began to develop new theories of women that reestablished the age-old claims that they were biologically unsuited for public and political life, but without having to rely on crude notions of inferiority and superiority.

From now on, the notion that women were inferior to men was replaced by the view that women were so completely different from men in all respects, and so obviously intended for childbearing and childrearing, that they could legitimately be excluded from ongoing political deliberations about who was entitled to vote or to take part in government. It is no coincidence that at just this time we see books appearing with titles such as Edward Thomas Moreau's *A medical question: Whether apart from genitalia there is a difference between the Sexes?* (1750) and Jakob Ackermann's *On the discrimination of sex beyond the genitalia* (1788). Numerous other medical writers begin to insist that women are different not just with respect to their genital organs but in their bone structure, their hair, their eyes, their sweat, their brains, and, indeed, as one writer insisted, "in every conceivable respect of body and mind." As Pierre Roussel put it in 1775, "the essence of sex is not confined to a single organ but extends through more or less perceptible nuances into every part." Ideas like this form the scientific background to Jean Jacques Rousseau's (1712–78) insistence in *Emile* (1762) that, "once it is demonstrated that man and woman are not and ought not to be constituted in the same way in either their character or their temperament, it follows that they ought not to have the same education" and, by a facile implication, that women ought not to be included in discussions of the political rights of man.

Philosophers such as Rousseau and Immanuel Kant (1724–1804) argued that moral action required the ability to reason abstractly: Since, they posited, women lacked this ability, their inferior moral sense was confirmed. Whereas women acted morally on the basis of emotion, men relied on reason to determine the appropriate moral response in a given situation. It was the heart that led women to acts of compassion, to tender nurturing, and to loving sacrifice. It was the mind that enabled men to develop just laws and a sense of social duty. Therefore, if the political sphere was to be rationally and scientifically constructed, women must continue to be excluded.

However, a somewhat contradictory view was simultaneously developing that posited the moral superiority of women. Since Enlightenment thinkers increasingly came to view religion as a nonrational (if not *irr*ational) endeavor, faith, revelation, and spiritual sensitivities were evidently more appropriate for women. Women were, therefore, more likely to acknowledge and obey the moral obligations arising from religious devotion than were men. Beginning in the eighteenth century and continuing into the nineteenth, society expected women to provide some kind of moral leadership on authority founded in religious experience. Thus, women established, organized, and led reform-oriented, benevolent associations, such as the Women's Christian Temperance Union and the British Society of Ladies for Promoting the Reformation of Female Prisoners. Those denying any moral superiority to women explained these forms of moral leadership as merely the extension of maternal feelings to those beyond the family circle. Both sides agreed that *real* institutional reform still necessitated the rational intervention of men using a scientific approach to the political, legal, and economic spheres.

Given the importance of the interests that ensured the perpetuation of this kind of sexual politics, it is hardly surprising that the new developments in the biological sciences of the nineteenth century were interpreted in such a way as to confirm these ideas. Charles Darwin's (1809–82) evolutionary theory, especially as extended and applied to humans in *The Descent of Man* (1871), seemed to corroborate earlier theories concerning the lower standing of women. All differences between males and females demonstrated for Darwin that the former were closer to perfection than the latter. Female traits resembled more closely those either of a child or of a lower species. The formation of the skull and the lack of facial hair, for example, placed women between the rank of children and the level of adult males—higher than the one but lower than the other. The emotional proclivities of women placed them lower than rational men but higher than the animals, which acted by instinct rather than by reason. Whereas in all races the female members were not as fully evolved as the males, Darwin was clear to point out that white women, while not as evolved as white men, were more highly evolved than men of other races.

The newly emerging science of psychology was also used to bolster traditional gender differences. In Greek Hippocratic medicine, hysteria had been defined as a woman's disease that resulted from a wandering uterus (the word *hysteria* is derived from the Greek word for uterus). Seventeenth- and eighteenth-century physicians had theorized that women's reproductive organs in general made women more susceptible to illness, and hysteria and many other mental conditions were classified as resulting primarily from their uterine condition. By the nineteenth century, the notion of nervous or psychological conditions was accepted, but women's vulnerability to these problems was still believed to be related to connections between the organs of reproduction and the central nervous system. The menstrual cycle was perceived to create an unstable condition in women, making them more easily overcome by internal and external stimuli that would leave men unaffected. Treatises on the prevention of nervous conditions in women, therefore, emphasized the need to eliminate excessive stimulation and to economize mental and physical energy in order to have sufficient resources to respond to the assaults in and on the person, including the rigors of childbearing.

Thus, advanced education was deemed acceptable only for single women, who, in choosing education over marriage, were thought to have picked the lesser alternative. For health reasons, therefore, women should not study too hard (especially at subjects that required a great deal of reasoning), try to perform masculine activities (such as working outside of the home or filling leadership roles), or exercise too much (e.g., by running rather than walking sedately). Using energy to perform such tasks put the woman at risk, since that energy was also needed to maintain stability and to respond to normal stimuli because of their innate weakness. Even when psychological theories changed with Sigmund Freud (1856–1939) and the introduction of psychoanalysis, cultural opinions reflected many of the earlier views. Moreover, the newer psychological theories still posited males as the norm. Freud's concept of penis envy assumed that females would recognize that the male anatomy was "normal" and that their own anatomy was, therefore, "abnormal." Such theories also reinforced traditional societal spheres for women as the natural spheres, limited, according to Freud, to *Kinder, Küche,* and *Kirche* (children, kitchen, and church).

In spite of the various changes in intellectual outlook from the Greeks, to the scientific revolution, through the Enlightenment, and on to the establishment of evolutionary biology and the major scientific achievements of the early twentieth century, the alleged incapacity of women for public life and high achievement remained so persistent as to be scarcely credible. The fact that the situation changed so considerably in the last two decades of the twentieth century undoubtedly owes more to the consciousness-raising efforts of recent feminism than it does to new developments in science. Furthermore, given the close alliance between science and religion through most of the period under discussion and the patriarchal nature of much traditional faith and practice (which merely reflected its cultural framework), it seems safe to say that religion played little or no significant part in the improved *scientific* understanding of female nature.

Women in Science

Scientific claims about the mental and physical inability of women to excel beyond the domestic sphere have been seen as one reason that women have been excluded from science and medicine and have, therefore, made only minor contributions to the history of science. But this is not the only reason behind women's lack of success in the history of science. The very fact that science has come to be seen in our culture as a masculine pursuit is another major factor, but even this does not cover all of the ground. Women's traditional absence from the history of science must be seen as one more example of the traditional exclusion of women from all but a few circumscribed aspects of social life. Until comparatively recently, women have been systematically excluded from the institutions of science and medicine, while their individual achievements have rarely been taken seriously.

Here again, however, the situation has begun to change. Recent work by feminist historians has done much to uncover the previously unnoticed history of women in science. In spite of a few pioneering efforts, beginning during the first feminist movement of the late nineteenth century, it is only since the 1980s that a feminist historiography, detailing women's contributions to science, has impinged in a significant way upon the consciousness of other historians. A major proportion of this work focuses upon individual heroines, women whose achievements are remarkable by any standards and all the more so given the barriers laid in their way by their own society. Sneers against women's achievements in science on the grounds that they are a long way from a Galileo, a Newton, or a Leibniz are silenced by the incidental details in the histories of women like Margaret Cavendish, Duchess of Newcastle (1623–73), and Anne, Viscountess Conway (c. 1630–79), in the seventeenth century; Émilie du Châtelet (1706–49) in the eighteenth; and Mary Somerville (1780–1872), Sofya Kovalevski (1840–1901), and Marie Curie (1867–1934) in the nineteenth. On reading their and other women's stories, one cannot help wondering what these women might have achieved had their society viewed them and their work differently or had their circumstances allowed them to pursue their work more single-mindedly. It cannot be denied, however, that such heroines are few. Accordingly, other feminist historians have preferred to look at the social history of women in science, the nature of the work they are allowed to do, the way they work and interact with male colleagues, and other patterns of their participation in science. This aspect of the feminist historiography of science also includes studies of the institutional context against which women all too often had to fight, such as a system of higher education that excluded women or provided them with a separate, more "suitable," education, or a system of scientific societies that excluded women, no matter what their achievements.

Some patterns are beginning to emerge from this historical research. Women have occasionally been able to colonize particular areas of science, such as botany in the eighteenth and nineteenth centuries or primatology in the twentieth century, but much work remains to be done before we can fully understand these unusual formations in the structure of science. It is also clear from contemporary research by historians and sociologists of science that women are increasingly entering science as a profession. In the last several decades of the twentieth century, women entered scientific careers at an unprecedented rate.

The inextricability of science and religion in the history of Western culture makes it inevitable that both must be considered together to understand the historical absence of women from science. David Noble has argued that Western science, because of its links to natural theology, "was always in essence a religious calling," and he has seen it as a "clerical culture." Margaret Wertheim, similarly, has suggested that "the priestly conception of the physicist continues to serve as a powerful cultural obstacle to women." Just as women were not permitted into the priesthood, so they were hindered from being priests of God's other book, the book of nature. It is easy to see, however, that, like their male counterparts, female scientists could have religious motivations for their interests. Anne Conway, for example, developed a vitalistic and monistic natural philosophy and used it to dismiss the traditional dichotomy between matter and spirit in order to counter the perceived atheism of dualistic mechanical philosophies. It seems safe to conclude, therefore, that religion, both in its alliance with science and in accordance with its own generally nonfeminist agenda, usually tended to accept the exclusion of women from science as much as from other areas of public life. In this respect, it merely reflected the broader culture.

Women and Science

A final and highly important aspect of feminist critiques of science has arisen in response to the perception that science itself is gendered and that its gender is masculine. Evelyn Fox Keller's 1978 article was primarily concerned with this "unexamined myth," which she saw as familiar and deeply entrenched in Western culture. Similarly, Carolyn Merchant's historical study of the scientific revolution, *The Death of Nature* (1980), was, in part, an attempt to understand the roots of the belief that science is a masculine pursuit. Other feminists have taken up this theme, pointing out that there was a prevailing assumption that women did not, indeed could not, think scientifically (notable exceptions being tacitly presumed to think like men). It is undoubtedly the feminist awareness of these claims that has led to the proliferation of historical studies of women's role in science, both as practitioners and as scientific subjects, but this awareness also has led to a profound reexamination of scientific epistemology. Rejecting the allegedly inherent sexism of current epistemologies propagated by men and believing that their own philosophies should bring some benefit to women, feminist philosophers have sought to develop new and more appropriate ways of knowing the world.

Some feminists simply believed that allegedly sexist and androcentric conclusions in science were merely the result of ideological distortion. The resulting errors arose because the truly "objective" scientific method had been insufficiently rigorously applied. It was their belief that proper vigilance against cultural bias and a more careful pursuit of the scientific method would lead to improved scientific knowledge. Implicit in these beliefs was a conviction that there was nothing innately masculine about science and its methodology, that science was not, in fact, gendered, but that scientists, predominantly male, were all too easily led astray by cultural pressures. This position was called "feminist empiricism" by Sandra Harding, a leading feminist philosopher of science.

Harding herself rejected this position and has tried to advocate a more ambitious approach, first signaled by Georg Wilhelm Friedrich Hegel (1770–1831) and, subsequently, by Marxist philosophers, called "standpoint epistemology." Originally developed in the social sciences, in which feminist practitioners became aware of the cultural biases in their questionnaires and other testing procedures, the standpoint approach takes it for granted that there is no one, privileged position from which value-free knowledge can be established. Assuming this, the standpoint theorist seeks to determine the best position for understanding the particular phenomena under investigation. Feminist sociologists, therefore, would valorize the perspective of the socially disadvantaged in the hope of learning something new about the social conditions of that group. Harding has tried to promote this approach in science, suggesting that women's perspectives may lead to an improved science.

The difficulty with this position, of course, is that it is not clear which women's perspectives would provide the best perspective. There are many different women, from different social, religious, or racial backgrounds, for example, who are all likely to have different standpoints on scientific issues. Similarly, should we take the standpoint of a female scientist, who nonfeminist critics might well claim has a rather masculine standpoint, or the standpoint of a woman far removed from scientific concerns, in which case it might legitimately be argued that her standpoint can hardly be considered the best available for understanding science? In spite of the formal difficulty of deciding upon this issue, feminist scholars have provided some excellent case studies to show just how women's perspectives have made major contributions to the improvement of our scientific understanding. Notable among them are studies of menopause related by Anne Fausto-Sterling and studies of primatology analyzed by Donna Haraway. There has, however, been a tendency among less careful feminist thinkers to suggest that women's "standpoint" allows greater recognition of nonrational, creative, and "intuitive" ways of thinking than masculine standpoints. But this, as other feminists have been quick to point out, is merely to accept the traditional male view of what women are supposed to be like (according to masculine science).

Another leading feminist philosopher, Helen Longino, has drawn upon recent work in the sociology of scientific knowledge to propose an alternative epistemology. Beginning from the traditional view that scientific objectivity, purged of cultural or political biases, is guaranteed by science's unique method (which relies upon repeated observations of the phenomena by different researchers and a thoroughly rational, even mathematical, analysis of the results), Longino reminds her readers of the work of N. R. Hanson (1924–67) and Thomas S. Kuhn (1922–96), who suggested that observations are theory laden, and Pierre Duhem (1861–1916) and W. V. O. Quine (1908–), who suggested that all theories are underdetermined by the data (i.e., not sufficiently grounded upon the data to ensure that no alternative theory is possible). She then goes on to develop, as nonfeminist sociologists of scientific knowledge had before her, a theory of scientific knowledge based on the consensus of scientific practitioners. "Scientific knowledge, on this view," she writes, "is an outcome of the critical dialogue in which individuals and groups holding different points of view engage with each other. It is constructed not by individuals but by an interactive dialogic community" (Longino 1993, 112). Longino calls this position "contextual empiricism." It is "contextual" because it acknowledges that scientific knowledge can be understood only by considering the context from which it emerged. It is "empiricist" in the same way as the feminist empiricists because it implies that there is nothing inherently masculine in scientific thinking, or method, merely that women's voices have thus far been excluded from the critical dialogue of scientific-consensus formation.

Debates about feminist epistemologies in science are continuing, but the literature on these matters thus far has paid no attention to religious or theological concerns. Similarly, in the ongoing debates about feminist theology, the major concern is to link feminist theology to other theories of liberation theology, and scant attention has been paid to scientific issues. It seems clear, however, that feminist theologians have as much right as anyone else to look at science from their particular standpoint (if it can be said that they have a single standpoint) or that female scientists who are theists have as much right to engage in the consensus formation of science as male theistic scientists or, for that matter, nontheistic scientists. Until a literature begins to emerge that specifically discusses these three related issues, it is worth noting the similarities between the treatment of women in science and Christianity. The Christian churches, like scientific theory and institutions, have often reflected cultural norms by depicting women as inferior to men. On the other hand, they have, on other occasions, elevated the status of women above conventional societal patterns (witness the medieval cult of the Virgin Mary or the idea of companionate marriage in the Protestant Reformation). While it is hardly surprising that feminist theologians have found Christian theology to be masculine in the way it has been gendered, just as feminist historians and philosophers have found science to be, perhaps one ought to caution against drawing facile generalizations that ignore the theological complexities of what is hardly a monolithic tradition, as well as the cultural conditioning of time and place. Essentialist and presentist approaches are as out of place here as they are in any historical endeavor.

BIBLIOGRAPHY

Abir-Am, Pnina G., and Dorinda Outram, eds. *Uneasy Careers and Intimate Lives: Women in Science, 1789–1979.* New Brunswick: Rutgers University Press, 1987.

Bem, Sandra Lapses. *The Lenses of Gender: Transforming the Debate on Sexual Inequality.* New Haven: Yale University Press, 1993.

Borresen, Kari E., ed. *Image of God and Gender Models in Judaeo-Christian Tradition.* Oslo: Solum Fodag, 1991.

Cadden, Joan. *Meanings of Sex Difference in the Middle Ages: Medicine, Science, and Culture.* Cambridge: Cambridge University Press, 1993.

Clatterbaugh, Kenneth. *Perspectives on Masculinity: Men, Women, and Politics in Modern Society.* Boulder, Colo.: Westview, 1990.

Fausto-Sterling, Anne. *Myths of Gender: Biological Theories about Women and Men.* New York: Basic Books, 1985.

Fox, Mary Frank. "Women and Scientific Careers." In *Handbook of Science and Technological Studies,* ed. Sheila Jasanoff, Gerald E. Markle, James C. Petersen, and Trevor Pinch. Thousand Oaks, Calif.: Sage, 1995, 205–23.

Gowaty, Patricia Adair, ed. *Feminism and Evolutionary Biology: Boundaries, Intersections, and Frontiers.* New York: Chapman and Hall, 1997.

Haraway, Donna J. *Primate Visions: Gender, Race, and Nature in the World of Modern Science.* New York: Routledge, 1989.

———. *Simians, Cyborgs, and Women.* New York: Routledge, 1991.

Harding, Sandra. *Whose Science? Whose Knowledge? Thinking from Women's Lives.* Ithaca: Cornell University Press, 1991.

Keller, Evelyn Fox. *Reflections on Gender and Science.* New Haven: Yale University Press, 1985.

———. "Gender and Science: Origin, History, and Politics." *Osiris* 10 (1995): 27–38.

Keller, Evelyn Fox, and Helen E. Longino, eds. *Feminism and Science.* Oxford: Oxford University Press, 1996.

Laqueur, Thomas. *Making Sex: Body and Gender from the Greeks to Freud.* Cambridge: Harvard University Press, 1990.

Laslett, Barbara, Sally Gregory Kohlstedt, Helen Longino, and Evelynn Hammonds, eds. *Gender and Scientific Authority.* Chicago: University of Chicago Press, 1996.

Longino, Helen E. *Science as Social Knowledge: Values and Objectivity in Scientific Inquiry.* Princeton: Princeton University Press, 1990.

———. "Subjects, Power, and Knowledge: Description and Prescription in Feminist Philosophies of Science." In *Feminist Epistemologies,* ed. L. Alcoff and E. Potter. New York: Routledge, 1993.

Lunbeck, Elizabeth. *The Psychiatric Persuasion: Knowledge, Gender, and Power in Modern America.* Princeton: Princeton University Press, 1994.

Merchant, Carolyn. *The Death of Nature: Women, Ecology, and the Scientific Revolution.* New York: Harper and Row, 1980.

Noble, David F. *A World without Women: The Christian Clerical Culture of Western Science.* New York: Oxford University Press, 1992.

Oreskes, Naomi. "Objectivity or Heroism? On the Invisibility of Women in Science." *Osiris* 11 (1996): 87–113.

Porter, Roy, and Mikulas Teich, eds. *Sexual Knowledge, Sexual Science: The History of Attitudes to Sexuality.* Cambridge: Cambridge University Press, 1994.

Rossiter, Margaret. *Women Scientists in America: Struggles and Strategies to 1940.* Baltimore: Johns Hopkins University Press, 1982.

———. *Women Scientists in America: Before Affirmative Action, 1940–1972.* Baltimore: Johns Hopkins University Press, 1995.

Schiebinger, Londa. *The Mind Has No Sex? Women in the Origins of Modern Science.* Cambridge: Harvard University Press, 1989.

———. *"Nature's Body": Gender in the Making of Modern Science.* Boston: Beacon, 1993.

———. *Has Feminism Changed Science?* Cambridge: Harvard University Press, 1999.

Scully, Diana, and Pauline Bart. "A Funny Thing Happened on the Way to the Orifice: Women in Gynecology Textbooks." *American Journal of Sociology* 78 (1974): 1045–49.

Terrall, Mary. "Émilie du Châtelet and the Gendering of Science." *History of Science* 33 (1995): 283–310.

Tuana, Nancy. *The Less Noble Sex: Scientific, Religious, and Philosophical Conceptions of Woman's Nature.* Bloomington: Indiana University Press, 1993.

Wertheim, Margaret. *Pythagoras' Trousers: God, Physics, and the Gender Wars.* New York: Random House, 1995.

29 The Social Construction of Science

Stephen P. Weldon

Since this essay was written, the discipline of history of science has been at the center of several widely publicized controversies, the so-called science wars. Both the terms *social construction* and *postmodernism* play major roles in these controversies. Although the terms are often conflated, the two essays on these topics in this volume treat them quite differently. Weldon employs *social construction* as a historiographical term to refer to specific methodological principles embraced by recent historians and scholars of science studies. *Postmodernism,* by contrast, refers to the post–World War II *Zeitgeist* that rejects modernist, Enlightenment-style thinking, and the essay about it considers the way in which postmodern ideas have affected the science-religion relationship.

The science-wars controversies have pitted scientists and others defending science against social scientists and humanists who study the practice of science. The two camps have been arguing over the understanding of how science works. On the one side, defenders of science adopt an essentially positivistic notion that science creates objective, neutral knowledge. By contrast, science-studies scholars claim that science is deeply influenced by the society and culture around it and that scientific knowledge is far from the objective, value-free creation that it is frequently claimed to be. *Social construction* has often misleadingly become the umbrella term used to characterize the whole discipline of science studies and to brand it as irrelevant, misguided, or even dangerous because it seems to threaten cherished beliefs about science.

Social construction is a scholarly method for getting at the complexity of the human activity of practicing science. Because social constructionists argue that social and cultural forces are inextricably tied to the practice of science and the creation of scientific knowledge, they do not assume that rational arguments alone—which are the focus of the work of practicing scientists—can fully determine the direction and outcome of scientific knowledge and understanding. A similar principle has long been employed in the scholarly study of religion: methodological atheism abjures explanations that rely upon God's action in the world, and many religious scholars use it in an effort to understand the social aspects of religion. In the same way, social construction brackets off the sciences' major explanatory system in order to see social, cultural, and psychological factors at work.

Stephen P. Weldon received his Ph.D. from the University of Wisconsin, Madison. He is currently a visiting scholar at Cornell University and is working on a history of science, skepticism, and secular humanism in America.

THE PHRASE "social construction of science" denotes the view that scientific knowledge is not autonomous or based on universal principles of rationality but, rather, tied directly to social interests and conditions. Science, in this view, is seen to be solely a human production that does not differ fundamentally from other human endeavors. By relativizing scientific knowledge in this way, social constructionism has had direct implications for the way in which one approaches the study of the relationship between science and religion in that it has forced scholars to stop privileging the scientific point of view over the religious.

The methodological orientation of social constructionism was spawned by Thomas Kuhn's classic analysis of the scientific enterprise, *The Structure of Scientific Revolutions* (1962). In it Kuhn (1922–96) argued that fundamental changes in scientific theories occurred through gestalt shifts in the way that communities of scientists perceived central problems of their field. By explaining certain basic theoretical transformations in terms of social and psychological factors, Kuhn deemphasized the role of rational thought in the establishment of scientific knowledge. This way of describing science ran counter to most prevailing conceptions of science that placed great weight on the autonomy of the scientific method. When Kuhn concluded that scientific theories were not independent of the social realm, he unleashed a theoretical current that radically redirected studies in the history of science.

Convinced by Kuhn's thesis, many scholars pursuing the sociology of knowledge undertook a research program to explore the socially contingent nature of scientific knowledge. Contrasting their position with the older sociology of science in the Mertonian tradition, they argued that, instead of correlating social factors with transformations in the institutional structures of scientific communities, the new sociology should investigate how social factors influenced the very content of scientific discovery. Termed the "strong programme" by its leaders at the University of Edinburgh, this new line of research held to several strict criteria for the investigation of science. David Bloor's treatise *Knowledge and Social Imagery* (1976, 2d ed. 1991) canonized this school of thought and proposed that research into science be highly empirical and avoid all attempts at what he called a teleological view of scientific developments. In essence, Bloor argued for the need to approach the study of science without any preconceptions regarding the truth or falsity of the knowledge itself. Methodological relativism thus formed the heart of his research proposals, which meant that sociologists studying knowledge would treat "accepted" knowledge and "rejected" knowledge symmetrically.

Previously, argued Bloor and his colleagues, students of scientific knowledge had treated what they knew to be false ideas quite differently from those they knew to be correct. The false ones were explained by sociological and psychological factors, whereas the true ones were seen to be merely the result of the unproblematic application of scientific method. For the new sociologists of knowledge who were influenced by Kuhn, this procedure was no longer considered to be viable because the very nature of rationality was the object under investigation. Their central project was to determine what caused people to think that a particular assertion was right.

The major premises outlined by the "strong programme" have been reiterated, expanded, and revised by other social constructionists. One alternative point of view comes from scholars who have used the methods of literary criticism in their study of the production of scientific knowledge. This group, led by Steven Woolgar, Bruno Latour, and others, believes that the best way to understand how knowledge comes into being is to pay attention to the rhetoric of scientists. Another group of researchers, who call themselves ethnomethodologists, have avoided such purely textual studies and attempted, instead, to learn about science's social features through participant observation in the laboratory. Finally, the philosopher Paul Feyerabend (1924–94) has likened scientific rationality to a performance. Science, he has contended, does not have any set methodology. Somewhat facetiously, he has claimed that scientific method is a method in which "anything goes": In essence, scientists do whatever they can in order to make their ideas convincing to others.

Out of this plethora of scholarship has come a substantially different picture of the nature and operation of scientific practice. Where science once appeared to be a universal source of knowledge about the world, the social constructionists see it as highly contextual and contingent upon local circumstances. According to these scholars, there is no single entity called "science."

Rather, each scientific discipline has its own methodologies, rules, and procedures that differentiate it from other fields. Furthermore, the line between science and other human endeavors threatens to disappear.

Theoretical principles that so clearly challenge the autonomy and rationality of science have had a pronounced effect on our understanding of the science-religion interaction. The relationship between social constructionism and the study of this interaction, however, varies considerably from case to case. Part of the reason for this is that, by and large, the major developments in social constructionism have subordinated the science-religion relationship to a secondary concern. By placing so much emphasis on the understanding of scientific knowledge, many social constructionists tend to regard religion as merely one social influence among many that affect the production of scientific knowledge. Hence, religion and science are no longer a focal point for analysis; instead, elements perceived to underlie both categories (such as linguistic factors, power relationships, and social hierarchies) have taken center stage.

New studies by social constructionists have shed light on the relationship between religion and science during the scientific revolution. When studying topics such as the origin of the mechanical philosophy or Robert Boyle's (1627–91) conception of science, scholars have shown that it is no longer clear where religion or irrationalism end and "pure science" begins. This is in marked contrast to earlier works by people like Robert Merton (1910–), who assumed a clear distinction between the two and sought to show the effect of one side upon the other. In other revisionist histories, political and cultural questions intrude on the standard story (as, e.g., in discussions of Galileo [1564–1642] and the church), thereby making the religion-science controversy of secondary importance.

In an important article published in 1981, the historian Martin Rudwick pointed out an asymmetry in the treatment of religious knowledge and scientific knowledge. Even among social constructionists, he contended, when the subject turned to modern religious views, far too many scholars still needed to depict a triumphant science and a defeated religion. Rudwick spelled out the implications of strictly adhering to the principles of the "strong programme" when treating religion-science interactions: The two ways of thinking, he argued, must be treated symmetrically. The study of Christian creationism has posed problems for the historian for this very reason. Creationists have presented their theory at times as a science and at other times as a religious position. All the while, however, they have argued their position using both physical and biblical evidences. Referring to Rudwick's article, the historian Ronald Numbers has explicitly used social constructionism to justify an even-handed, unbiased treatment of creationist ideas, calling for the need to treat them with the same seriousness and rigor as other historians of science have treated the views of evolutionists.

Social constructionism has not gone unchallenged among science-studies scholars. One of the most damning criticisms asserts that the insights of social construction are, in fact, not new. The notion that science is a social enterprise and that the knowledge it produces is prone to the same errors and problems as any other human activity should not, argue some, surprise scientists themselves. Furthermore, one need not invoke social constructionism merely to justify a rigorously historicist perspective, one that treats discoveries and failures according to local and historical contexts. The fact that many historians uninterested in questions about the contingent nature of rationality have produced thoughtful and fair-minded studies of science and religion suggests that social constructionism is not as influential in this regard as some suppose. The discipline of the history of ideas has long demonstrated the fluidity with which beliefs move between scientific and religious contexts.

One recent work that has (perhaps inadvertently) tested the limits of social constructionism insofar as it relates to the science-and-religion question is John Brooke's monumental synthesis, *Science and Religion: Some Historical Perspectives* (1991). In this book, Brooke invokes a "complexity thesis" to replace the old conflict and harmony models of the relationship between the two enterprises. This complexity thesis stemmed, in part, from Brooke's view that religion and science can no longer be viewed in broad universal terms. Much of the recent literature in the history of science upon which Brooke has drawn has demonstrated the need to understand both science and religion according to locally contingent factors. In this sense, Brooke's thesis finds much in common with social construction. Nevertheless, one social constructionist has taken Brooke to task for not going far enough in this direction because the very terms *science* and *religion* work against the constructionist enterprise. Any historian who uses those terms, claims Brooke's critic, needs to be aware that the meanings of the words are themselves constructed and may not be useful in understanding a particular situation. In other words, the late-twentieth-century categories of the historian interfere with our understanding of the social and intellectual categories of the period being studied.

Interestingly, David Bloor's 1991 afterword to the second edition of his

Knowledge and Social Imagery makes explicit mention of the historical study of religion, pointing out that the same kinds of arguments currently being waged over the social construction of science were, a century before, waged over the study of religion—namely, could religious dogma still be maintained even when beliefs were subject to a probing analysis that deprivileged them? The revelation of this parallel between nineteenth-century religious studies and twentieth-century science studies illustrates something not only about the changing relationship between science and religion in Western culture but also about the role of the investigator who studies science and religion. Bloor's point seems to be that the debate over the more radical claims of social constructionism will not vanish quietly but will continue to inform all areas of science studies, including the history of science and religion.

BIBLIOGRAPHY

Berger, Peter L. *The Sacred Canopy: Elements of a Sociological Theory of Religion*. New York: Doubleday, 1967.

Bloor, David. *Knowledge and Social Imagery*. 2d ed. Chicago: University of Chicago Press, 1991.

Brooke, John Hedley. *Science and Religion: Some Historical Perspectives*. New York: Cambridge University Press, 1991.

Brooke, John, and Geoffrey Cantor. *Reconstructing Nature: The Engagement of Science and Religion*. Edinburgh: T&T Clark, 1998.

Brooke, John Hedley, Margaret J. Osler, and Jitse M. van der Meer, eds. *Science in Theistic Contexts: Cognitive Dimensions*. *Osiris* 16 (2001).

Cole, Stephen. *Making Science: Between Nature and Society*. Cambridge: Harvard University Press, 1992.

Collins, H. M., and T. J. Pinch. *Frames of Meaning: The Social Construction of Extraordinary Science*. Boston: Routledge and Kegan Paul, 1982.

Desmond, Adrian. *The Politics of Evolution: Morphology, Medicine, and Reform in Radical London*. Chicago: University of Chicago Press, 1989.

Feyerabend, Paul K. *Against Method*. New York: Verso, 1993.

Gieryn, Thomas F. "Relativist/Constructivist Programmes in the Sociology of Science: Redundance and Retreat." *Social Studies of Science* 12 (1982). 279–97.

Hacking, Ian. *The Social Construction of What?* Cambridge: Harvard University Press, 1999.

Hess, David J. *Science in the New Age: The Paranormal, Its Defenders and Debunkers, and American Culture*. Madison: University of Wisconsin Press, 1993.

Levitt, Norman. *Prometheus Bedeviled: Science and the Contradictions of Contemporary Culture*. New Brunswick: Rutgers University Press, 1999.

Lindberg, David C., and Ronald L. Numbers, eds. "Introduction." In *God and Nature: Historical Essays on The Encounter between Christianity and Science.* Berkeley: University of California Press, 1986, 1–18.

Kaye, Howard L. *The Social Meaning of Modern Biology: From Social Darwinism to Sociobiology.* New Haven: Yale University Press, 1986.

Kuhn, Thomas S. *The Structure of Scientific Revolutions.* Chicago: University of Chicago Press, 1962.

Latour, Bruno, and Steve Woolgar. *Laboratory Life: The Social Construction of Scientific Facts.* Beverly Hills, Calif.: Sage, 1979.

Moore, James. "Speaking of 'Science and Religion'—Then and Now." Review of *Science and Religion* by John Hedley Brooke. *History of Science* 30 (1992): 311–23.

Myers, Greg. *Writing Biology: Texts in the Social Construction of Scientific Knowledge.* Madison: University of Wisconsin Press, 1990.

Numbers, Ronald L. *The Creationists.* New York: Knopf, 1992.

Pickering, Andrew, ed. *Science as Practice and Culture.* Chicago: University of Chicago Press, 1992.

Rudwick, Martin. "Senses of the Natural World and Senses of God: Another Look at the Historical Relation of Science and Religion." In *The Sciences and Theology in the Twentieth Century,* ed. A. R. Peacocke. Notre Dame: University of Notre Dame Press, 1981, 241–61.

Ruse, Michael. *Mystery of Mysteries: Is Evolution a Social Construction?* Cambridge: Harvard University Press, 1999.

Segerstrele, Ullica, ed. *Beyond the Science Wars: The Missing Discourse about Science and Society.* Albany: State University of New York Press, 2000.

Shapin, Steven. "History of Science and Its Sociological Reconstructions." *History of Science* 20 (1982): 157–211.

Shapin, Steven, and Simon Schaffer. *Leviathan and the Air-Pump: Hobbes, Boyle, and the Experimental Life.* Princeton: Princeton University Press, 1985.

Slezak, Peter. "The Social Construction of Social Constructionism." *Inquiry* 37 (1994): 139–57.

Wallis, Roy, ed. *On the Margins of Science: The Social Construction of Rejected Knowledge.* Keele: University of Keele, 1979.

30 Postmodernism

Stephen P. Weldon

Ever since the end of World War II, forces critical of science have been building in many segments of society. In the academy, critical philosophical arguments attacking the dominant influence of Enlightenment rationalism have arisen in various departments, from literary criticism to newly minted disciplines like culture studies. The various forms of these arguments constitute philosophical postmodernism, which has been both influential and enormously controversial in recent years.

By posing questions that undercut the traditional foundations of Western culture, especially science and Judeo-Christian religion, postmodern theory tends to be radically destabilizing: both rationality and the notion of universal truths come under attack. Hence, postmodernism tends to be excoriated by most religious adherents as well as by defenders of science. Nonetheless, it has attracted some theologians, most often on the liberal side of the spectrum, where so-called postliberal theology has blossomed, although even some evangelical thinkers have begun to explore "postfoundational" theology.

Recently, postliberal writers trying to understand religion and science have dealt with serious problems arising from the ramifications of postmodernism. When religion is seen as distinct communities of belief and universal truths are disallowed, the question of how to hold a conversation among different traditions becomes paramount. Since each community holds to its own conceptions of truth, how is interaction possible—either with other religious traditions or with nonreligious traditions like science? The upshot seems to be a longing for dialogue and urgent efforts to reestablish it. Until some of those dialogues have begun in earnest, it remains unclear what kinds of postmodern interactions are possible among the many religions and sciences.

Stephen P. Weldon received his Ph.D. from the University of Wisconsin at Madison. He is currently a visiting scholar at Cornell University working on a history of science, skepticism, and secular humanism in America.

POSTMODERNISM is a chameleon-like word that refers variously to the artistic and cultural production of the late twentieth century, to the philosophical or critical orientation of Western scholars in this period, or to specifically Christian theological positions that distinguish themselves from religious modernism and that may or may not draw on the views of secular critical theorists. This essay discusses the philosophical and theological postmodernisms. As a critical orientation to philosophical problems, postmodernism has no single school or line of thought. Instead, its perspective surfaces in numerous areas of modern scholarship. The orientation is explicitly antagonistic toward several principles deemed to have dominated Western philosophy since René Descartes (1596–1650) and, according to some scholars, since the emergence of metaphysics in ancient Greece.

Postmodernist Theory

In general, postmodernist intellectuals have waged a war against "totalizing" systems or perspectives. Reacting to conditions of modernity that they find inimical to freedom—namely the bureaucracy, technocracy, and rationalism of twentieth-century capitalist societies—the postmodernists have developed methods of analysis and discourse aimed at breaking down those monolithic systems and have done so in the cause of heterogeneity and pluralism. They offer what often turns out to be a despairing view of the human condition, one that depicts people as trapped in webs of language, social structures, cultural conventions, and economic forces so constraining that individual freedom and autonomy become virtually impossible. Language, in particular, has drawn the attention of postmodernists because of the extreme dependence of people on it. Hence, understanding the limitations of language is essential for comprehending the ineluctability of the human condition.

A survey of a few key tenets espoused by a majority of postmodern scholars provides an insight into the nature of their understanding of the world. First of all, postmodernism asserts that there are no foundations for ethical principles or knowledge claims and that morality and knowledge are grounded only in particular circumstances of history and culture. This means that, for all human endeavors (including religion and science), there can be no transcendental or transcultural truths. Language and culture constrain behavior and thought, making the world appear differently to people in different times and places. However cogent these appearances, postmodernists warn, they must not be mistaken for universal truths, since such universals do not exist.

Second, postmodernists have generally repudiated all representational theories of language, which means that words do not derive their meaning by referring to objects in the world but, rather, take their meanings from specific contexts in which they are found. Focusing on the relationships between texts, linguistic theorists, using a method called deconstruction, have turned their attention to the act of reading a text because it is only in that act itself that the meanings of the words take shape. Deconstruction completely undermines the idea that there is any stability in language. There are no permanent structures or rules governing language use; everything is in constant flux. As a result, meaning becomes so slippery that no specific text can be said to have a single correct interpretation. Furthermore, deconstructionists have claimed that any text can be shown to exhibit radical discontinuities; every text contains elements that undermine its basic assumptions, thereby rendering meaning completely fragmentary and ephemeral. Drawing out the implications for human beings, deconstructionists have posited that individuals are products of the language they use; even thoughts about one's self are constrained by language. This means that, given the fragmented nature of language, human self-conceptions cease to have any unity or coherence. The individual becomes "decentered."

All of this theorizing has significant implications for the understanding of rationalism, which postmodernists also attack. In a word, they deny its autonomy and, hence, its legitimacy as a privileged mode of finding knowledge and solving problems. One of the principal expositors of this view, Michel Foucault (1926–84), has argued that knowledge and power are inextricably interrelated to the extent that knowledge is impossible without power. By describing knowledge in this way, Foucault radically undermined the notion that rational thought produces any privileged perspective. That which is called rational, Foucault asserted, is as socially mediated as any other claim to knowledge.

Science and Religion

Perceptions of the relationship between religion and science have been substantially affected by postmodernism. In the first place, many religious apologists have embraced the academic attacks on rationalism and humanism, finding common cause with the postmodernists in fighting the hegemony of contemporary secular culture. Postmodernists have denounced humanism for its misguided view of human beings and its naively optimistic ideas about the capability of human control over the natural and social worlds. Humanism is not warranted, according to postmodernists, because the constraints imposed by society, language, and culture limit the ability of people actually to achieve any real measure of freedom. Religionists, too, have often upbraided humanism for precisely this reason: its arrogant appraisal of mankind's status on this earth. Even Pope John Paul II (b. 1920, p. 1978–) has expressed himself as being in agreement with critical postmodern scholarship on certain points. On the whole, however, most traditional religionists have only limited use for postmodern theory, since postmodernism can as equally undermine the foundations of traditional religion as it can scientific humanism. For their part, the humanists, with their strong faith in the power of science and the scientific method, find postmodernism dangerous because it provides a legitimation of irrationality. Although some humanists have found a way to incorporate the postmodern critique, as a group they tend to distrust it and decry it as a serious threat to human progress.

Despite the wide differences between traditional religion and academic postmodernism, a number of theologians have designed theologies that they explicitly label postmodern. Of these various theologies, two are especially important with regard to perceptions of the science-religion relationship. First, theologians like David Griffin have asserted that transformations in science and culture necessitate a transformation in theology. In particular, developments in ecology, quantum mechanics, and psychology have altered the face of Western science so that it is no longer the mechanistic and positivistic study that it once was. In fact, Griffin's reading of science now admits a place for spiritual values. In this account, then, postmodern theology has less to do with rejecting the scientific worldview than with recognizing a new renaissance in which science and religion can once again be found compatible partners. For Griffin, this reconciliation takes the form of a Whiteheadian or Hartshornian process theology. Theologians of this sort use the adjective *postmodern* primarily to contrast their theology with early-twentieth-century religious modernism, as their postmodern theology has little to do with the academic version of postmodernism discussed above. In fact, in some respects, it differs little from its modernist precursor in the basic notion that religion must be made compatible with a scientific understanding of the world; the difference lies in the nature of what each considers scientific and the way in which the reconciliation takes place.

Following an altogether different direction, Mark C. Taylor's postmodern theology derives its insights directly from the deconstructionist wing of academic postmodernism. Like the secular academics, Taylor has called into question all forms of foundationalism (the idea that human knowledge must rest on a foundation of axiomatic beliefs). His attempt to create a new religious system has led him to what he calls an "a/theology," in which he explores the space between atheism and theism. He has endeavored to read the Bible in a deconstructive mode that reveals, in a surprising fashion, basic Christian moral tenets. His practice of deconstruction has led him to establish a religious outlook that comes close to the negative mystical tradition in which God is defined only by that which he is not. Indeed, this parallelism is even indicated in the writings of Jacques Derrida (1930–), one of the founders of deconstructive criticism.

Critiques of Postmodernism

Aspects of secular postmodernism have come under heavy scrutiny by many scholars who vehemently disagree with its assumptions and use, as well as by partisans of one or another school of postmodern thought itself. Some scholars have asserted that postmodernism actually sets up a straw man, attacking an antiquated idea of modernism. This positivistic and reductionistic view of modernism, these critics argue, is more or less a caricature of eighteenth-century Enlightenment views and not a serious twentieth-century standpoint. This critique often raises the question of whether postmodernism is really a radical departure from modernist thought or merely an extension of it.

In defending the latter interpretation, analysts have pointed to the fact that postmodern theory has deep roots in the literary modernist tradition and has been strongly influenced by Karl Marx (1818–83), Friedrich Nietzsche (1844–1900), and Martin Heidegger (1889–1976), among others. These roots demonstrate continuity in terms of both a general Romantic sentiment against antihuman social conditions and a similarity of ideas. On a political level, postmodernists are seen by some to be making unjustified claims about the subversiveness of their radicalism when they are, in fact, merely playing meaningless intellectual games. Critics have also pointed out logical inconsistencies in postmodernist views: The statement, for example, that there can be no truth claims is itself a truth claim. In essence, there seem to be certain foundational elements hidden beneath all of the rhetoric of antifoundationalism. Whether or not the postmodernists have adequately addressed these apparent contradictions in their work is a matter of great debate. Regardless of the outcome, the relationship between religion and science in late-twentieth-century thought cannot be fully understood without accounting for the postmodernist point of view.

BIBLIOGRAPHY

Allen, Diogenes. "The End of the Modern World." *Christian Scholars Review* 22 (1993): 339–47.

Betty, Philippa, and Andrew Wernick, eds. *Shadow of Spirit: Postmodernism and Religion.* New York: Routledge, 1992.

Eagleton, Terry. *Literary Theory: An Introduction.* Minneapolis: University of Minnesota Press, 1983.

Erickson, Millard J. *Postmodernizing the Faith: Evangelical Responses to the Challenge of Postmodernism.* Grand Rapids, Mich.: Baker, 1998.

Ferry, Luc, and Alain Renaut. *Heidegger and Modernity.* Trans. Franklin Philip. Chicago: University of Chicago Press, 1990.

Feyerabend, Paul K. *Against Method.* New York: Verso, 1993.

Foucault, Michel. *The Order of Things: An Archaeology of the Human Sciences.* New York: Vintage Books, 1970.

Gregersen, Niels Henrik, and J. Wentzel van Huyssteen, eds. *Rethinking Theology and Science: Six Models for the Current Dialogue.* Grand Rapids, Mich.: Eerdmans, 1998.

Grenz, Stanley J., and John R. Franke. *Beyond Foundationalism: Shaping Theology in a Postmodern Context.* Louisville, Ky.: Westminster / John Knox Press, 2001.

Griffin, David Ray. *God and Religion in the Postmodern World: Essays in Postmodern Theology.* Albany: State University of New York Press, 1989.

———. "Green Spirituality: A Postmodern Convergence of Science and Religion." *Journal of Theology* 96 (1992): 5–20.

Griffin, David Ray, William A. Beardslee, and Joe Holland. *Varieties of Postmodern Theology.* Albany: State University of New York Press, 1989.

Habermas, Jürgen. *The Philosophical Discourse of Modernity: Twelve Lectures.* Cambridge: MIT Press, 1987.

Harvey, David. *The Condition of Postmodernity: An Enquiry into the Origins of Cultural Change.* Cambridge: Blackwell, 1989.

Hollinger, David A. "The Knower and the Artificer, with Postscript 1993." In *Modernist Impulses in the Human Sciences, 1870–1930,* ed. Dorothy Ross. Baltimore: Johns Hopkins University Press, 1994, 26–53.

Jameson, Fredric. "Postmodernism, or the Cultural Logic of Late Capitalism." *New Left Review* 146 (1984): 53–92.

Kuhn, Thomas S. *The Structure of Scientific Revolutions.* Chicago: University of Chicago Press, 1962.

Lyotard, Jean-François. *The Postmodern Condition: A Report on Knowledge.* Trans. Geoff Bennington and Brian Massumi. Minneapolis: University of Minnesota Press, 1984.

McGowan, John. *Postmodernism and Its Critics.* Ithaca: Cornell University Press, 1991.

Murphy, Nancey, and James Wm. McClendon Jr. "Distinguishing Modern and Postmodern Theologies." *Modern Theology* 5 (April 1989): 191–214.

Norris, Christopher. *What's Wrong with Postmodernism: Critical Theory and the Ends of Philosophy.* Baltimore: Johns Hopkins University Press, 1990.

Nuyen, A. T. "Postmodern Theology and Postmodern Philosophy." *Philosophy of Religion* 30 (1991): 65–76.

Rorty, Richard. *Philosophy and the Mirror of Nature.* Princeton: Princeton University Press, 1979.

Stewart, Jacqui A. *Reconstructing Science and Theology in Postmodernity: Pannenberg, Ethics, and the Human Sciences.* Burlington, Vt.: Ashgate, 2000.

Tilley, Terrence W. *Postmodern Theologies: The Challenge of Religious Diversity.* New York: Orbis Books, 1995.

van Huyssteen, J. Wentzel. *Duet or Duel? Theology and Science in a Postmodern World.* Harrisburg, Pa.: Trinity Press International, 1998.

Index

Academia Linceorum, 252
academic freedom, university and, 43–44
accommodation theory, 20, 121–22
activism, 262
Agassiz, Jean Louis Rodolphe, 184
agnosticism, 220
Albert the Great, 67–68
alchemy, 127, 150
Alpher, Ralph, 317–18
American Scientific Affiliation, 270
Anglican Church, 127; geology and, 192
Anthropic Principle, 339–40
anthropocentrism, 345–53
anthropological naturalism, 322, 323
antievolution crusade, 289–90
apocalyptic theology, 189–90
Aquinas, Thomas. *See* Thomas Aquinas
Arianism, 150, 156
Aristotelianism: in Byzantine Empire, 37; claims of natural impossibility in, 40–41; commentary tradition and, 34, 36–38; determinism of, 65–66; domestication of, 67; as dominant intellectual system, 42; in early modern period, 42–44; historical development of, 37; Islam and, 37; in Middle Ages, 38, 65; naturalism of, 65–66; old logic and, 65; opposition to, 42; propositions condemned, 39–40, 68–70; questions format of, 38–39; rationalism and, 66; rejection of, 143–44; rival natural philosophies of, 42; scientific revolution and, 43; theological implications of, 65–70; theological objections to, 39–40; universities and, 38, 66–69; University of Paris and, 39–40, 68–69
Aristotle, 164, 324, 336; christianizing of, 68; commentaries on, 34, 36–38; conception of God of, 35–36; metaphysics and, 34–35; natural philosophy and, 34–35, 38–39, 40, 65–70; soul and, 197; theology of, 35–36; treatises of, 34–35; view of human soul of, 36, 65; view of women of, 361–62
Arius, 156
Ascent of Man (Drummond), 225
Ash'ari, al-, 85
Ash'ari philosophical theory, 89
astrology, 127
Astronomia Nova (New Astronomy) (Kepler), 101
astronomy, 90, 250; of Galileo, 107
Astronomy and General Physics Considered with Reference to Natural Theology (Whewell), 240, 305
atheism, 124
atheistic materialism, 151
a / theology, 385
atom, 146
atomism, 133, 145, 166
Augustine of Hippo, 60, 324; on Creation, 53–54; natural philosophy and, 51–53
autonomy of science, 256
Aveling, Edward, 216
Averroës, 85–88, 134
Avicenna, 78

Bacon, Francis, 120
Bacon, Roger, 66, 67
Barth, Karl, 172
Basil of Caesarea, natural philosophy and, 50–51
behaviorism, 328

Behe, Michael, 342–43
Bellarmine, Cardinal Robert, 108, 111–13, 114
Benedict, Saint, 351
bestiary, 200
Between Science and Religion (Turner), 23
Bible: American evangelical appropriation of, 266–68; higher criticism and, 267; metaphorical and figurative expressions in, 111; natural history and, 198; Protestant church and, 118
biblical exegesis, 61, 108; Galileo and, 110–11
biblical literalism, 54, 293–97
biblical prophecy, Newton and, 157–60
biblicism, 262
big bang theory, 255–56, 307–8, 314, 317–20
biology: design argument in, 338–39; gender and, 365–67
Biruni, al-, 82–83
Bloor, David, 376, 378–79
Bohm, David, 311
Bondi, Hermann, 318–19
Boscovich, Rodger Joseph, 303–4
Boyle, Robert, 124, 138, 148–49, 155; publications of, 165–66, 326
Brahe, Tycho, 100
Brewster, David, 269
Briggs, Charles A., 268
Brooke, John, 378
Browne, Thomas, 201
Bruno, Giordano, 102
Bryan, William Jennings, 274–75, 280–81; Scopes trial, 291–98
Büchner, Ludwig, 305
Buckland, William, 181, 183–85, 186, 187; hyena-den theory, 183
Buffon, Georges Louis Leclerc, Comte de. *See* Leclerc, Georges Louis (Comte de Buffon)
Bugg, George, 272
Butler, Samuel, 228
Butterfield, Herbert, 17–18
calculus, 154
Calvin, John, 118
Calvinism, 126
Cartesianism, 135–36
Cassiodorus, 61
cathedral school, 63; rationalistic turn of, 64
causation, 130–41; al-Ghazali's rejection of, 133; in Christian orthodoxy, 134–35; emanationist hierarchy of secondary causes in, 134–35; naturalistic theories and, 131–41; nature of, 130; nature of providence and, 138; physics and, 141; role of God in, 131–41
celestial mechanics, 303
Chain of Being, 186, 197, 200
Chalmers, Thomas, 269, 274, 304–5
Chambers, Robert, 240
Christianity: as abstraction, 58; Aristotelianizing of, 68; church fathers and natural philosophy in, 49–53; intellectual tradition of, 47–49; patristic period of, 47–55, 59–61, 199; treatment of women in, 367, 372; use of pagan Greek learning in, 39
church, as abstraction, 58
Cicero, 336
Clarke, Samuel, 138–39, 167–68
Clavius, Christopher, 102–3
Clement of Alexandria, 60
clockwork metaphor, 168–69, 326
Cohen, Chapman, 317
complex specified information, 341–43
complexity thesis, x, 23–26
concept of force, 150
conflict thesis, ix, 3–11, 14–19, 57–58; continuing influence of, 21–23; criticism of, 24; distorted view of disputes in, 9; ethics and, 6; Huxley, Thomas Henry and, 6–7, 10; issues of contention in, 4–7; methodology and, 5–6; as obscuring variety of response, 9; other relationships between science and religion and, 7–8; as positivist, Whiggish historiogra-

phy, 8–9; reaction to, 17–19; reasons for endurance of, 10–11; reevaluation, ix; science and religion operating in close alliance and, 8; scientific naturalism and, 5–7; scriptural geologist and, 5; social power and, 6–7; weaknesses of, 7–9
Congregationalism, 266
consciousness, 332
conservation principles, 348
contextual empiricism, 371
contextualist approach, 13–14
conversionism, 262
Conybeare, William Daniel, 183–84
Copernican revolution, 94–104; Catholic Church and, 99, 102–4, 112–13, 248–50; heliocentric planetary theory in, 42; initial reception of, 98–99; later reception of, 100–102; Protestant church and, 100–102, 121; scriptural texts and, 99; University of Wittenberg and, 98–99
Copernican Revolution (Kuhn), 18
Copernicus, Nicholas, 4–5, 95–98; publications of, 95, 96–100, 102, 109
corpuscular philosophy, 148–49
cosmogony, 234–43; cosmology distinguished, 315; Genesis and, 242; meaning of, 315
cosmology: big bang–steady state debate in, 317–19; cosmogony distinguished, 315; Einstein and, 315–16; expanding universe in, 316–17; meaning of, 315; observational, 315–16; religious implications of, 317–20; theoretical, 315–16; twentieth-century, 314–20
Council of Trent, 108, 110, 248
Counter Reformation, 107, 248
Creation Research Society, 285
creation science, 230, 271, 273
Creation Science Research Center, 286
creationism, 220–21, 229–30, 271; antievolution movement and, 279–84; Darwinian debates and, 279; influence of, 287; popularity of, 278; progressive creationists in, 278; public-school textbooks and, 283–84, 285–86; revival of, 284–87; since 1859, 277–87; strict creationists in, 278
crucicentrism, 262
cultural relativism, xiii
Cuvier, Georges, 182–83

Dana, James Dwight, 270
Darrow, Clarence, Scopes trial and, 291–98
Darwin, Charles, xi, 140, 180, 208–17, 222–24; autobiography of, 215; death of, 216–17; education of, 209–10; private life of, 213–16; Protestant responses to, 23; publications of, 171, 180, 213–15, 222, 225; reaction to, 170, 213, 327, 338–39; religious views of, 210–13, 215, 216; voyage of, 210. *See also* evolutionary theory
Darwin, Erasmus, 221
Darwiniana (Gray), 227
Davies, Paul, 319–20
Dawson, John William, 270, 277–78, 279
day-age exegesis, 184, 190
De revolutionibus orbium coelestium (On the Revolutions of the Heavenly Bodies) (Copernicus), 95, 96–100, 102–3
death, before Fall of man, 186–87
Death of Adam (Greene), 19
Decisive Treatise on the Harmony between Religion and Philosophy (Averroës), 86
deconstruction, 383, 385
deep ecology movement, 347
deism, 125, 139, 151
Deluc, Jean André, 191
Deluge. *See* Flood
Derham, William, 202, 348–49
Descartes, René, 119, 135–36, 147–48, 165; cosmogony and, 235
Descent of Man (Darwin), 171, 214–15, 225
design argument, 335–43; Anthropic Principle and, 339–40; biology and, 338–39; criticisms of, 337–39; historical development of, 335–37; Intelligent Design and,

design argument (*continued*)
341–43; mechanical philosophy and, 336–37; metaphysical commitment to design and, 336; recent incarnations of, 339–43
Determination of Coordinates of . . . Cities (al-Biruni), 83
determinism, 65–66, 134
Dialogo (Galileo), 103
Dialogue Concerning the Two Chief World Systems (Galileo), 114
Dialogues Concerning Natural Religion (Hume), 168–69, 338
Diderot, Denis, 221
Digges, Thomas, 100
Discourse on the Studies of the University of Cambridge (Sedgwick), 189
Discourse on Two New Sciences (Galileo), 106, 115
dispensational premillennialism, 263–64
Disquisition about the Final Causes of Natural Things (Boyle), 165
Disquisitions Relating to Matter and Spirit (Priestley), 304
dissenter, 127
divine will, 122–23
doctrine of papal infallability, 15–16
Draper, John William, ix, 4, 10, 15–16
Draper-White thesis. *See* conflict thesis
dualism: naturalism and, 329–30; weaknesses of, 330
Duns, John, 270
Dyson, Freeman, 311

earth history: anthropocentric design in, 189, 190; prehuman, 180; reconciliation scheme and, 184–85; vs. world (or human) history, 181, 183
Eastern philosophy, 332
ecology, 345–53; economy of nature and, 346; interconnectedness of, 352; meaning of, 345–46; theology and, 353
Eddington, Arthur, 316–17
education: gender and, 366–67; in Islam, 78–79, 89–90; medieval, 63–64
egalitarianism, gender and, 364–65
Einstein, Albert, 306–7, 310–11; cosmology and, 315–16; general theory of relativity of, 307; grand unified theory and, 307; religious beliefs of, 306–7
Elements of Philosophy (Hobbes), 147
eliminativism, 328–29
Empedocles, 323
empiricism, 120, 138
Enquiries into . . . Vulgar and Common Errors (Browne), 201
environment, 345–53; Christianity ecologically culpable, 349–51; White thesis and, 349–51
environmental management, 34
Epicurean hypothesis, 337–38
Epicureanism, 143, 145
Epicurus, 144–45, 146
epistemological naturalism, 322
epistemology, 4, 369–71
Époques de la nature (Buffon), 180
Erigena, John Scotus, 62
essentialism, xi
ethical naturalism, 328
ethics, 328; conflict thesis and, 6
evangelicalism: American, 262; differentiated, 264; evangelical scientists in, 269–71; evolution and, 274–75; history of, 263; intellectual consequences of, 262–63; larger meanings of science in, 273–75; meaning of, 262–63; narrowly evangelical science in, 271–73; natural theology and, 274; science and, 264–68
Eve, 362
Evelyn, John, 348
evolutionary theory, 140, 171–72, 210–14, 219–31, 278; antievolution movement and, 279–84; antiselectionist arguments and, 227; Catholic Church and, 225, 227, 228–29, 231, 257–58; creationism and, 277–87; Darwin and, 23, 171–72, 180, 213–15, 220, 222–24, 225; design in na-

ture and, 226–29; early evolutionism and, 220–22; evangelicalism and, 274–75; fundamentalism and, 274–75; gender and, 366; human origins and, 224–26; human spirit in, 225; Lamarckism, 221, 226, 228, 229; materialistic aspects of, 220; modern creationist backlash to, 220; modern Darwinism and, 229–31; moral reaction against, 228; moral values and, 229–30; non-Darwinism mechanisms of, 226; ongoing sources of conflict for, 220; on origin of species, 139–40; Pope Pius XII and, 254–55; Protestant responses to, 23; purposeful process in, 230–31; reaction to, 170, 213, 327, 338–39; religious debate about, 223–24; soul and, 225; Teilhard and, 253–55; theistic evolutionism and, 227; threats of, 219
Exhaustive Treatise on Shadows (al-Biruni), 83
Expanding Universe (Eddington), 317
experimentalism, 138
Exposition du systeme du monde (Explanation of a World System) (Laplace), 239, 303

faith, 118
Faraday, Michael, 269
Farish, William, 269
feminism, xii
feminist empiricism, 370
feminist historiography of science, 368
feminist theology, 371–72
final cause, 131, 164
Finney, Charles G., 266
first or primary cause, 130, 131
Fleming, John, 269
Flood, Noah's, 182–85, 271, 273; natural history and, 204–5
Force and Matter; or, Principles of the Natural Order of the Universe (Büchner), 305
form, Aristotelian concept of, 134
fossil record, 139, 169; death and, 186–87; degeneration in, 190; design argument and, 187
Foucault, Michel, 383
foundationalism, 385
Francis of Assisi, 351
Free Enquiry into the Vulgarly Received Notion of Nature (Boyle), 326
freethinker, 214, 215
Freud, Sigmund, 226, 367
Froidmond, Libert, 103
From the Closed World to the Infinite Universe (Koyré), 18
functionalism, 328
fundamentalism: differentiated, 263, 264; evolution and, 274–75; history of, 263; intellectual life and, 263–64; meaning of, 263; science and, 264

Gaia hypothesis, 347
Galen, 132; on women, 361–62
Galileo affair, 247–48; biblical dispute in, 110–12; central issues in, 105; concerns for religious orthodoxy in, 110; condemnation of Copernicanism in, 112–13; Decree of 1616 against Copernicanism of, 112–13, 114; Galileo's published writings in, 109–10, 112–13; Galileo's trial in, 113–15; historical background of, 106–8; lack of proof of Copernicanism and, 109; methodology and, 106; Pope Urban VIII in, 113–14; *Revolutions of the Celestial Spheres* (Copernicus) and, 109; scientific dispute in, 108–10; supposed injunction in, 112–13
Galileo Galilei, xi, 42–43, 102, 105–15, 121–22, 145, 257; astronomy of, 107; biblical exegesis and, 110–11; Catholic Church and, 249; heliocentric system and, 111; physics and, 106; publications of, 103, 106, 109–10, 112–13, 114, 115
Gamow, George, 317–18
gap theory, 279
Gassendi, Pierre, 146–48

gender, 359–72; biology and, 365–67; education and, 366–67; egalitarianism and, 364–65; evolution and, 366; meaning of, 359–60; medicine and, 364–67; psychology and, 366; scientific epistemology and, 369–70; scientific revolution and, 363–64; women according to science, 361–67; women in science, 367–69
general theory of relativity, 307
Genesis: cosmogony and, 242; natural history and, 198; Scopes trial and, 291; traditional reading of, 277, 278
Genesis Flood (Whitcomb and Morris), 284–85
genetics, 229, 253
geocentric worldview, 95, 108–9
geology, 19, 25, 139–40, 169, 170, 179–92, 221; Anglican Church and, 179, 192; in Britain, 190–91; on Continent, 191–92; dating in, 180–82; death and, 186–87; eschatological ends of, 188–90; eternalism and, 181; Genesis and, 179; national context of, 190–92; schemes to reconcile Genesis with, 184; species extinction and, 185–87
Geology and Mineralogy Considered with Reference to Natural Theology (Buckland), 187
Gerbert (Pope Sylvester II), 62
Ghazali, al-, 85, 87, 88, 133
Glas, John, 269
God: omnipotence of, 41, 132, 137–38; ongoing supernatural activity of, 325–26; ordained power of, 137–38; reason of, 123; will of, 123
God and Nature (Lindberg and Numbers), 25
God and the Universe (Cohen), 317
Gold, Thomas, 318–19
good works, 118
Gosse, Philip, 272
Gould, Stephen Jay, 22–23
gravitation, 150–51
Gray, Asa, 227
Greek philosophy, 47–55; attitudes toward, 59–61; beliefs about nature of, 49; Islam and, 79, 133
Green, Henry, 268
Griffin, David, 384
Grosseteste, Robert, 67
Guyot, Arnold, 277–78, 279

Hale, Matthew, 348–49
handmaiden formula, 53, 61, 62–63, 67, 324
Harding, Sandra, 370
Hartshornian process theology, 384
Hawking, Stephen, 307–8
Hegel, Georg Wilhelm Friedrich, 370
Heisenberg, Werner, 308, 309, 310
Heisenberg's uncertainty principle, 140
heliocentric system, 42, 96, 119; aesthetic considerations in, 97; Catholic Church and, 248–50; Galileo and, 111; initial reception of, 98–99; later Protestant reception of, 100–102; observational evidence in, 96–97; thesis development of, 96–97
Henry, Joseph, 270
Herman, Robert, 317–18
Hermeticism, 120
Historia animalium (History of Animals) (Gesner), 200
history, cyclic view of, 20
History of the Conflict between Religion and Science (Draper), ix, 4, 10, 15–16
History of the Warfare of Science with Theology in Christendom (White), ix, 4, 10, 15–16
Hitchcock, Edward, 190
Hobbes, Thomas, 147
Hodge, Charles, 267
holism, 332
House of Wisdom, 77
Hoyle, Fred, 318–19
Hubble, Edwin, 315, 316
Humani generis, 254–55
Hume, David, 151, 168–69, 338

Hutchinson, John, 271–72
Hutton, James, 170
Huxley, Julian, 229
Huxley, Thomas Henry, 213, 224–25, 229; conflict thesis, 6–7, 10

Ibn Rushd, 85–88, 134
Ibn Sina, Abu 'Ali, 78
imam, religious law and, 81
Incoherence of the Incoherence of the Philosophers (Averroës), 86
Incoherence of the Philosophers (al-Ghazali), 85
indeterminacy, 309
Institute of Creation Research, 286
Intelligent Design, 341–43
interconnectedness, 352
internalism, orthodoxy of, 22
Irving, Edward, 272
Isidore, bishop of Seville, 61–62
Isis (Rupke), 19
Islam: Abbasid revolution in, 75; appropriation and naturalization of science in, 76–80; Arabic in, 75; Aristotelianism and, 37, 78, 79–80; attitudes toward "foreign" sciences of, 80–88; basis of authority in, 74; causation and, 133–34; characterized, 73; conflict between political and religious elites in, 75; consensus of religious scholars in, 87; continual dialectic with competing worldviews of, 87; court and private patronage in, 77–78; decline of science in, 88–90; doctrinal and institutional diversity in, 86–87; education and, 78–79, 89–90; Greek philosophy and, 79, 133; lack of central hierarchical religious institution in, 87; medicine and, 78; medieval, 73–90; paper availability in, 77; patronage patterns in, 89; philosophical theology and, 81, 83–84; profane sciences in, 82; as religion, empire, and civilization, 73–76; religious law and, 81; sacred sciences in, 82; science in the service of, 79; secondary causation and, 37; translation and, 77; unitary vision of, 75–76; worldview of, 79–80
Isma'ili, 80

Jeans, James, 317
Jesuit order, 253–55
justification by faith alone (*sola fide*), 125–26
Justin Martyr, 60

Kant, Immanuel, 169, 304–5, 338
Keller, Evelyn Fox, 360
Kepler, Johannes, 100–101, 121
Kindi, al-, Abu Ya'qub, 77
Kirkwood's law, 241
knowledge, 83; power and, 383; sources of, 120
Knowledge and Social Imagery (Bloor), 376, 378–79
Koran, religious law and, 81

labor, manual, 20
Lagrange, Louis, 303
Lamarck, Jean Baptiste, 221
Lamarckism, 221, 226, 228, 229
Lammerts, Walter E., 285
language: in postmodernism, 382, 383; translation of, 77
Laplace, Pierre Simon, 151, 303, 327; cosmogony and, 239–43; nebular hypothesis and, 239–43
Latin classics, 63–64
law of independent assortment, 253
law of segregation, 253
law of the excluded middle, 131
law of the pendulum, 106–7
Le Monde (The World) (Descartes), 235
Le Phénemène humain (The Phenomenon of Man) (Teilhard), 255
Leclerc, Georges Louis (Comte de Buffon), 180, 181–82, 197, 221; cosmogony and, 237–39
Lectures on Revivals of Religion (Finney), 266

Leibniz, Gottfried Wilhelm, 138–39, 167–68

Lemaître, Georges, 255–56, 257, 316–17, 318

Les Epoques de la nature (The Epochs of Nature) (Buffon), 238

Linnaeus, Carolus, 346

Literal Meaning of Genesis (Augustine), 54

Locke, John, 151

Longino, Helen, 371

Lucretius, 323

Luther, Martin, 118

Lyell, Charles, 183–84, 210, 211

madrasa, 87

Maestlin, Michael, 100–101

Maimonides, Moses, 134, 325

Malebranche, Nicolas de, 136

Maliki religious scholar, 87

Ma'mun, al-, 77

manual labor, 20

Mästlin, Michael, 121

Maragha observatory, 88

marginality thesis, 88

materialism, 329

mathematical formalism, 141

Mathematical Principles of Natural Philosophy (Newton), 154, 155, 160, 167, 236

mathematics, 150, 250; mathematical demonstration, 120

matter, infinitely divisible, 147

mechanical philosophy, 119–20, 143–51; advocates of, 145–48; challenge of, 122–23; characterized, 143–44; doctrine of primary and secondary qualities and, 144; historical development of, 144–45; later developments in, 148–51; matter and motion in, 143; soul and, 146–47

medicine: early Christian attitudes toward, 54–55; gender and, 364–67; Greco-Roman medicine, 54–55; in Islam, 78

Melanchthon, Philip, 121

Mendel, Gregor, 252–53

Mersenne, Marin, 326

Merton, Robert, 125

Merton thesis, 125–28; criticisms of, 125; puritanism and, 125–27; Restoration and, 127

Metaphysical Foundations of Modern Physical Science (Burtt), 16

metaphysics: Aristotle and, 34–35; principles of, 120

Metaphysische Anfangsgrunde der Naturwissenschaft (Metaphysical Elements of Natural Science) (Kant), 304–5

methodological naturalism, 322–23, 331–32

methodology, 23–24, 57–59; conflict thesis and, 5–6; Galileo affair and, 106; pluralism and, 331–32

microscope, 120, 166

Middle Ages, 33, 57–71; Aristotelianism and, 38, 65; early, 61–63; eleventh- and twelfth-century renewal, 63–64; Islam in, 73–90; later, 65–71; medieval theologian–natural philosophers in, 41–42; natural history and, 196, 199; preservation and transmission of scientific knowledge in, 63; science and religion in, ix

military metaphor. *See* conflict thesis

millenarianism, 189–90

Miller, Hugh, 170, 269

Milner, Isaac, 269

mind, Descartes's concept of, 147

miracle, 122, 326

Mivart, St. George Jackson, 227

Monist, 305–6

monkey trial, 230

monotheism, 132

moral values, evolution and, 229–30

More, Henry, 136

Morris, Henry M., 284–87

Mosaical geologists, 182, 186

Muhammad, 74

Mutakallims, 133

Mysterious Universe (Jeans), 317

natural history, 195–206; Bible and, 198; characterized, 196; deluge and, 204–5; Genesis and, 198; Greek rationality and, 197; mechanistic cosmological model and, 203–4; in Middle Ages, 196, 199; natural philosophy distinguished from, 196; nonteleological approach to, 203; in patristic period, 199; pre-Darwinian, 195–205; professional natural historians and, 203; scientific naturalism and, 203; secularizing historicization of, 204
natural philosophy, 41–42; Aristotle and, 34–35, 38–39, 40, 65–70; Basil of Caesarea and, 50–51; characterized, 49; church fathers and, 49–53; handmaiden status of, 53; meaning of, 59; medieval, 43–44, 59; natural history distinguished from, 196; physics and, 301; St. Augustine and, 51–53; Tertullian and, 47, 48, 49–50, 51; University of Paris and, 39
natural reason, 163
natural science, 59
natural selection, 171–72, 211, 219, 220, 223; antiselectionist arguments and, 227; branching model of relationships in, 223; materialistic implications of, 226; models of theistic evolution in, 172; radical materialism of, 223; theodicy and, 171, 172
natural theology, 139, 163–73; characterized, 163–64; Darwinian challenge to, 171–72; design arguments of, 169–70; different styles of, 170; diversification of, 170; evangelicalism and, 274; functional adaptation in, 188; functionalist design argument and, 179; higher profile of, 166–67; models of theistic evolution in, 172; order and design in nature and, 164–66; paleontology and, 188; presuppositions of, 167–69; social pressures and, 166; survival of, 169–70; in twentieth century, 172
Natural Theology (Paley), 187, 205, 220
naturalism, 5–6, 65–66, 322–32; before 1900, 323–27; dualism and, 329–30; strengths of, 329; in twentieth century, 327–32
nature: change in worldview of, 119–20; early Christian attitudes toward, 47–55; as female, 347; mechanistic model and, 347–48; metaphors of, 345–49; as nonliving creation of rational God, 20; organic view of, 20; as organism, 346–47; theory of oscillating universe and, 20–21
Nature of the Universe (Hoyle), 318
nebula, nature of, 314, 315
nebular hypothesis, 239–43, 327; in America, 241–43; confirmations of, 241–43; religious opposition to, 240
New Age, 332, 347
New Astronomy (Kepler), 121
New Geology (Price), 283
new learning: organization and assimilation of, 65; utility of, 67
New Paganism, 347
Newton, Isaac, 124–25, 138–39, 149, 153–61, 166–68, 326; Arianism of, 156–57; biblical prophecy and, 157–60; cosmogony and, 235–37; cyclical pattern of human history and, 159–60; education of, 153–54; one true religion and, 159–60; theological studies of, 154–60; at Trinity College, 153–54, 157
Newtonian natural theology, 302
Nicholson, Max, 350
Noah, 159–60
nominalism, 325
non-Whiggish history, 23–24
nuclear physics, 317–18

Observations on the Prophecies (Newton), 157–58
occasionalism, 133–34, 135–36
On the Aims of the Philosophers (al-Ghazali), 85
Opticks (Newton), 124, 150–51, 154, 155, 167

organicism, 346–47
Origen of Alexandria, 60
Origin of Species (Darwin), 180, 213, 222
Origins of Modern Science (Butterfield), 17
Osborn, Henry Fairfield, 225–26
Ostwald, Wilhelm, 305–6
Owen, Richard, 170, 189

pagan philosophy, 59–61
pagan schooling, 47–48
paleontology, 187–88, 221; dating in, 180–82; ecological criterion and, 188–89; natural theology and, 188; Noah's Flood and, 182–85
Paley, William, 170, 187, 205
Pannenberg, Wolfhart, 308
papal infallability, 15–16
patristic period, 47–55, 59–61, 199; natural history, 199
Pentateuch, 185, 192
Perfect Cosmological Principle, 318–19
philosophical theology in Islam, 81, 83–84
philosophy, meaning of, 59
Philosophy and the Physicists (Stebbing), 317
philosophy of mind, 328
physicotheology, 201
physics, 150, 301–11; causation and, 141; classical, 302–6; Galileo and, 106; meaning of, 301; modern, 306–11; modern field theories of, 308; natural philosophy and, 301
Physiologus, 199
pietism, 262
Pius XII, Pope, 254–55
Planck, Max, 308–9
planet, motion of, 238
plant-breeding experiment, 253
Plateau, Joseph, 241
Plato, 165, 197, 323–24, 335–36; on women, 361
plenitude, 186
Pliny the Elder, 197–98
politics, social-contract theories of, 364
Pontifical Academy of Sciences, 251–52
positivism, xii
Post-Darwinian Controversies (Moore), 23
postmodernism, xii, 381–86; critiques of, 385–86; language and, 382, 383; meaning of, 382; rationalism and, 383; religion and science in, 383–85; theology and, 384–85; theory of, 381–82
Powell, Baden, 222
power, knowledge and, 383
Presbyterian Church, 265–66, 268
presentism, xi, 14, 26
Price, George McCready, 282–83
Priestley, Joseph, 170, 304
Principia (Philosophiae naturalis principia mathematica) (Newton), 154, 155, 160, 167, 236
Principia philosophiae (Principles of Philosophy) (Descartes), 135, 147, 235
principle: indeterminacy or uncertainty, 309; universal gravitation, 150–51
Principles of Geology (Lyell), 189
printing, 201
process theology, 173, 384
progressivism, 189
progressivist synthesis, time and, 189
Protestant church: Bible and, 118; Copernican revolution and, 100–102, 121; direct access to God and, 118; early modern, 117–28; scientific revolution and, 119–20; substance of science and, 121–25
Protestant Reformation, 107, 118
providence, 147–48; nature of, 131
psychology, gender and, 366
Ptolemaic theory, 96–97, 108–9
public-school textbooks, 283–84, 285–86
puritanism: definition of, 125–26; Merton thesis and, 125–27; science and, 125

quantum mechanics, 308–11

radical Aristotelian, 68
rationalism, 120, 122; Aristotelianism and, 66

Ray, John, 201–2
reason: faith and, 68; God's, 123
reductionism, 327–28
religion: changeability of, xiii; conflict with science of, 3–11; meanings of, 25–26; science as handmaiden of, 53, 61, 62–63, 67, 324
Religion and the Rise of Modern Science (Hooykaas), 19–20
Religion of Geology and Its Connected Sciences (Hitchcock), 190
religious law in Islam, 81
Reliquiae Diluvianae (Diluvial Remains) (Buckland), 183
Restorationist movement, 267
retrospective fallacy, 23–24
revelation, meaning of, 163
Revival of the Religious Sciences (al-Ghazali), 82, 84
revivalism, 266
Revolutions of the Celestial Spheres (Copernicus), 109, 112
Rheticus, Georg Joachim, 5, 97, 99, 121
Riley, William Bell, 281
Rimmer, Harry, 281–82
Roman Catholic Church: autonomy of science and, 256; clerical science (1600–1800) and, 250–51; Copernican revolution and, 99, 102–4, 248–50; evolution and, 225, 227, 228–29, 231, 257–58; Galileo and, 249; heliocentric system and, 248–50; modern science and, 256–58; papal patronage of science in, 251–52; two truths doctrine and, 256–57
Rudwick, Martin, 377–78
Russell, Bertrand, 329

sacred chronology, 180–82
Safavid state, 89
Sandeman, Robert, 269
Schroedinger, Erwin, 308, 309–10
science: as abstraction, 58; Calvinistic origin of, 20; changeability of, xiii; Christian foundations of, 19–21; conflict with religion, 3–11; evangelicalism and, 264; feminist critiques of, 369; feminist historiography of, 368; as gendered, 369; as handmaiden of theology, 53, 61, 62–63, 324; ideological distortion in, 370; meanings of, 25–26, 59, 264; medieval attitudes toward, 43; medieval science characterized, 58–59; professionalized, 7; puritanism and, 125; religion's relationship to, x; social construction of, 374–79; treatment of women in, 367, 372
Science and Creation (Jaki), 20–21
Science and Religion: Some Historical Perspectives (Brooke), 25, 26, 378
Science and Religion in Seventeenth Century England (Westfall), 18–19
Science and the Modern World (Whitehead), 16
science wars, xii
scientific biblicism, 266
scientific creationism, 277, 286–87
scientific knowledge, theory of, 371
scientific naturalism, 322–33; adaptation to, 330–31; conflict thesis and, 5–7; critique of, 331; meaning of, 322–23; reinterpreting religious beliefs and, 330–31; rejection of, 330
scientific revolution, 17, 90; Aristotelianism and, 43; gender and, 363–64; image of nature and, 348; Protestant church and, 119–20; sexual metaphors of male dominance in, 363; social constructionism and, 377
Scopes, John Thomas, 283, 289–91
Scopes trial, 230, 283, 289–98; arguments in, 292–97; Bryan in, 291–98; Darrow in, 291–98; Genesis and, 291; influence of, 297–98; as media circus, 290; national interest in, 292–93
Scottish Enlightenment, 269
scriptural geology, 25, 272; conflict thesis and, 5
scriptural interpretation, 61, 108, 110–11

secondary causation, 64; attributing natural efficacy to, 139; in Islam, 37; rise of, 139–41
Sedgwick, Adam, 183–84
Segraves, Nell J., 286
shape of Earth, 54
Shia, 74; religious law and, 81
Siger of Brabant, 68
Silva: A Discourse of Forest Trees and the Propagation of Timber in His Majesty's Dominions (Evelyn), 348
social constructionism, xii, 374–79; contextual and contingent, 376–77; criticisms of, 378; methodological orientation of, 375; relativizing scientific knowledge in, 375; religion and, 377; scientific revolution and, 377; "strong programme" of, 375–76
social Darwinism, 224
social power, conflict thesis and, 6–7
solar system, origin of, 234–43
soul: Aristotle's view of, 36, 65, 197; evolution and, 225; mechanical philosophy and, 146–47
species extinction, geology and, 185–87
specified complexity, 341–43
Spencer, Herbert, 225
spontaneous generation, 221
standpoint epistemology, 370–71
steady state theory, 314, 318–19
Stebbing, L. Susan, 317
Steward, Balfour, 305
stewardship, 348–49, 351
Stoicism, 336
Stokes, George, 270
Structure of Scientific Revolutions (Kuhn), xii, 375
Summa Theologiae (Thomas Aquinas), 164
Sunni, 74; religious law and, 81; theological philosophy and, 89
Syntagma philosophicum (Gassendi), 146
Systematic Theology (Hodge), 266
Tait, Peter Guthrie, 305
Taylor, Mark C., 385
Teilhard de Chardin, Pierre, 229, 231, 253–55; evolution and, 253–55
telescope, 107, 120
Tertullian, 59–60; natural philosophy and, 47, 48, 49–50, 51
Thales of Miletus, 323
theistic evolutionism, 227
theodicy, 168; natural selection and, 171, 172
Theologiae gentilis origines philosophicae (The Philosophical Origins of Gentile Theology) (Newton), 158–61
theological rationalism, 123
theology: as abstraction, 58; ecology and, 353; greening of, 353; postmodernism and, 384–85; transformation in, 384–85
Theoria philosophiae naturalis (Theory of Natural Philosophy) (Boscovich), 303–4
theory of information, 342
Thomas Aquinas, 67–68, 134, 137, 164, 325
Thomist synthesis, 144
Timaeus (Plato), 63–64, 323–24
time, progressivist synthesis and, 189
Tipler, Frank, 320
Toynbee, Arnold, 350
translation, 64; Islam and, 77
Treatises of the Brothers of Sincerity, 80
trinitarianism, 156–57, 158
Tusi, al-, Nasir al-din, 88
Twelver philosophical theology, 88–89
two truths doctrine, 256–57

Ummayad dynasty, 75
uncertainty, 309
universe, history of, 307–8
university, 33, 64; academic freedom and, 43–44; Aristotelianism and, 38, 66–69; arts masters in, 44; faculty rights and privileges in, 44
University of Paris: Aristotelianism and, 39–40, 68–69; natural philosophy and, 39

University of Wittenberg, 98–99
Unmoved Mover, 35–36

Vatican Observatory, 251–52
Venerable Bede, 62
Vestiges of the Natural History of Creation (Chambers), 222, 240
Victoria Institute, 269, 270, 317
void, existence of, 70, 146
voluntarism, 20, 123, 124, 147–48
von Humboldt, Alexander, 346

Wallace, Alfred Russel, 225
warfare model. *See* conflict thesis
Western civilization, x
Whewell, William, 15, 170, 240, 304–5
Whig historiography, 17; reaction against, xi
Whig Interpretation of History (Butterfield), 17
Whitcomb, John C., Jr., 284
White, Andrew Dickson, ix, 4, 10, 15–16
White, Ellen G., 282
White, Lynn, Jr., 349
White thesis, 349–51
Whitehead, Alfred North, 173
Whiteheadian process theology, 384
Wickramasinghe, Chandra, 319
William of Ockham, 138, 325
Wisdom of God Manifested in the Works of Creation (Ray), 166, 220
Wollaston, Francis, 269
women: Aristotle and, 361–62; early Christian view of, 362; Galen and, 361–62; Greek attitudes toward, 361–63; Hebrew attitudes toward, 362; moral superiority of, 365; perspectives useful for science of, 370–71; Plato and, 361; in science, 367–69; Western culture's view of, 361–67
World's Christian Fundamentals Association, 281
Wright, George Frederick, 270

Zuñiga, Diego de, 102